Crystallization in Foods

The author has made every effort to ensure the accuracy of the information herein. However, appropriate information sources should be consulted, especially for new or unfamiliar procedures. It is the responsibility of every practitioner to evaluate the appropriateness of a particular opinion in the context of actual clinical situations and with due considerations to new developments. The author, editors, and the publisher cannot be held responsible for any typographical or other errors found in this book.

Aspen Publishers, Inc., is not affiliated with the American Society of Parenteral and Enteral Nutrition.

Library of Congress Cataloging-in-Publication Data

Hartel, Richard W., 1951-
Crystallization in foods/Richard W. Hartel.
p. cm.—(Aspen food engineering series)
Includes index.
ISBN 0-8342-1634-5
1. Food. 2. Crystallization. 3. Food texture. I. Title. II. Food engineering series (Aspen Publishers)
TP370.5.H38 2001
664—dc21
00-068991

Editorial Services: Erin McKindley
Library of Congress Catalog Card Number: 00-068991
ISBN: 0-8342-1634-5

Printed in the United States of America

1 2 3 4 5

To my wife, Paula, and my children, AnnaKate and Scott.

Contents

feedback about much of this work. Of particular importance are the numerous industry people with whom I have had discussions on the practical (and sometimes not-so-practical) aspects of crystallization, which have certainly helped me think about the importance of this area. The publishers at Aspen (and Chapman and Hall before that) also deserve credit for keeping me moving on this project. Finally, I would like to thank my family for supporting me throughout the years of seemingly endless writing.

CHAPTER 1

Introduction

Crystallization is a term that describes several different phenomena related to the formation of a crystalline lattice structure. The four steps in crystallization typically include: (1) generation of a supersaturated state, (2) nucleation—the formation of crystalline lattice structure from solution or melt, (3) growth—subsequent growth of nuclei until equilibrium is attained, and (4) recrystallization—a reorganization of the crystalline structure to a lower energy state, generally without any further change in the amount of crystalline phase volume.

First, in order to form a crystal, the liquid phase must exceed the equilibrium condition and become supersaturated (solution) or supercooled (melt). This state can be accomplished by either cooling below the melting point of a compound (eg, water) or by somehow producing a concentration in solution greater than some solubility concentration (sugars, salts, etc). Under these nonequilibrium conditions, the molecules in the liquid state can organize and form into crystal lattice structures. The conditions of supersaturation or subcooling in foods are governed by formulation factors and processing conditions.

Once the solution or melt has been supersaturated, there is a thermodynamic driving force for crystallization. That is, the molecules tend toward the crystalline state to lower the energy level of the system. During nucleation, the molecules in the liquid state rearrange and eventually form into a stable cluster that organizes into a crystalline lattice. The ordered arrangement of molecules in the lattice involves a release of latent heat as the phase change occurs. Nucleation also depends on the formulation (types and concentrations of nucleating promoters or inhibitors) and on the processing conditions (heat and mass transfer rates). In many systems, however, there may be conditions, particularly at extremely high supersaturations, when nucleation is severely inhibited and the system converts into a glassy state. This highly supersaturated, viscous fluid may have solid-like characteristics desirable in certain food products.

Nuclei that form can grow to larger size based on the available supersaturation in the solution. The extent of this growth depends on the magnitude of supersaturation remaining in solution after nucleation has occurred. Growth continues until all of the available supersaturation has been depleted and the system approaches an equilibrium in phase volume, which depends on temperature and composition of the system. For example, if a highly concentrated sugar syrup is allowed to crystallize, the concentration of the remaining solution phase decreases until it reaches the equilibrium concentration

Table 1–1 Typical Food Products Where Controlling Crystallization Is Important to Quality and Shelf Life

Component	*Product Categories*	*Product Examples*
Ice	Frozen desserts	Ice cream, frozen yogurt
	Frozen foods	Meats, fruits and vegetables, entrees
	Freeze-dried products	Instant coffees, dried fruits, rehydratable meals
Sugars	Confectionery products	Grained caramel, fudge, fondant, panned goods, chewing gum
	Cereals	Frosted cereals, glazed cereals
	Dairy products	Whey and milk powders
	Other	Icings, frostings
Sugar alcohols (ie, sorbitol)	Confectionery products	Chewing gum, sugar-free products
Lipids	Spreads	Butter, margarine
	Confectionery products	Chocolate, compound coatings, toffee
	Baked goods	Croissants, puffed pastries, cookies, pie crusts
	Other	Peanut butter, frosting
Salts	Table salt	Sodium chloride (NaCl), salt substitutes
	Dairy products	Milk, cheese
Starches	Baked goods	Breads, cookies, cakes, crackers
	Extruded products	Snacks, crackers
Proteins	Additives	Lysine
Organic acids	Powdered additives	Citric acid
Emulsifiers	Additives	Lipid-based products, dressings

excellent melt-down and cooling properties. When these ice crystals grow larger during storage (recrystallization), the product becomes coarse and unacceptable. Similar concerns apply to sugar crystals in fondant and frostings and to lipid crystals in chocolate, butter, and margarine.

The texture of different types of caramels provides an interesting example of the effect of sugar crystals on product characteristics. When sugar crystals are produced in a caramel, the stretchy texture caused by the protein matrix is interrupted by the large number of small sugar crystals, and the grained product has a "short" texture. Thus, when one bites into a grained caramel, there is a relatively smooth break. In contrast, a caramel with no small sugar crystals has a "chewy" texture, and the uninterrupted protein strands produce a "stretch" to the product. In this case, the caramel is chewy and sticks to the teeth. Each of these products (grained and ungrained caramels) is desirable in certain circumstances, and both formulation and processing conditions must be controlled to obtain the desired texture. Another example of the impact of sugar crystals on texture of a confectionery product is the grained mint. Figure 1–2 shows an optical micrograph of sugar crystals in a grained mint. It is the presence of these numerous, small sugar crystals that gives this product the desired crumbly texture. Techniques for observing and measuring crystalline microstructure will be discussed in Chapter 3.

Sometimes the shape of a crystal is important to the texture. Smooth, rounded crystals flow across one another easily and give a smooth texture, whereas jagged edges and rough

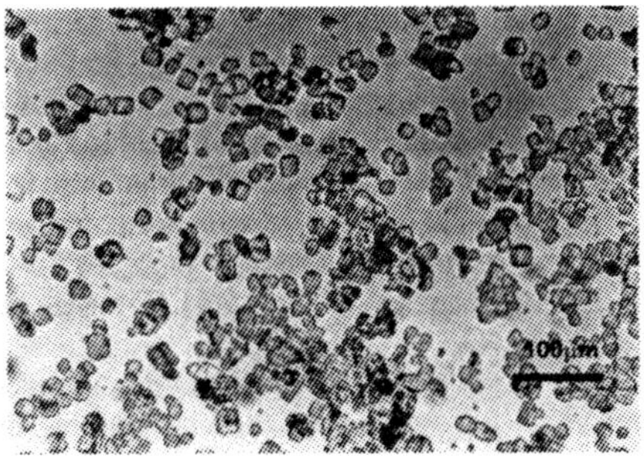

Figure 1–2 Photomicrograph of Commercial Grained Mints Taken by Optical Microscopy with Crossed Polarizers. Scale bar is 100 μm.

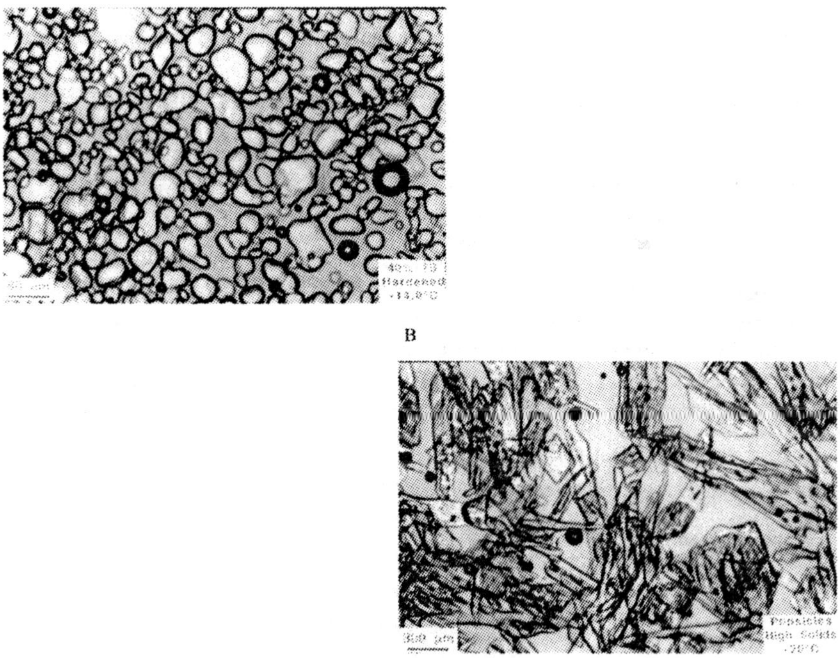

Figure 1–3 Ice Crystals in (A) Ice Cream and (B) Popsicles.

surfaces on crystals may cause them to flow unevenly as the product is sheared during consumption. Contrast the ice crystals in popsicles (large and uneven) with those found in ice cream (Figure 1–3). These differences in ice crystals between popsicles and ice cream are due to the combination of differences in formulation and processing conditions between the two products. In lipids, the correct polymorphic form is required since this can affect texture and shelf life. When the fats in margarine form into undesirable polymorphs, the product often becomes grainy and undesirable to the consumer. Careful control of each of these crystalline elements (number, size, shape, and polymorph) is necessary to produce the highest quality products.

In the previous examples, the crystals provided an integral aspect of the food product. However, sometimes crystals separated from the crystallizing matrix are the desired products (Table 1–2). Refined sugars (lactose, sucrose, glucose, fructose), salts (NaCl, etc), organic acids (citric acid, etc), and proteins are examples of these products. In addition, the process of fat fractionation is widely used to separate components of natural fats into more usable ingredients. Fats that are fractionated in this way include tallow, palm oil, palm kernel oil, and milk fat. In this situation, crystallization is controlled to produce the desired number, size, and shape of crystals that are most appropriate for efficient separation. Typically, this requires smaller numbers of crystals that have grown to larger sizes. In this case, it is important to maintain uniform crystal size and shape to ensure optimal filtration.

Another concern for controlling crystallization is when it occurs during storage in an undesired and uncontrolled fashion. This can affect product quality and may signal the end of shelf life in some food products. Table 1–3 provides examples of products where crystallization during storage is undesirable. In general, systems that can form metastable solutions are those that can undergo crystallization at some later time. For example, hard candies, ungrained caramels, and the unfrozen phase of ice cream are all unstable (or metastable) to sugar crystal formation. If sugar crystallizes in these products during storage, the product is defective, and this generally signals the end of shelf life. Although one can eat a "sandy" ice cream, one where lactose crystals have grown to detectable size, the experience is less satisfying than when consuming a smooth product with no lactose crystals.

Once crystals have been formed in a product and the equilibrium phase volume has been attained, there still may be changes in the crystalline structure as the system seeks

Table 1–2 Examples Where Crystallization Is Applied as a Separation Process

Component	*Product Categories*
Sugars	Refined sugars (sucrose, lactose, fructose, glucose)
Sugar alcohols	Refined polyols (sorbitol, lactitol, etc)
Salts	Refined salts (NaCl, etc)
Fats	Fat fractionation (palm oil, palm kernel oil, milk fat, etc)
Ice	Freeze concentrates (beer, vinegar, citrus juices, etc)
Organic acids	Refined powders (citric acid, etc)

Table 1–3 Examples of Food Products Where Crystallization during Storage Causes Product Defects

Component	*Product Categories*	*Product Examples*
Sugar	Confectionery products	Hard candy, ungrained caramels
–Lactose, sucrose	Dairy products	Ice creams, milk powders
–Fructose	Baked goods	Soft cookies
Salt		
–Potassium bitartrate	Grape juices	Wine
–Calcium lactate	Dairy products	Cheese

a lower energy state. This may either be a reduction of surface free energy or internal energy. In fact, recrystallization may be defined as "any change in the number, size, shape, orientation or perfection of crystals following initial solidification."[1(p 203)] Table 1–4 lists examples of products where recrystallization is important. One of the most common examples of recrystallization in foods is the coarsening of ice crystals in frozen foods during storage. Over time, the ice crystals rearrange, with smaller crystals disappearing and larger ones growing. This results in larger ice crystals. Subsequently, the product either feels coarse during consumption (ice creams) or has a distinct loss of texture upon thawing (frozen fruits or meats, etc). This recrystallization is enhanced by the warm and fluctuating temperatures during storage in our frost-free home freezers. Figure 1–4 shows how ice crystals in ice cream change during storage.[2] Note that in Figure 1–4, the same amount of ice (phase volume) is present before and after storage. The ice cream stored at fluctuating temperature (between –10 and –15°C) for several days has fewer and larger ice crystals due to recrystallization (discussed in detail in Chapter 8).

CONTROLLING CRYSTALLIZATION

In order to control crystallization during processing and storage, several pieces of information are required. First, the equilibrium conditions of the system must be known. One of the most important pieces of information required to control crystallization is the

Table 1–4 Examples of Recrystallization during Storage of Food Products

Component	*Product Categories*	*Manifestation*
Ice	Ice creams and frozen desserts Frozen foods	Increased mean crystal size and range of sizes
Sugars	Fondants, creams	Increased mean crystal size and range of sizes
Lipids	Margarines Chocolates Compound coatings	Polymorphic transformation—grainy texture or fat bloom
Starches	Breads, baked goods	Staling

A

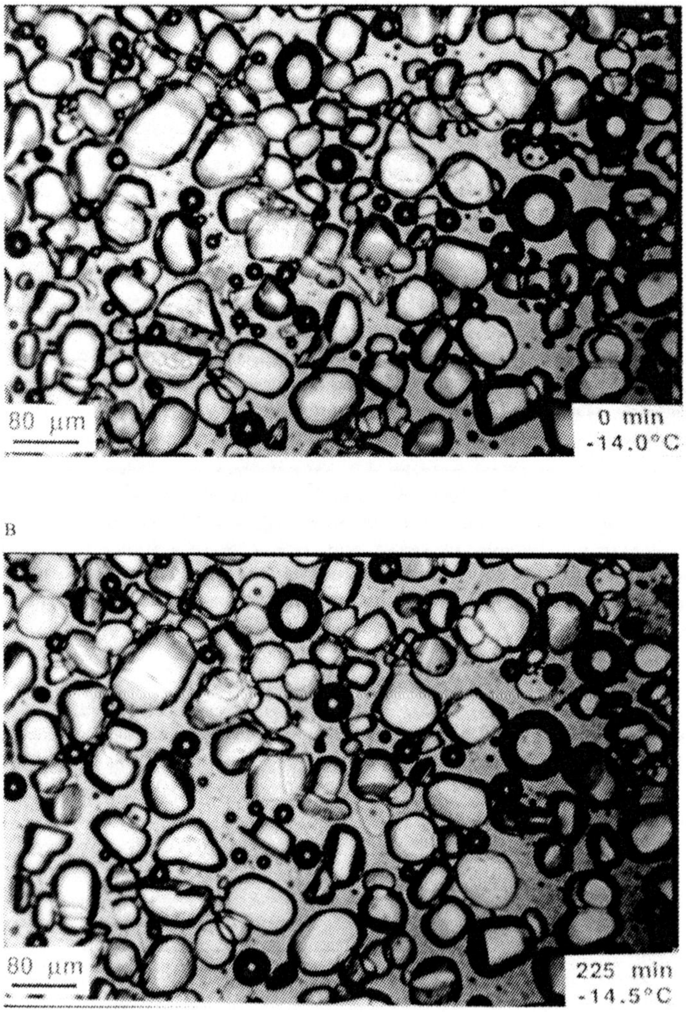

Figure 1–4 Typical Change in Ice Crystals during Storage. (A) Ice crystals initially present in ice cream. (B) Ice crystals after storage with temperature fluctuations between –10 and –15°C.

phase diagram of the system. In simple systems, the phase diagram may be quite familiar. For example, the solubility curve of sucrose in water is quite well known, as is the freezing point depression curve for water in the presence of salts and sugars. Thus, the phase diagram for the simple binary mixture of sucrose and water contains only the solubility curve and the freezing-point depression curve. In complex food products, however, the influence of all other components on the phase behavior of the crystallizing species must be known, and this may be a difficult task to determine in complex products. For example, the solubility concentration of sucrose or lactose in a caramel formulation is not easily found since the milk salts also influence solubility and the high viscosity of the system makes it difficult to perform equilibrium studies.

The effects of operating conditions on crystallization kinetics also must be known to control crystallization. The kinetics of nucleation, crystal growth, and glass formation must be known to control formation of the correct number, size, shape, and polymorph of crystals. Rates of heat and mass transfer that exist during manufacture of a food product govern the conditions under which crystallization takes place and may dramatically influence the nature of the crystalline phase. These kinetics are dependent on formulation, since many ingredients influence nucleation and growth rates.

Finally, storage conditions influence the rate of recrystallization that takes place after crystals have been formed and the product goes through the storage and distribution system. Ideal control of crystallization during processing can be quickly destroyed by improper storage conditions. Temperature of storage, magnitude of temperature fluctuations, and relative humidity all can affect the changes in crystalline structure during storage. The rates of these changes also may depend on the composition of the product. For example, stabilizers added to ice cream inhibit ice recrystallization during storage.

SUMMARY

Controlling crystallization involves careful choice of ingredients to promote the desired structure, proper processing conditions to either promote or prevent crystallization, and proper storage conditions to either maintain the desired crystalline structure or prevent undesired crystallization from taking place. In many food products, controlling crystallization is not an easy task since our understanding of how parameters influence crystal formation and growth is limited. The complex nature of most foods has led to the trial-and-error nature of controlling crystallization in foods. Through a better understanding of the fundamentals of crystallization and glass transition, better control of product quality should result.

REFERENCES

1. Fennema OR, Powrie WD, and Marth EH, eds. *Low-Temperature Preservation of Foods and Living Matter*. New York: Marcel Dekker; 1973.
2. Donhowe DP. *Ice Recrystallization in Ice Cream and Ice Milk*. Madison, WI: University of Wisconsin; 1993. PhD thesis.

CHAPTER 2

Structure of the Solid State

Crystals occur in a wide variety of shapes and sizes throughout nature. Rocks and minerals, biologic materials (eg, cholesterol), salts, sugars, lipids, water (or ice), proteins, and many drugs and chemicals all can be found in the crystalline state. The characteristics or physical properties of many materials depend on the crystalline nature, either of the entire system or of a component part. For example, the ice in a glacier has different characteristics than that in fresh snow, even though both are formed from the same material. In the same way, the ice crystals in a popsicle take a slightly different form than those found in ice cream (Figure 1–3); this leads, at least in part, to the different textures between the two products. In foods, the crystalline structure plays an important role in product texture and shelf life.

When a crystal forms, the molecules orient themselves in a regular pattern, or lattice structure (Figure 2–1). In contrast, molecules in gases and liquids are more randomly oriented and move about freely as dictated by external forces. Freedom of motion is restricted greatly in a crystal lattice, as each molecule interacts with its nearest neighbors. Crystalline lattices are composed of an ordered layering of molecules, with the ordering dependent on the nature of the forces and the interactions between individual molecules. The molecular organization of the lattice gives rise to the macroscopic structure of the crystals. However, looking at a crystal of refined table sugar gives no indication of the ordering of the individual sucrose molecules within the lattice. A diagram of a typical sucrose crystal—an ordinary crystal from a bag of table sugar—is shown in Figure 2–2. The sucrose molecules are arranged in a well-defined pattern within the lattice that makes up this crystal. As seen in Figure 2–3, different faces of the crystal have different conformations of the sucrose molecule apparent at the surface. These slight differences between faces within a single crystal impact the physical properties and growth characteristics of this crystal.

Solid-like materials also occur in which there is no crystalline lattice structure; rather, the molecules are "frozen" into some random orientation, as in an amorphous glass (Figure 2–1). Glasses are not true solids; they may be more accurately defined as highly viscous liquids. However, for practical time scales, amorphous glasses can be considered to have solid-like characteristics. The main difference between glasses and crystals is the extent of "long-term" order, or regularity in spatial dimension. In a crystal lattice, long-term order is apparent, whereas there is essentially no long-term order in glass. Only "short-term" order

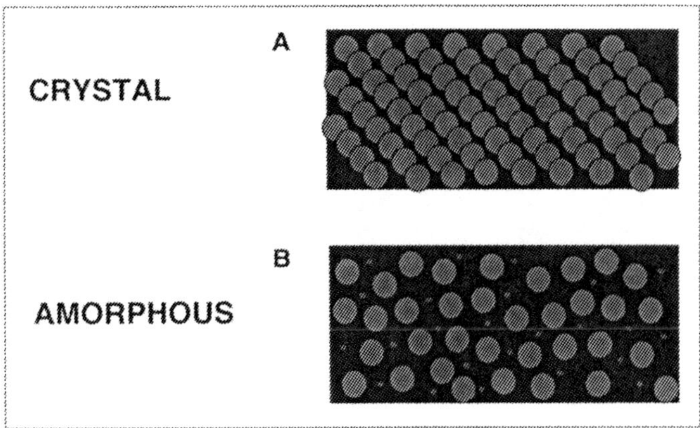

Figure 2–1 Schematic Diagram of Molecular Arrangement in (A) Crystalline and (B) Amorphous Materials.

is apparent in amorphous glasses, indicating that the molecules have ordering over a small range in spatial dimension. This degree of order is typically measured by X-ray diffraction, which gives characteristic diffraction spectra based on the structure of the solid material. Crystals diffract X-rays in a very uniform pattern to produce a well-defined scattering spectrum, whereas glasses diffract X-rays diffusely due to the random orientation of the molecules and produce only a broad X-ray diffraction spectrum. Glasses also can be distinguished from crystals through their melting profiles and transition energies. That is, the energy of transformation from glass to liquid is far less than the energy of transformation from crystal to liquid. Thus, glasses also can be distinguished from crystals based on their melting profiles in a calorimeter.

Amorphous materials also occur regularly in nature; for example, window (silica) glasses, polymers, sugars, gums, and starches all take amorphous forms. Window glass is probably the most common example of an amorphous material, although many plastics are amorphous as well. Many food products, including hard candies and milk powders, are amorphous sugar glasses. The texture and overall quality of these products depend on the formation, structure, and stability of these amorphous glasses. Further discussion of the glassy state is reserved for Chapter 4.

CRYSTAL SYSTEMS AND COORDINATES

The nature of crystal lattices follows the general rules of crystallography, or the study of crystal structures. Crystallography is a well-studied field, and entire books are devoted to the subject. Only a rudimentary understanding of these principles is needed for

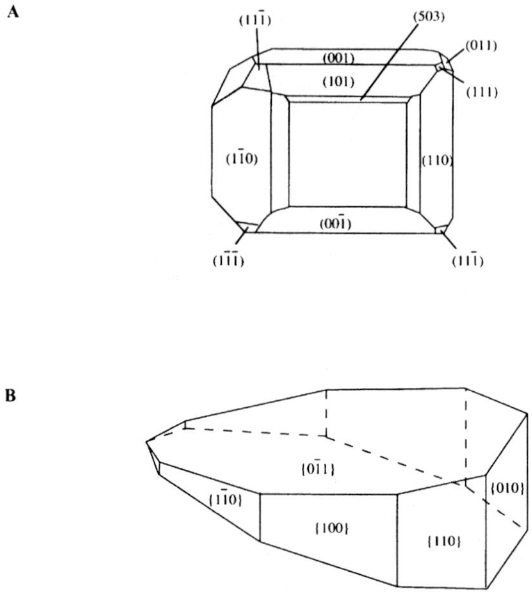

Figure 2–2 Diagram of Typical (A) Sucrose and (B) Lactose Crystals Showing Faceted Structure.

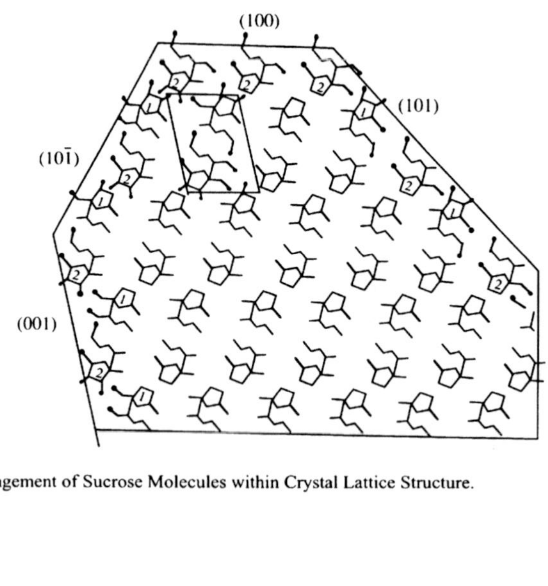

Figure 2–3 Arrangement of Sucrose Molecules within Crystal Lattice Structure.

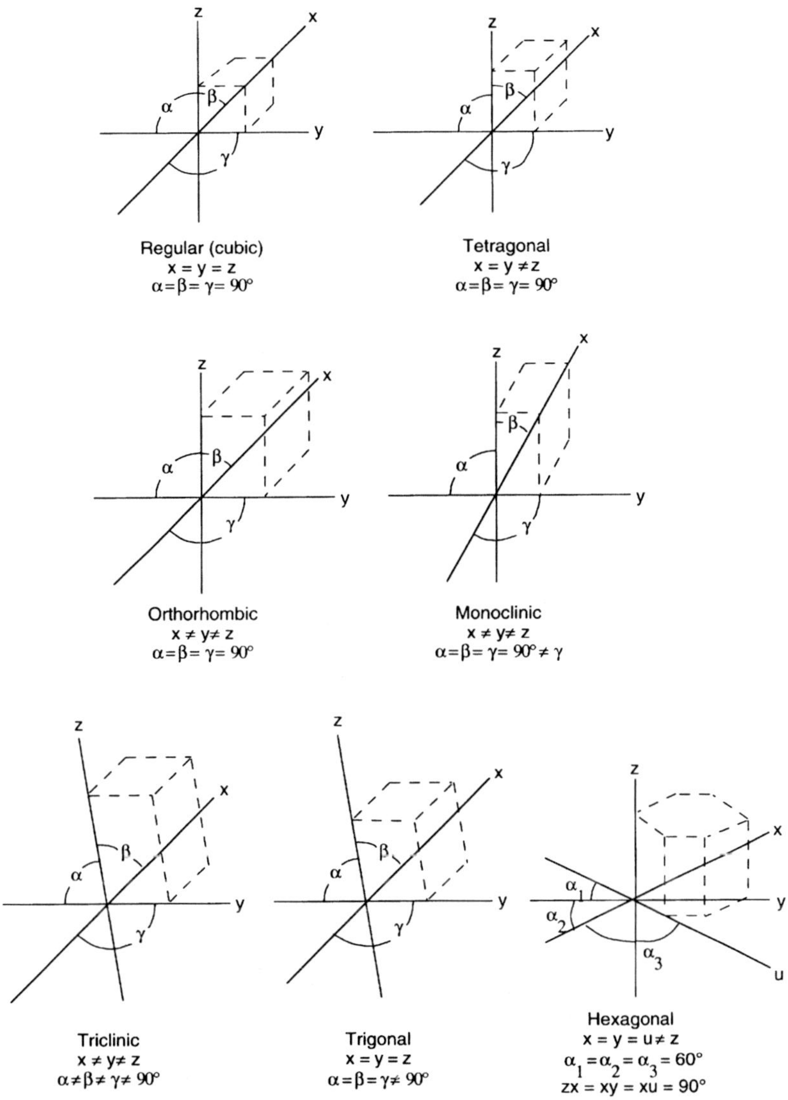

Figure 2–4 Crystallographic Axes and Unit Cells for the Seven Crystal Systems.

understanding most crystallization processes in foods, however. Seven recognized crystal systems describe crystallographic structure and crystal symmetry. All but one of these crystal systems can be described using a coordinate system with three dimensions; the remaining system requires a fourth dimension (Figure 2–4). Differences in coordinate systems arise due to differences in the angles between axes and the length scale of each axis. Table 2–1 provides some typical examples of crystals that fall into each category. For more information about crystallographic systems, see Bennett et al[1] and Mullin.[2]

Unit cells, which are formed by the molecular arrangement in the lattice and are the basic building blocks of crystals, for each of the seven crystal systems are shown in Figure 2–4 (dashed lines). Crystals are made up of a series of these unit cells stacked in three dimensions. These unit cells, their particular arrangement within a crystal, and the resulting macroscopic appearance of the crystal are dependent on the nature of the ions,

Table 2–1 The Seven Different Crystallographic Systems with Some Examples

System	*Example*
Regular (cubic)	Sodium chloride
	Potassium chloride
	Diamond
Tetragonal	Nickel sulfate
	Urea
Orthorhombic	Triacylglycerols (β′-polymorph)
	Diacylglycerols, asymmetric
	Monoacylglycerols (β′-polymorph)
	Silver nitrate
	Citric acid
	Fructose (D)
	Glucose
	Raffinose
	Mannitol (D)
Monoclinic	Sucrose
	Lactose
	Oxalic acid
	Tartaric acid
	Monoacylglycerols (β-polymorph)
	A form of starch
Triclinic	Triacylglycerols (β-polymorph)
	Diacylglycerols, symmetric
Trigonal	Sodium nitrate
	Calcium carbonate
	Calcium chloride
Hexagonal	Ice
	Potassium nitrate
	Silver iodide
	Triacylglycerols (α-polymorph)
	Monoacylglycerols (α-polymorph)
	B form of starch

atoms, or molecules that make up the crystal. Within each unit cell, there is a specified arrangement of the molecules or atoms. This is defined by the space lattice. Within each of the seven crystal systems, there may be multiple space lattices, depending on the nature of the constituent ions, atoms, or molecules and the type of bonding that occurs between them in the crystal. For example, in the cubic crystal system, the unit cell may be either cubic, body-centered cubic, or face-centered cubic depending on the arrangement of molecules within the cubic form.[1-3]

The nature of the crystallizing material dictates the type of bonding between neighbors in the crystal lattice. Bonding can be ionic, covalent, molecular, or metallic. In ionic crystals, such as sodium chloride (NaCl) and other salts, the ions are bound in the lattice by electrostatic forces. Covalent bonds, where outer electrons of neighboring atoms are shared, bind the atoms of covalent crystals, as typified by diamonds. Organic materials, such as sugars, lipids, and organic acids, are bound by weak molecular forces or hydrogen bonds. In metals, the cations weakly share their outer electrons, which are free to migrate through the crystal lattice and result in metallic properties of heat and electrical conduction. Semiconductors fall somewhere between covalent and metallic structure.

MOLECULAR COMPLEXITY IN LATTICE ORGANIZATION

There are several cases important to food applications where complexity of the molecular structure may lead to differences in the crystal lattice structure. Such cases may be due to: (1) crystallization of different enantiomers of a molecule present in solution; (2) crystallization of hydrates, where one or more molecules of water are incorporated into the crystal lattice along with the molecule of the crystallizing species; or (3) polymorphism, where complex molecules crystallize in more than one molecular arrangement depending on conditions.

Isomers

Some substances may take different molecular forms in solution depending on the direction of side groups off the main molecule. Some saccharides (eg, lactose, glucose, fructose) and organic acids (eg, lactic, tartaric), in particular, show this behavior.

Enantiomers

One type of configurational isomer of importance in foods is the positional enantiomer. Enantiomers are molecules that are mirror images of each other.[4] In this case, each asymmetric atom is transposed to give the opposite enantiomer. For example, D-glucose and L-glucose represent enantiomeric forms. In nature, however, typically the D forms of simple sugars are found, and these are the forms of interest in crystallization in food materials.

Anomers

Configurational isomerism that occurs due to the configuration of hydroxyl groups at the anomeric carbon (the asymmetric carbon created by the ring structure of the sugars) of simple sugars plays an important role in crystallization. These configurational isomers are labeled α and β depending on the orientation of the hydroxyl (OH) group at this carbon. For example, the difference between α- and β-lactose is in the orientation (up or down) of the OH group at the reducing end (Figure 2–5). Both forms exist in solution, with an equilibrium maintained between them depending on such factors as pH and temperature. The least soluble form at a given temperature is the form that crystallizes even though both forms are present in solution. Thus, for lactose, it is α-lactose that crystallizes below 93.5°C, whereas the temperature must exceed 93.5°C to obtain the β crystal form. Further details of the influence of anomers on crystallization are presented in Chapter 4.

Hydrates

In many systems, hydrated crystals can form where a certain number of water molecules are incorporated within the crystalline lattice. This depends on the level of hydrogen bonding of the molecule and the arrangement of molecules within the lattice. In lactose, the α form crystallizes as a monohydrate (one water molecule incorporated for each lactose molecule), whereas the β form crystallizes in the anhydrous (no water incorporated) state.

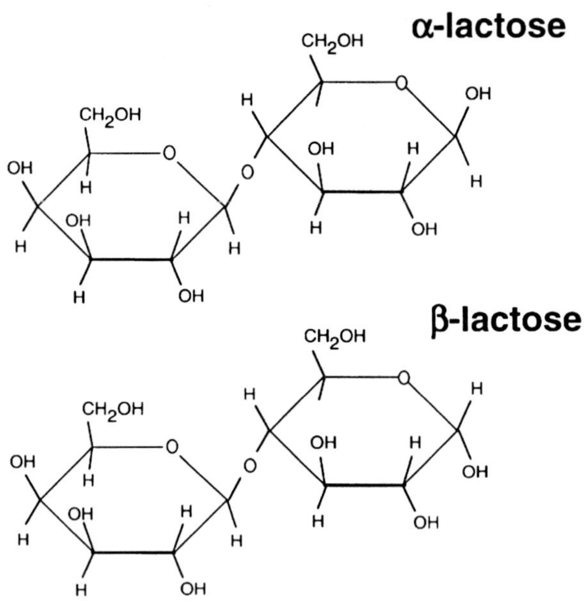

Figure 2–5 Chemical Structures of α- and β-Lactose.

Conversion between enantiomeric and hydrated crystalline forms may occur depending on the conditions. Figure 2–6 demonstrates some of the many forms and interconversions possible in the lactose system. A lactose solution can crystallize into either α-lactose monohydrate at low temperatures (< 93.5°C) or β-lactose (anhydrous) at high temperatures (> 93.5°C). If a lactose solution is concentrated and cooled rapidly, an amorphous lactose glass can be formed. Transformations between these forms also may occur depending on the environmental conditions (primarily temperature and relative humidity). In addition, an anhydrous α-lactose crystal can be formed, either by drying α-lactose monohydrate crystals or by direct crystallization from nonaqueous solvent. Even greater complexity is found in other systems such as fructose and glucose, which can take multiple crystalline forms. For example, fructose can crystallize in the dihydrate, hemihydrate, or anhydrous forms, depending on conditions. On the other hand, sucrose (a nonreducing sugar) is a very simple system since it takes only a single crystalline form.

Polymorphism

Some materials exhibit the ability to crystallize into multiple crystal structures depending on the arrangement of the molecules. Carbon, for example, can crystallize in several forms including graphite (hexagonal) or diamond (cubic). Some sugar alcohols (eg,

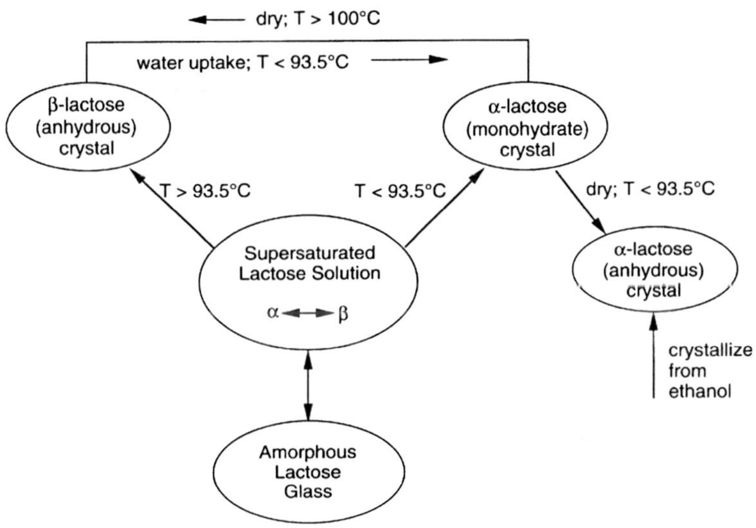

Figure 2–6 Different Forms of Lactose.

sorbitol) also take different polymorphic forms. Sorbitol, a widely used ingredient in many food products, forms four polymorphs (α, Δ, β, and γ). As with lipids, sorbitol exhibits monotropic polymorphism[5] where the least stable polymorph (α) gradually transforms to the most stable form (γ). The different polymorphs of sorbitol are characterized by different melting temperatures, as shown in Table 2–2.

Lipids

The polymorphic variability of lipids (triacylglycerols in particular) also is quite diverse. The polymorphism of most fats can be simply classified according to three main forms, usually given as α, β', and β,[6] although in some cases there may be additional classifications (a very low stability form γ) and subclasses within a main polymorphic type (β_1 and β_2). For example, tristearin crystallizes into four different polymorphs, denoted as γ, α, β', and β.[6] Formation of these different polymorphs for a given lipid material depends mainly on the temperature at which crystallization takes place,[7] as shown figuratively in Figure 2–7, although cooling rate and agitation rate also may affect polymorphism.

Different polymorphs typically are characterized by their distinct X-ray diffraction spectra, which quantify the different crystallographic forms possible. The arrangement of the lipid molecules (angle of orientation and packing arrangement) in the crystal lattice structure is different for each polymorph. Triacylglycerol molecules, which look something like chairs or tuning forks, can be oriented in either double or triple packing modes, as seen in Figure 2–8. Double packing means that the fatty acid groups of adjacent molecules overlap, and the long spacing is equivalent to two lengths of the fatty acid residue. A triple packing arrangement means that the fatty acids of adjacent molecules do not overlap, and the long spacing is equal to three lengths of fatty acid residues. The short spacings between adjacent molecules in a layer depend on the nature of the lipid molecule (length and degree of saturation) and the angle of orientation.

Differences in physical properties between lipid polymorphs arise due to these differences in molecular arrangement.[7] The less stable forms have lower density, lower melting point, and lower heat of fusion than the more stable forms, since the molecules in the less stable forms are not arranged in their lowest energy conformation. These less stable forms are prevalent when crystallization takes place rapidly so that the molecules do not have a chance to order in their lowest energy state. Table 2–3 shows some differences between polymorphic forms of monoacid triacylglycerols. In general, both melting temperature

Table 2–2 Characteristics of Sorbitol Polymorphs

Polymorphic Form	*Melting Point (°C)*
Delta	86.7–88.3
Alpha	90.0–91.7
Beta	94.4–95.0
Gamma	98.9–101.7

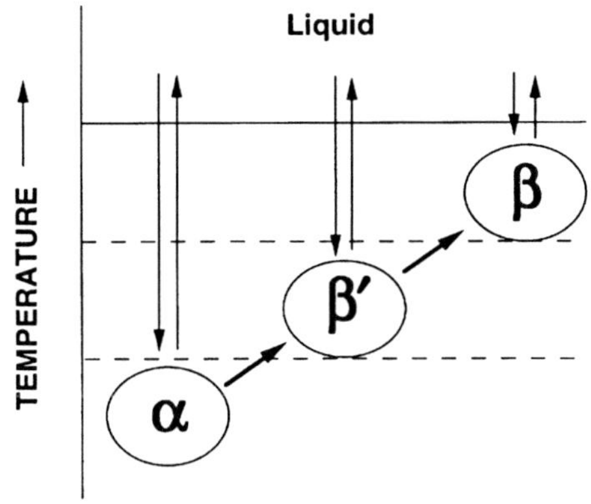

Figure 2–7 Monotropic Polymorphism of Lipids.

and heat of fusion increase with increasing fatty acid chain length. Unsaturated monoacid triacylglycerols have a lower melting temperature and heat of fusion than the corresponding saturated fatty acid when in the *cis* conformation, although the *trans* conformation has values closer to those of the saturated fatty acids.

Polymorphs are typically distinguished by differences in their X-ray spectra. According to Hernqvist,[6] the following criteria can be used to distinguish lipid polymorphs based on their X-ray diffraction pattern.

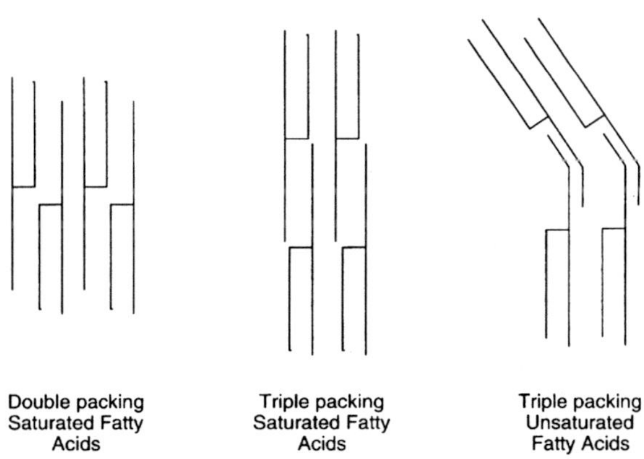

Figure 2–8 Packing Arrangements of Triacylglycerol Molecules in the Crystal Lattice.

Table 2–3 Melting Points and Latent Heats of Fusion (ΔH) for Different Polymorphs of Some Monoacid Triacylglycerols (TAG)

	Melting Temperature (°C)			*ΔH (kJ/mole)*		
TAG	α	β′	β	α	β′	β
LLL	14.0–15.2	30.0–40.0	46.0–46.5	60.3–72.0	81.2	114.3–116.4
MMM	31.0–32.8	41.0–51.0	56.0–58.5	78.7–85.0	100.5	135.2–136.9
PPP	44.7–46.0	53.0–56.6	66.0–66.4	93.4–103.0	131.5	163.3–165.8
SSS	54.7–55.0	61.0–64.0	73.0–73.5	109.3–112.2	142.8	188.4–191.8
OOO (*cis*)	(−32.0)–(−37.0)	−5.0	(−2.5)–5.5			95.6–113.1
OOO (*trans*)	15.0–16.6	37.0	41.0–42.0			133.5–148.3

L, lauric acid; M, myristic acid; P, palmitic acid; S, stearic acid; O, oleic acid.

- α-polymorph: One strong short-spacing line near 4.15 Å
- β′-polymorph: Two short-spacing lines near 3.80 Å and 4.20 Å (also shows a doublet near 720 cm^{-1} of the infrared absorption spectrum)
- β-polymorph: A form that does not satisfy the above

D'Souza et al[8] summarize the X-ray short spacings for various natural and commercial solid fats.

Lipid polymorphs also may be distinguished by differences in melting point (found by differential scanning calorimetry). Nuclear magnetic resonance, infrared, and Raman spectroscopy also have been applied to distinguish polymorphs based on differences in magnetic spin, infrared absorption, and vibrational energies, respectively.[9] In some cases, optical microscopy can be used to distinguish polymorphs based on their different crystal habit, although the range of shapes observed makes it difficult to distinguish polymorphic form based solely on appearance.[10] More recently, molecular modeling programs have been used to characterize polymorphic structures. These models, which orient the molecules so as to minimize lattice energies, may be used to predict polymorphic structures under certain conditions. Typically, X-ray spectroscopy is used to verify and augment the molecular modeling results.[11,12]

Lipids exhibit monotropic polymorphism, where the less stable polymorphs form first and then transform, either by solid-liquid or solid-solid transformations, to the more stable polymorphs (Figure 2–7). Since each polymorph has a unique melting point, the polymorph that forms first depends on the nature of the molecular interactions for that species and the temperature conditions that exist during crystallization. Rapid cooling generally results in formation of less stable polymorphs since the more stable polymorphs require longer times to allow proper orientation of the molecules into the most stable lattice structure. Thus, the less stable polymorphs (γ, α, β′) generally crystallize before the most stable β-polymorph. After crystallization has taken place, the less stable polymorphs then transform to the more stable forms. Eventually, the most stable polymorph will form, although this may take many months in food products.

In some situations, there may be more than one subtype within the main polymorph category. For example, cocoa butter, a relatively simple natural fat, may take up to six different polymorphic forms (Table 2–4). There is still some discussion over nomenclature of these forms, although most people use the simple Greek letter nomenclature as above. Two β′ and two β forms have been identified for cocoa butter, although it is still widely debated that only one of each is a true polymorph.[13,14] These subforms have slightly different melting points and X-ray spectra, although the X-ray spectra still fit the main polymorphic type classification. In the Greek nomenclature, the most stable polymorph within a classification is given the subscript 1; all other polymorphs are ordered in decreasing melting temperature. It also is common to see a hyphenated number following the Greek letter—usually 2 or 3. This number refers to the long-chain packing arrangement in the crystal form, whether there is a double or triple packing arrangement (Figure 2–8). The polymorphic forms of cocoa butter remain somewhat a matter of debate as recent articles by Loisel et al[15] and van Malssen et al[16] indicate.

As mentioned above, cocoa butter may be considered a relatively simple natural fat with only a handful of triacylglycerol species present at substantial concentration. It is for this reason that cocoa butter exhibits such complex polymorphism. Milk fat, on the other hand, is one of the most complex natural fats, with over 250 triacylglycerol species present.[17] Due to this molecular complexity, milk fat does not crystallize into the β form and the β′-polymorph is the most stable form found in milk fat. Table 2–5 shows the most stable polymorph found in various natural fats.[18]

Starch

Starches also form multiple crystalline patterns (polymorphs or allomorphs) depending on the nature of the molecule that crystallizes. In a starch granule, there are regions of amorphous (noncrystalline) and crystalline regions in an orderly layered pattern. The crystalline regions come from an organized pattern of amylopectin molecules, where the regions of linear polymer in amylopectin form the crystalline regions; the branched section of the amylopectin, along with amylose, makes up the amorphous region. A very uniform pattern of crystallinity and amorphous regions occurs in the starch granule, depending on the type of amylopectin present and the amount of amylose.[20]

Table 2–4 Polymorphic Forms of Cocoa Butter

		Melting Temperature (°C)	
Form		*Wille and Lutton*[13]	*Davis and Dimick*[19]
I	γ	17.3	13.1
II	α	23.3	17.7
III	β'_2	25.5	22.4
IV	β'_1	27.5	26.4
V	β_2	33.8	30.7
VI	β_1	36.3	33.8

Table 2–5 Most Stable Polymorph Found in Various Natural Fats

β-Forming	*β′-Forming*
Cocoa butter	Milk fat
Lard	Palm oil
Palm kernel oil	Cottonseed oil
Sunflower oil	Menhaden oil
Canola oil	Rapeseed oil
Coconut oil	Tallow
Olive oil	
Peanut oil	
Safflower oil	
Soybean oil	

Generally speaking, there are three different forms of starch crystal: A, B, and V. There is an intermediate C form,[20] although some[21] consider the C form to be distinct from the A and B forms. Typically, cereal starches (rice, corn, wheat, etc) take the A form and tuber starches (potato, etc) take the B form, although there are exceptions to this rule. Differences between the A and B types are due to differences in water content (the A type has lower water content within the crystal structure than the B type) and the packing configuration of the molecules.[20,22] Another form is recognized for starch, the V (or Vh) form, although this is due to an amylose-complex formation with compounds like aliphatic alcohols and fatty acids.[21,23]

CRYSTAL SHAPE

Crystal shape, sometimes called habit or morphology, is an important technological parameter in many circumstances. Crystal shape can influence the texture and organoleptic characteristics of a product. Smooth, rounded crystals flow across each other easily, whereas crystals with jagged edges do not flow nearly as well and may result in a stiffer, more brittle texture. Crystal shape also influences separation efficiency during refining or fractionation. For optimal separation, all crystals should be about the same shape and size with no liquid inclusions.

The individual faces apparent on a macroscopic crystal are labeled according to their relation with some standard or parametral plane based on the crystal system coordinates (see Figure 2–2). The Miller indices of a crystal face are inversely proportional to the intercepts of the plane on the crystal coordinate system. The standard parametral plane is defined to intercept the axes at the equivalent of a unit cell of crystal structure and has a Miller index of (111), indicating a unit cell on the x, y, and z coordinates, respectively. Further details on crystallography and Miller indices may be found in Mullin[2] and Bennett et al.[1]

The shape of the final crystal, with individual faces and facets, depends not only on the crystal system classification, but also on the conditions under which the crystal was formed and grew. A look at a handful of table sugar crystals reveals the wide diversity of individual crystal shapes. Most are approximately the same shape, but there is a wide variety of slightly different shapes. In fact, over 1,000 different crystal shapes have been identified for sucrose.[24]

Crystal shape can vary from tabular (thin disks) to prismatic to acicular (needles), as shown figuratively in Figure 2–9. Each crystal system has a characteristic habit that occurs when crystals are grown slowly under near-equilibrium conditions that allow the molecules to form into their most stable arrangement. Lactose crystals, for example, form a standard wedge or tomahawk crystal when grown slowly, whereas sucrose crystals grow as prisms (seen in Figure 2–2). However, both sucrose and lactose can appear as needle-shaped crystals under certain circumstances. The different forms appear when crystals are grown under different conditions. The factors that affect crystal habit include: speed of crystal growth, temperature, and the presence of even microlevels of certain impurities. For example, when lactose crystals are grown at high supersaturation, they appear as needle-shaped crystals instead of the tomahawk shape,[25] as shown schematically in Figure 2–10. The tomahawk shape only is found when lactose crystals are grown slowly.

When some crystals are grown under extremely rapid crystallization conditions, the crystals form in the shape of dendrites. These tree-like formations involve growth of a main trunk from which crystal branches extend into the solution, as shown schematically in Figure 2–11. Systems that have high latent heats and low thermal conductivities, including ice and metals, are likely to form dendrites. Frost growth (ice crystals formed from water vapor) on cold windows and freezing of puddles (under low temperature conditions) provide common examples of dendritic crystal forms. The shape of dendrites depends on the growth conditions. For example, ice crystals grown from 10% sucrose solution on a microscope slide (Figure 2–12A) have different structure than those grown under the same conditions but in a 20% sucrose solution (Figure 2–12B). Differences in heat and mass transfer rates due to the solute cause these variations in dendritic growth pattern.

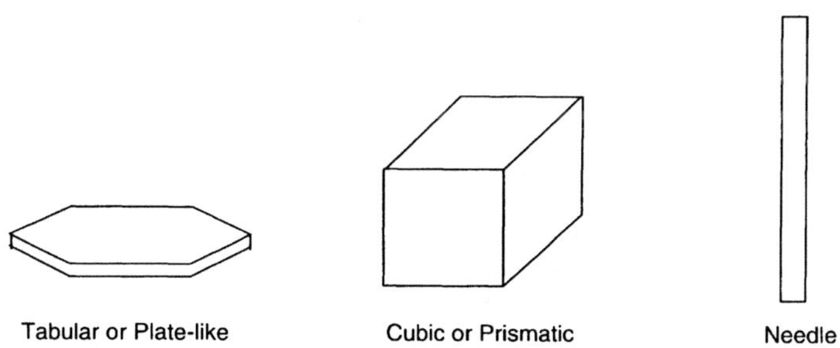

Figure 2–9 Typical Crystal Shapes from Needle (Acicular) to Plate-like.

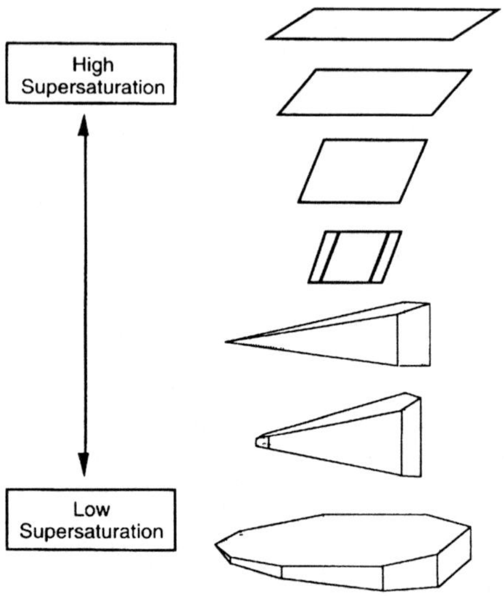

Figure 2–10 Shape of Lactose Crystals Grown at Different Supersaturations.

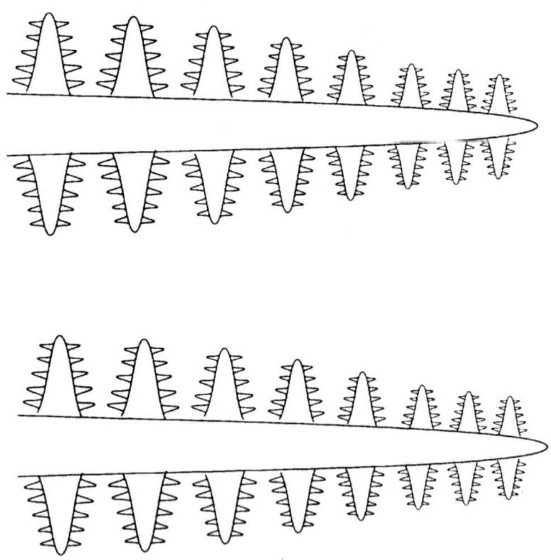

Figure 2–11 General Form of Dendritic Crystals.

A

B

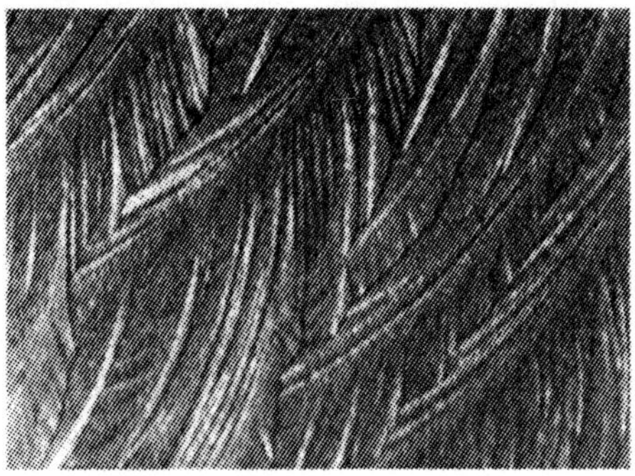

Figure 2–12 Forms of Dendritic Crystals of Ice Grown in (A) 10% Sucrose Solution and (B) 20% Sucrose Solution. Images taken by optical microscopy.

Different lipid polymorphs also have different shapes.[26] The β′-polymorph typically takes the shape of spherulitic crystal that looks something like a pin cushion, whereas the more stable β polymorph crystal often (but not always) forms plate-shaped crystals. Even though they are composed of the same molecules, lipid polymorphs rearrange into more stable crystal lattice structures, with a subsequent change in lattice geometry, which results in different crystal shapes. This can have important consequences to product texture. For example, formation of a β crystal in margarine, where β′ is the desired form, causes the margarine to have a grainy texture.[27] The larger, plate-shaped crystal of the β-polymorph melts in the mouth more slowly; this produces the perception of a grainy texture.

LATTICE IMPERFECTIONS

Due to the nature of molecular forces, crystal lattices generally contain imperfections. That is, lattice sites may be skipped over during crystal growth or sometimes low levels of impurity molecules may be incorporated into the lattice that disturb its subsequent structure. These imperfections generally occur to a greater extent when crystals are grown rapidly and in the presence of impurities. Typically, crystals contain numerous imperfections, and these imperfections play an important role in crystal growth, as will be discussed in Chapter 6.

Several types of lattice imperfections occur, defined either as point, line, or surface defects. Point defects include site vacancies or substitutional impurities, where either lattice sites are unfilled during growth or the lattice site is filled with a competitive substance (or impurity). Line defects, generally either edge or screw dislocations, are formed under the influence of surface or internal stresses during growth. These stresses may be due to either thermal or mechanical influences. Screw dislocations are particularly important to crystal growth as they often control how rapidly a crystal face grows. Further details on imperfections in crystal lattices can be found in Bennett et al.[1]

COMPOSITE CRYSTALS

In many systems, composite crystals form rather than single crystals of uniform type or species. Crystal aggregates, agglomerates, and conglomerates are composite crystals formed under certain conditions. Kidney stones, for example, are aggregates of calcium oxalate crystals (combined sometimes with other mineral crystal species). Composite crystals also can occur due to twinning, which appears as one crystal growing out of another crystal.[2]

Complex triglyceride systems, particularly mixtures of natural fats, crystallize in complex ways depending on the interactions of chemical and physical properties.[28] Natural fats generally contain a wide range of different triacylglycerols that differ in chain length and degree of saturation. For example, milk fat contains the saturated fatty acids ranging from butyric acid (C4:0) to stearic acid (C18:0), and it also contains several unsaturated fatty acids such as oleic (C18:1) and linoleic (C18:2) acids. These fatty acids are placed almost randomly on the glycerol molecule (by the cow) to create many unique triacylglycerols in milk fat. In fact, researchers[17] have identified over 250 different triacylglycerols in milk fat based solely on the 11 most important fatty acids. None of these triacylglycerols are present at concentrations greater than about 4% (molar basis). Thus, there is no dominant molecular species that crystallizes out of milk fat, but rather the similarity in molecular structure between many of the triacylglycerols causes a group of them to crystallize together as a composite crystal. This wide range of molecular species prevents milk fat from forming the most stable β-polymorph, except under extreme conditions.

Recent work[29] on fractionation of milk fat clearly shows the differences in crystalline characteristics due to the different components present in milk fat. Figure 2–13 shows milk fat crystals produced at different stages in the fractionation process. Crystals formed at 30°C, under agitation, from intact milk fat are spherulitic in nature with individual crystalline needles emanating from the central core (Figure 2–13A). These crystals are composed primarily of the highest-melting triacylglycerols present in the natural milk fat. Removal of these high-melting crystals by filtration and cooling of the liquid to 20°C for further crystallization results in formation of crystals with a slightly different shape (Figure 2–13B). This second crop of crystals is still spherulitic with needles emanating from a central core, but the crystals are significantly more dense. These crystals have a more diverse triacylglycerol composition than the original crystals. Subsequent crystallization of the liquid fat from the second stage at 15°C gives a completely different crystal shape (Figure 2–13C). These crystals, which are composed of the low- to middle-melting triacylglycerols, are spherical in nature with smooth surface and strong birefringence in polarized light. This shape is due to the wide diversity of molecular species that make up these crystals plus the substantial amount of liquid fat entrained within the crystal.

Formation of composite crystals in lipids may have significant impact on the quality, texture, and shelf life of lipid-based food products, as the potential for phase separation during storage exists. More dissimilar triacylglycerols may actually migrate out of a composite crystal during storage at certain conditions, such as may occur at elevated and fluctuating temperatures. This recrystallization may result in rearrangement of the crystalline structure and cause quality changes. It has been hypothesized that this phase separation is one of the mechanisms of fat bloom formation during long-term storage of chocolate.

LIQUID CRYSTALS

Some types of molecules form a "fourth state of matter," an intermediate state between liquid and crystal called liquid crystals. Typically, molecules that form liquid crystals have different chemical characteristics over different portions of the molecule and have a rod or disk shape. Emulsifiers and surfactants form liquid crystals under certain circumstances, although they do not always do so when incorporated into foods. These are molecules that have a strongly polar group attached to an aliphatic group, giving them both hydrophobic and hydrophilic characteristics. It is this difference between components within the same molecule that results in liquid crystal formation.

Liquid crystals may play some role in stabilizing food systems, particularly in products with emulsions. There is some evidence that emulsifiers surrounding an oil droplet may actually stabilize the droplet, in part by forming a liquid crystal layer around the droplet. The appearance of an oil-in-water emulsion, with an emulsifier at the surface, under polarized light shows a birefringence characteristic of a liquid crystal. It also has been suggested that emulsifiers in ice cream form a thin layer of liquid crystal around the emulsion droplet, although there has been some debate about this issue.[30]

Recently, Savage and Dimick[31] suggested that phospholipids form liquid crystals in some natural fats. This study suggests that the phospholipid liquid crystal provides the seed for nucleation of triglycerides. Evidence comes from the chemical analysis of

A

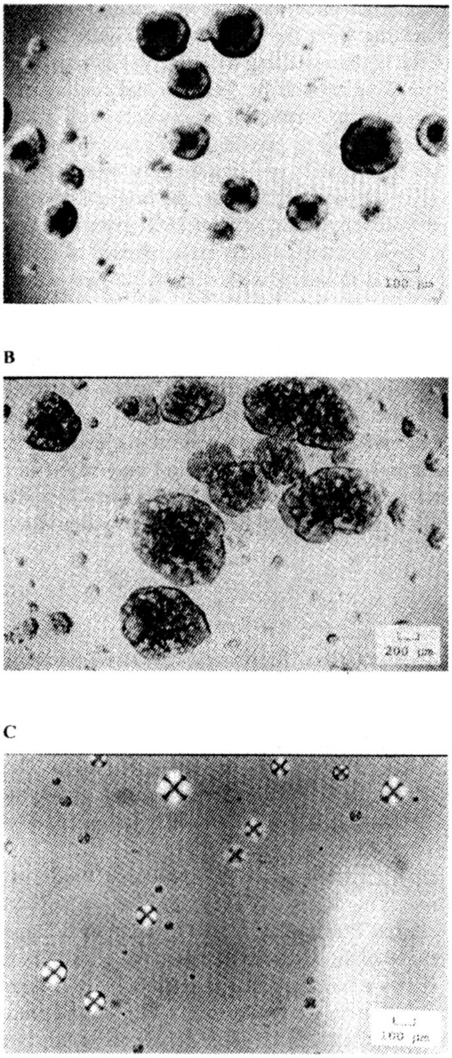

Figure 2–13 Composite Crystals of Milk Fat Formed by Sequential Crystallization at (A) 30°C, (B) 20°C, and (C) 15°C.

the initial crystals that form in cocoa butter. These nuclei are significantly higher in phospholipid content than subsequent crystals formed at later times. In addition, removal of phospholipids from cocoa butter by refining the oil leads to significantly slower crystallization and different crystalline habits, suggesting that the phospholipids are needed to provide seeds for cocoa butter crystallization.

There are actually two types of liquid crystals: those formed by temperature variation (thermotropic mesomorphism) and those caused by the presence of solvent (lyotropic mesomorphism). Within each classification of liquid crystals, there are many subtypes, depending on the arrangement of the molecules.

Thermotropic Mesomorphism

Thermotropic mesomorphs may take several forms: smectic, nematic, or isotropic. The smectic forms are uniform layers—either single or double—of molecules with the polar and aliphatic groups aligned in specific directions. The smectic groupings have organized polar head groups with aliphatic groups that may be frozen or melted and arranged either perpendicularly or at some defined angle. These molecules may be ordered in either single or double layers. The smectic A form has melted aliphatic groups that are perpendicular to the layer of the plane (Figure 2–14A). Smectic B has more rigid aliphatic chains (frozen) that are organized hexagonally (Figure 2–14B). Smectic C has melted aliphatic chains that are not perpendicular to the layer (Figure 2–14C).

In nematic liquid crystals, the molecules are arranged side by side with the aliphatic group in the frozen state, but the molecules are not arranged in specific layers (Figure 2–14D). In the isotropic state, there is very little specific molecular ordering although there is evidence of clustering of molecules (short-range order) in the liquid.

Lyotropic Mesomorphism

Probably the more important form of liquid crystals in foods is that induced by solvent molecules (lyotropic mesomorphism). In this case, interactions between the solvent and the hydrophilic group of the molecule determine the ordering in solution, depending on the concentration of the molecule forming the liquid crystal (Figure 2–15). At low levels of emulsifier in solvent (eg, lecithin in water), the aliphatic groups of the molecules interact to form a micellar solution. Within the micelle are the aliphatic groups of the surfactant, and the solvent is excluded. At slightly higher concentrations of surfactant, the molecules form a hexagonal arrangement. This is essentially a tubular arrangement of the surfactant with these tubes arranged in a hexagonal pattern. Again, solvent is excluded from the interior of the tubes. At even higher concentrations of surfactant, the lamellar liquid crystal phase forms. This is an ordered layering of the surfactant molecules, with aliphatic groups oriented toward each other and the solvent excluded from within the layer. Finally, at very high surfactant concentrations (low solvent content), the molecules organize in a reversed hexagonal form. Here, the molecules again form tubes, but this time with the polar groups

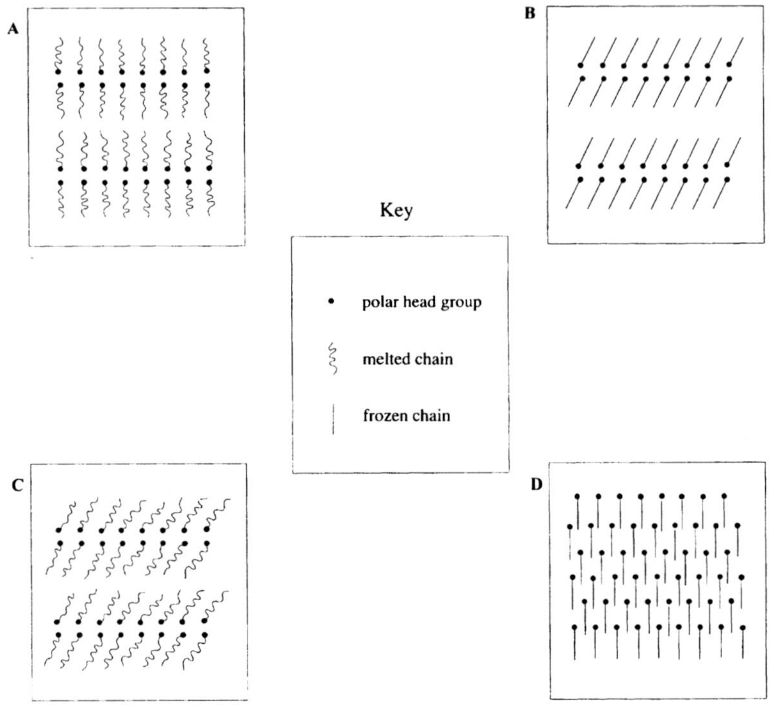

Figure 2–14 Various Forms of Thermotropic Liquid Crystals of Type (A) Smectic A, (B) Smectic B, (C) Smectic C, and (D) Nematic.

heading in and the solvent molecules contained within the micelle. Again, the tubes are oriented hexagonally.

This discussion only shows the simplest forms of liquid crystals. For more information about liquid crystals, see Small.[32]

SUMMARY

The nature of the solid state is important to controlling crystallization in foods. The manner in which molecules orient in crystalline, liquid crystalline, or amorphous materials may impact such aspects as quality, texture, and organoleptic properties of foods as well as their storage stability. Thus, it is necessary to have some understanding of how molecules arrange into the crystalline lattice.

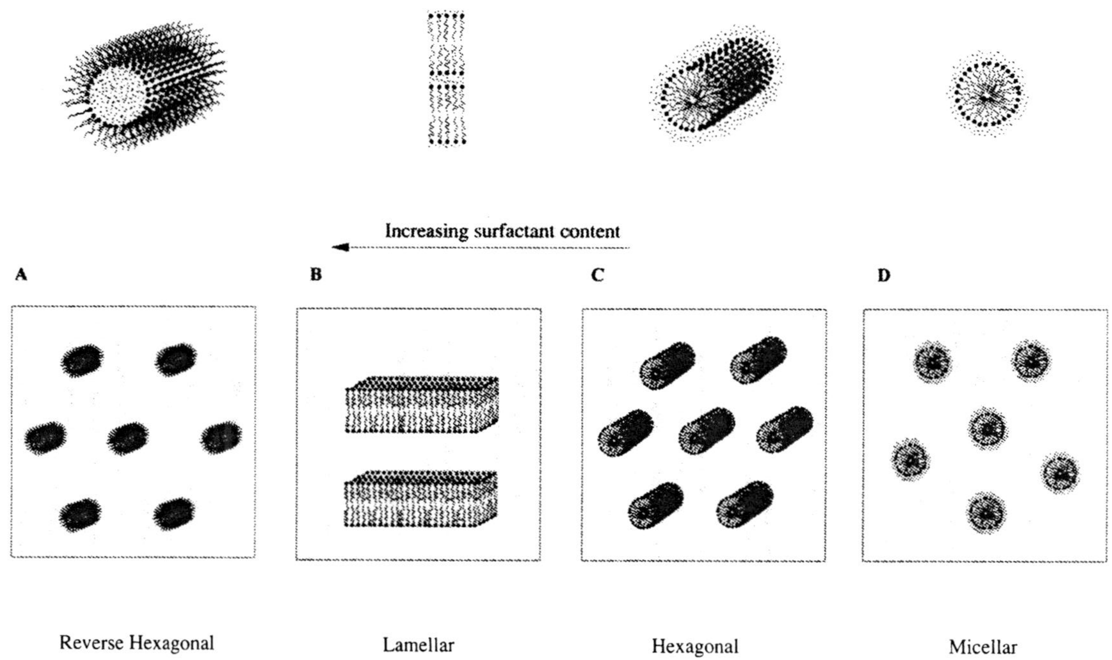

Figure 2–15 Various Forms of Lyotropic Liquid Crystals of Type (A) Reverse Hexagonal, (B) Lamellar, (C) Hexagonal, (D) Micellar.

REFERENCES

1. Bennett A, Hamilton D, Maradudin A, Miller R, Murphy J. *Crystals: Perfect and Imperfect*. New York: Walker and Co; 1965.
2. Mullin JW. *Crystallization*. 2nd ed. London: Butterworth; 1993.
3. Strickland-Constable RF. *Kinetics and Mechanism of Crystallization*. London: Academic Press; 1968.
4. Shallenbarger RS, Birch GG. *Sugar Chemistry*. Westport, CT. AVI; 1975.
5. DuRoss JW. Modification of the crystalline structure of sorbitol and its effects on tableting characteristics. *Pharm Technol*. September 1984; 8:42–53.
6. Hernqvist L. Crystal structure of fats and fatty acids. In: Garti N, Sato K, eds. *Crystallization and Polymorphism of Fats and Fatty Acids*, New York: Marcel Dekker; 1988:97–138.
7. Hagemann J. Thermal behavior and polymorphism of acylglycerides. In: Garti N, Sato K, eds. *Crystallization and Polymorphism of Fats and Fatty Acids*. New York: Marcel Dekker; 1988:9–96.
8. D'Souza V, deMan JM, deMan L. Short spacings and polymorphic forms of natural and commercial solid fats. *J Am Oil Chem Soc*. 1990; 67(11):835–843.
9. Kobayashi M. Vibrational spectroscopic aspects of polymorphism and phase transition of fats and fatty acids. In: Garti N, Sato K, eds. *Crystallization and Polymorphism of Fats and Fatty Acids*. New York: Marcel Dekker; 1988: 139–188.
10. Kellens M, Meeussen W, Reynaers H. Study of the polymorphism and the crystallization kinetics of tripalmitin: a microscopic approach. *J Am Oil Chem Soc*. 1992;69(9):906–911.
11. van de Streek J, Verwer P, de Gelder R, Hollander F. Structural analogy between β′ triacylglycerols and n-alkanes: toward the crystal structure of β′-2 p.p+2.p triacylglycerols. *J. Am Oil Chem Soc* 1999;76(11):1333–1341.
12. van Langevelde A, van Malssen K, Sonneveld E, Peschar R, Schenk H. Crystal packing of a homologous series β′-stable triacylglycerols. *J Am Oil Chem Soc*. 1999,76(5).603–610.
13. Wille RL, Lutton ES. Polymorphism of cocoa butter. *J Am Oil Chem Soc*. 1966;43:491–496.
14. Schlichter-Aronhime J, Garti N. Solidification and polymorphism in cocoa butter and the blooming problems. In: Garti N, Sato K, eds. *Crystallization and Polymorphism of Fats and Fatty Acids*. New York: Marcel Dekker; 1988:363–394.
15. Loisel C, Keller G, Lecq G, Bourgaux C, Ollivon M. Phase transitions and polymorphism of cocoa butter. *J Am Oil Chem Soc*. 1998:75(4):425–439.
16. Van Malssen K, van Langevelde A, Peschar R, Schenk H. Phase behavior and extended phase scheme of static cocoa butter investigated with real-time X-ray powder diffraction *J Am Oil Chem Soc*. 1999;76(6):669–676.
17. Gresti J, Bugaut M, Maniongui C, Bezard J. Composition of molecular species of triacylglycerols in bovine milk fat. *J Dairy Sci*. 1993;76:1850–1869.
18. O'Brien RD. *Fats and Oils*. Lancaster, PA: Technomic; 1998.
19. Davis TR, Dimick PS. Solidification of cocoa butter. *Proceedings of the Pennsylvania Manufacturing Confectioners' Association (PMCA) Production Conference*. Center Valley, PA: PMCA; 1986:104–108.
20. Oates CG. Towards an understanding of starch granule structure and hydrolysis. *Trends Food Sci Technol*. 1997; 8(11):375–382.
21. Zobel HF. Starch crystal transformations and their industrial importance. *Starch/Starke*. 1988;40:1–7.
22. Buleon A, Le Bail P, Colonna P, Bizot H. Phase and polymorphic transitions of starches at low and intermediate water contents. In: Reid DS, ed. *The Properties of Water in Foods · ISOPOW 6* Glasgow: Blackie; 1998: 161–178.
23. Godet MC, Bouchet B, Colonna DJ, Gallant DJ, Buleon A. Crystalline amylose–fatty acid complexes: morphology and crystal thickness. *J Food Sci*. 1996;61(6):1196–1201.
24. Vavrinecz G, *Atlas of Sugar Crystals*. Berlin: Verlag Dr. Albert Bartens; 1965.
25. Nickerson TA. *Lactose: Fundamentals of Dairy Chemistry* Westport, CT: AVI; 1974.
26. Sato K. Crystallization of fats and fatty acids. In: Garti N, Sato K, eds. *Crystallization and Polymorphism of Fats and Fatty Acids*. New York: Marcel Dekker; 1988: 227–266.
27. Watanabe A, Tashima I, Matsuzaki N, Kurashige J, Sato K. On the formation of granular crystals in fat blends containing palm oil. *J Am Oil Chem Soc*. 1992;69(11):1077–1080.

28. Walstra P. Fat crystallization. In: Blanshard JMV, Lillford P, eds. *Food Structure and Behaviour* New York: Academic Press; 1987:67–86.

29. Grall DS, Hartel RW. Kinetics of butterfat crystallization. *J Am Oil Chem Soc*. 1992;69(8):741.

30. Euston SR. Emulsifiers in dairy products and dairy substitutes. In: Hasenhuettl GL, Hartel RW, eds. *Food Emulsifiers*. New York: Chapman and Hall; 1997.

31. Savage CM, Dimick PS. Influence of phospholipids during crystallization of hard and soft cocoa butters. *Proceedings of the Pennsylvania Manufacturing Confectioners' Association (PMCA) Production Conference* Center Valley, PA: PMCA; 1995:54–59.

32. Small DM. *Physical Chemistry of Lipids From Alkanes to Phospholipids*. New York: Plenum Press; 1986.

CHAPTER 3

Measurement of Crystalline Structure in Foods

There are many techniques that can be used to characterize the crystalline microstructure and its development in foods. Important measurements on crystalline content in foods include the following:

- *Verification of the presence and amount of crystals.* The presence of crystals in a food product can often be distinguished by the existence of certain textural properties or by appearance. For example, a grained caramel has a characteristic "short" texture as the sugar crystals disrupt the protein matrix. Ungrained caramel, in contrast, has a chewy texture that depends on the rheological properties of the matrix. However, sometimes the presence of crystals may not be so obvious from textural changes. Other techniques, such as microscopy and spectroscopy, also can be used to monitor the presence of crystals in foods.

  It is often difficult to quantify the precise amount of crystalline material present. The total amount of crystalline material may be determined using techniques sensitive to total crystal presence (eg, X-ray or thermal analysis) or by quantifying the number and size distribution of crystals in the product.
- *Quantification of the number and size of crystals present.* The number of crystals present and the crystal size distribution (CSD) are important parameters for control of crystallization in foods. The number and size distribution lead to either the desired product quality and texture or the efficiency of separation during refining operations. The CSD can be measured quantitatively in a number of ways (eg, microscopy or light scattering techniques). These techniques are described in greater detail in a later section.
- *Characteristics of crystals present.* The shape and polymorphic form of crystals in foods also may impact texture and stability of the product. Shape plays an important role in texture, especially in products where crystals interact with one another. For example, crystals in ice cream are extremely close to one another (within 4 to 10 μm). When ice cream is consumed, the movement of ice crystals across each other influences product texture. As shown in Figure 1–3A, ice crystals in ice cream are typically quite smooth and "roll" across each other during consumption. On the other hand, ice crystals in popsicles (Figure 1–3B) are long and dendritic (needle-like) in nature and result in a completely different texture.

The polymorphic form of a crystal (ie, in lipids, organic acids, etc) also influences product characteristics and shelf life. Transformation from one polymorph to another during storage is often accompanied by a loss of product quality, as experienced during bloom formation in chocolates or graininess in margarines.

- *Type of crystals present.* In some complex foods, there may be more than one component that can crystallize. In caramels and caramel toppings, for example, the solution phase may be supersaturated in both lactose and sucrose. Which of these components crystallizes depends on the equilibrium phase behavior and crystallization kinetics. The ability to distinguish which component crystallizes may be important to controlling undesired crystallization. However, it often is difficult to distinguish the type of crystals present.
- *Rate of crystallization.* To control the crystalline structure in foods, it is important to control how and when crystals form during processing. In general, techniques for measuring aspects of crystal structure can be used to characterize the rate of their formation, although it is often very difficult to employ these techniques for use as process control indicators during processing. It is more common to characterize rates of crystallization in laboratory systems that are more easily controlled and then extend that information to control of the commercial process.

Measurements of the rate of formation and specific characteristics of crystals in foods often are made difficult by the presence of other structural elements of the product. Other structural elements, such as air cells, fat globules, and particulate dispersions, often impede our ability to monitor changes in crystalline microstructure. Common techniques for measuring crystal characteristics in foods include microscopy, thermal analysis, spectroscopic techniques, and light scattering. Other techniques, such as rheological or dielectric measurements, also have been used to study crystals in foods. More recently, people have begun to use molecular modeling programs to predict certain characteristics of the crystalline phase in foods. The choice of technique for measuring crystalline characteristics depends on many factors including type of product and the nature of the crystalline microstructure being measured, the processing conditions, and other environmental factors.

IMPORTANCE OF CRYSTAL SIZE DISTRIBUTIONS

The spread and distribution of sizes of a population of crystals play a crucial role in final product quality, no matter the application. Virtually every crystallization process requires careful control of the CSD. In many foods, product quality is defined by not only the total amount of crystalline material, but also the characteristics of the crystalline phase. It is the number of crystals and the range in sizes, shapes, and even polymorphic stability, as well as the arrangement of that crystalline material with the other elements of the food (ie, network structures), that lead to the mechanical/rheological properties (Figure 1–1). On the other hand, when crystallization is employed as a separation process during processing of foods, different requirements for the size distribution are important. Crystals with larger size and more uniform size distribution typically are more efficiently filtered, resulting in a separated product with less liquid entrained and higher purity. Thus, in both cases during processing of foods, it is critical to understand the required CSD and to know how to obtain that CSD through proper processing conditions.

For example, ice cream contains numerous ice crystals, most of which must be below some threshold detection size (about 50 μm) or else the ice cream is perceived as coarse. The same applies to fondants, which contain numerous small sugar crystals. In fact, the palate can detect sugar crystals larger than about 15 to 20 μm, so a fondant must contain crystals primarily smaller than this size. Other foods containing sugar crystals where control of CSD is crucial to texture include fudge, caramel, and panned confections. Similar concerns occur in products in which the lipid crystallizes such as butter, margarine, and chocolate. The texture and snap of chocolate depend on controlling cocoa butter crystallization during tempering. The goal of tempering is to produce a large number of very small cocoa butter crystals with the proper polymorphic form in the final chocolate. This provides a smooth internal structure that snaps sharply. In contrast, poorly crystallized chocolate has a large-grained internal structure that does not snap sharply. In products containing lipid crystals, the polymorphic form also may be important to product texture. For example, the formation of the β-polymorph in margarines results in a grainy texture since this crystal typically forms as a larger, plate-like crystal compared with the more spherulitic form of the β′-polymorph.

There are several food processes where controlling crystallization is important to the purification or separation of components. In sugar refining (ie, sucrose, lactose, glucose, and fructose), the CSD is crucial for adequate separation of the crystalline material from the mother liquor (or saturated solution). A uniform distribution of large crystals with similar size is desired for efficient centrifugation of the crystal mass. Small crystals, formed during the later stages of the crystallization process, decrease the efficiency of centrifugation by trapping liquid in the interstitial spaces between larger crystals. The shape or habit of the crystals also is important in filtration or centrifugation efficiency. In particular, needle-shaped crystals severely limit the rate of filtration.

Fractionation of fats provides another example of a purification process resulting from crystallization of a selected component. Fractionation of fats (such as tallow, palm oil, palm kernel oil, and milk fat) is a relatively common technique to produce fats and oils with desirable properties. During fractionation, the liquified fat is cooled to a crystallization temperature that allows only a portion of the fat to crystallize. The crystal slurry, with typically less than 30% phase volume, is filtered to produce solid (stearin) and liquid (olein) fractions. The ability to separate these fractions depends on the size distribution of crystals formed during the fractionation process. Typically, crystallization is controlled to produce relatively large and uniform crystals. If the CSD is too broad, with too large a range from small to large crystals, filtration is impeded and the separation is less efficient. Figure 3–1 shows schematically the differences in filtration and melting point between several different size distributions obtained during fractionation of milk fat.[1] The best separation, as indicated by shortest filtration time and highest melting point of the solid fraction, arises when crystallization is controlled to produce a narrow distribution of

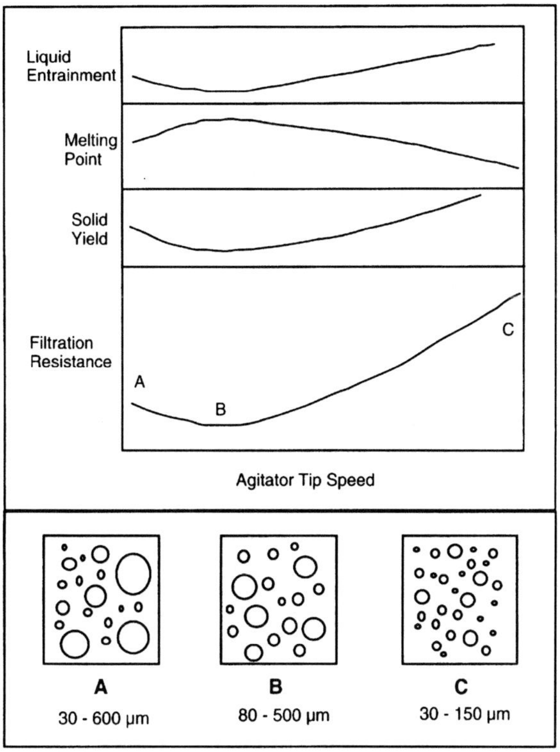

Figure 3–1 Schematic Drawing Showing the Effect of Particle Size Distribution on Filtration of Milk Fat.

relatively large crystals. Either a wider distribution or small crystals result in more difficult and less efficient separation.

CRYSTAL CHARACTERIZATION TECHNIQUES

There are numerous techniques available to measure the size, amount, or type of crystals present in a food. These techniques are based on various physical properties of the crystals. The choice of measurement technique most suitable for a specific application must be made with consideration for the type of results desired and the practicalities of the situation. For example, crystals embedded into a solid matrix, as in ice cream, are not easily amenable to separation. It is extremely difficult to separate these crystals from the matrix and count them using sophisticated automatic particle counters. However, it may be possible to do this for some types of materials, such as the sugar crystals present in chocolate.

Measurement of crystalline characteristics, especially the CSD, is necessary for optimizing and controlling the product and the process, whether it be for separation efficiency or product quality. However, an accurate CSD is not always easy to obtain, particularly in many complex food products. Incorrect sampling techniques can result in inconsistent size distributions that have little resemblance to the actual CSD. It is very important that sampling techniques provide a representative distribution of the crystals in the product and not induce undesired changes, either during sample preparation or during analysis. Each measurement technique also implies certain assumptions and/or simplifications that may result in inaccurate or biased CSD measurements.

To ensure an accurate characterization of crystalline microstructure in foods, several different measurement techniques should be used. Multiple analytical techniques ensure that the true nature of the crystalline phase is determined and that the biases inherent in each measuring technique do not give a misleading representation. Often, it is best to use some microscopic technique to verify the nature of the crystals present.

There also are many techniques with which to follow or characterize the rate of crystallization. Typically, these techniques are used on model systems or on food products being produced in a simulated environment. It is usually quite difficult to measure crystallization rate and extent directly in a processing environment, although some indirect techniques may be used in some situations. In the following discussion, techniques for both characterizing crystalline microstructure and measuring crystallization kinetics will be described. Often, a technique may be used for both measurements in slightly different ways.

Desupersaturation

An indirect measure of crystallization processes often can be obtained by evaluating the change in solution characteristics. As crystallization progresses in a batch system, the supersaturation decreases as equilibrium is approached. The rate at which the equilibrium condition (solubility concentration) is approached gives an indirect measure of crystallization rate. Desupersaturation typically exhibits a curved shape as shown in Figure 3–2. If there are no crystals present initially, supersaturation does not begin to change until massive nucleation occurs. Although there may be a few small nuclei present earlier, it is not until many crystals form and start to grow that a decrease in supersaturation can be detected. This is especially true when measuring concentration of solution, due to the errors in this measurement. A more accurate means of measuring desupersaturation in many solutions is by using refractive index. Once substantial crystallization (combined nucleation and growth) occurs, the supersaturation decreases rapidly. As the system approaches equilibrium (solubility concentration), the rate of decrease in supersaturation tapers off until eventually crystallization ceases as equilibrium between crystal phase volume and solution concentration is attained.

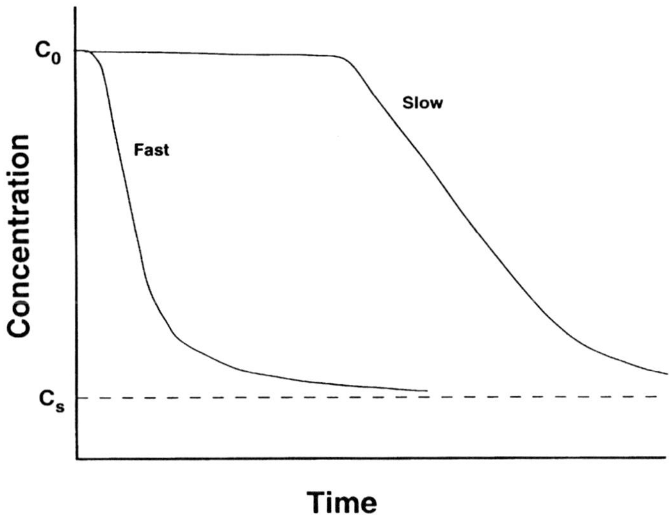

Figure 3–2 Typical Desupersaturation Curve Showing How Solution Concentration Decreases with Time during Either Fast or Slow Crystallization. Initial concentration is C_0, and solubility concentration is C_s.

The shape of the desupersaturation curve can be used to estimate crystallization rates. The induction time required for onset of nucleation can be measured as the time required for massive crystallization to occur. The rate of crystallization can be taken as the slope of the desupersaturation curve (usually taken at the inflection point). However, the technique, as described, does not discriminate between nucleation and growth. Thus, these parameters are not truly quantitative measures of nucleation or growth rates, and other techniques must be employed to obtain more fundamental information. Nevertheless, comparisons between different systems are easily and quickly performed with this technique; they can provide substantial information on the effects of different components on crystallization rates. This technique by itself also does not provide information on the characteristics of the crystalline dispersed phase.

Sieves

One of the oldest methods for sorting, sizing, and counting particles is sieving.[2] A mesh screen of controlled and well-known size opening is used to segregate particles on screens of decreasing size. After shaking for sufficient time, the weight of crystals remaining on each screen is measured and correlated to the average mesh size of each screen. Weight distributions according to crystal size are generated in this way. Sieving works well for powders with particles above about 60 μm. It is more difficult to sieve particles much smaller than this accurately, although wet-sieving techniques have been developed that can be used for smaller sizes (above 10 to 20 μm). In this case, a particulate suspension is applied to the top screen and allowed to drain through the stack of screens. As the liquid carrier flows through the screens, the particulates are sieved according to the particle size.

The most commonly available sieves are those made up of a wire mesh to produce a square opening for particles to pass through. Sieves are characterized by their mesh size rather than by the size of the opening on the screen. The mesh size refers to the number of wires per linear inch that make up the screen. For example, a 635-mesh screen refers to a screen size of 20 μm, a 400-mesh screen refers to a screen size of 37.5 μm, and a 5-mesh screen refers to a screen size of 125 μm. Electroformed micromesh sieves are available to smaller sizes (about 5 μm). These sieves are formed by photo-etching of a metal sheet.

As a particle is vibrated on the top of a screen, the speed at which it passes through the opening depends on its size relative to the screen size. If the particle is much smaller than the screen size, it passes through quite quickly. Of course, if the smallest square dimension of the particle is larger than the screen opening, it will never pass through. When the particle is slightly smaller than the opening, the time to pass through the screen depends on how long it takes to orient in the proper dimension during vibration. The particle will remain on top of the screen until its smallest square dimension orients correctly with the screen opening. Thus, for accurate screening, sufficient vibration times are needed to ensure that all particles of appropriate size have passed through the screen.

The results of a sieve analysis are displayed as a percentage of the total weight that remains on each screen. Thus, the weight percentage of particles smaller than a certain size, based on the nominal sizes of the screen openings, can be plotted to give the size distribution of a sample.

Sieving techniques were widely used in the past for analysis of size distribution in the sugar refining industry, and they are generally applicable to any powdered material. However, sieves are not as widely used these days in the food industry, since more sophisticated techniques are available that are easier to use and provide more detailed information about the size distribution.

Sedimentation

Sedimentation devices also have been used to size and count crystals.[2,3] The simplest form of this technique uses a sedimentation column, which allows the crystals to settle, in a suitable liquid carrier of known viscosity (μ), according to Stokes' law.

$$v_s = \frac{g}{18}\frac{d_p^2\left(\rho_p - \rho_f\right)}{\mu} \tag{3.1}$$

Here, v_s is settling velocity, g is the gravitational acceleration, d_p is the particle diameter, and ρ_p and ρ_f are densities of the particle and fluid phases, respectively. The size obtained from these measurements is sometimes called the Stokes' diameter since it is calculated from Stokes' equation based on the time required for a crystal to settle through a certain distance in the liquid column. This equation assumes that the particle has attained its maximum settling (or terminal) velocity while in laminar flow. Corrections to Stokes' law are available for nonspherical particles and for other force fields (centrifugal sedimentation). See Allen[2] for more details.

The simplest sedimentation techniques generally rely on weighing of suspension within a certain sedimentation zone at specified time intervals. These give a weight fraction of sample related to the Stokes' diameter for particles falling through that zone (column height) after a certain time of sedimentation. Size measurements down to a few microns are possible in this way. However, the sample must be adequately diluted and dispersed in the sedimentation fluid for accurate measurements. Sedimentation rates also can be measured by light attenuation or light-scattering techniques. That is, the density of particles at a given sedimentation depth can be found from the attenuation of scattered light passing through that zone. X-rays also have been used to characterize the particle density during sedimentation to measure particle size distribution.

More sophisticated versions of the sedimentation technique utilize inertial forces to separate crystals of different sizes. Disk centrifuge devices may be used to measure extremely small crystals of submicron size. The movement of a particle in a centrifugal field is determined by the balance of the drag force holding it in and the centrifugal force pushing toward the outer ring of the disk. Then, particle size is characterized from the attenuation of scattered light in the disk plane.

Although sedimentation techniques may be used more widely in other fields, it is rare for sedimentation techniques to be used to analyze crystalline material in the food industry. This is primarily due to the difficulties in separating crystalline material from the food matrix and the relative simplicity of other techniques.

Electrozone Sensing

Electrozone sensing devices are a common method of automatically measuring particle size for crystals that can be dispersed in an electrolyte solution. The principle behind this technique is that changes in electrical resistance as a particle passes through an orifice of specified dimension can be related to the size of the particle.[2] When a crystal passes through the orifice, as shown in Figure 3–3, it displaces a certain amount of electrolyte, which changes the electrical resistance between two electrodes on opposite sides of the opening. The magnitude and duration of the voltage impulse are related to the size of the particle, and the electronics of the device correlate and sort them into the appropriate size channel. Particle sizes between 0.5 μm to about 1,000 μm can be measured, depending on orifice size. However, radio-frequency noise often effectively limits the lower size range to 2 to 3 μm unless extreme precautions are taken to eliminate external noise.

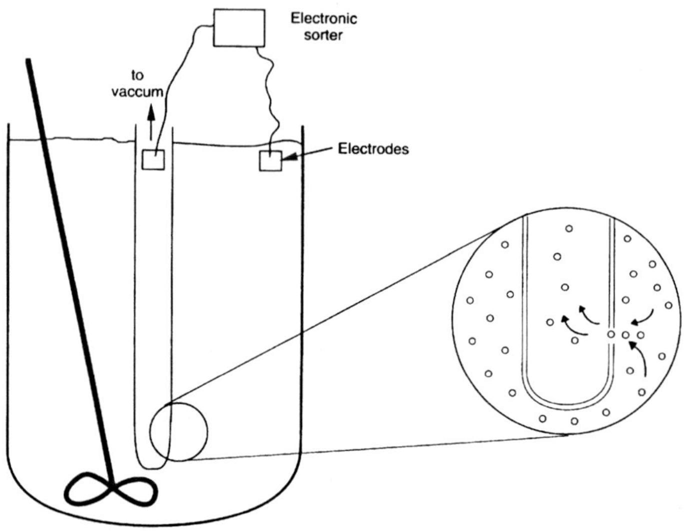

Figure 3–3 Principle of Electrozone Sensing.

One primary advantage of this technique is that thousands of particles can be counted in a very short time, thereby allowing an accurate and statistically significant size distribution to be determined quite easily. However, sampling procedures are very important. The particles must be dispersed in an electrolyte solution that is essentially inert to the particles. That is, the solvent must be chosen so that particles do not either grow or dissolve during the time of sampling. Also, the particles must be sufficiently dispersed so that only one particle passes through the orifice at a time. In these units, the level of coincidence, where multiple particles are passing through the orifice at one time, is characterized by the shape of the electronic pulses. If coincidence is unacceptably high, a warning is given.

Electrozone sensing devices, like the Coulter Counter™ (Coulter Electronics, Hialeah, FL), have been used to analyze a variety of crystals formed during processing of foods. For example, sugar crystals produced during refining operations are easily separated and analyzed in these devices.[4] Choice of solvent is typically an aqueous-based solution of salts and sugars that provides sufficient viscosity to suspend the crystals and is saturated in the sugar being studied so that no growth or dissolution occurs. The Coulter Counter also has been used to study protein crystallization. However, precautions must be taken to ensure that the shear forces during transit through the aperture do not break up the fragile protein crystals and cause errors in the measured size distribution. The Coulter Counter may be used to study ice crystals; however, precautions must be taken to ensure that no melting or growth of ice occurs when operating at freezer temperatures.

The use of electrosensing devices for monitoring crystalline structures within food products is more limited. This would require separation of the crystalline solids in a food and dispersion in a stable electrolyte. Conceivably, one could use this technique for analyzing sugar crystals in caramel, for example, by dissolving away the caramel matrix with an appropriate organic solvent to leave a slurry of sugar crystals for analysis. However, no studies of this type have been reported in the literature.

Microscopy

Microscopy is probably the most widely used technique for characterizing structure, especially crystalline content, in foods.[5–7] Under the right circumstances, particle sizing through image analysis also can be performed on the images obtained by microscopy. Thus, microscopy has the potential both to document the nature of the crystalline structure in a food visually and to give quantitative information regarding the size distribution. Microscopy also can be used to evaluate rates of nucleation and growth directly. It therefore is an extremely valuable and versatile tool for analyzing crystals and crystallization rates in foods.

There are numerous types of microscopy, based on the type of electromagnetic radiation used and the nature of the optical system.[8] Standard optical microscopy using white light is a simple technique that allows visualization of crystals in foods under certain conditions. The primary advantage of using optical microscopy is that it allows visual observation of the size and shape of crystals that influence texture and quality. However, crystals must be greater than about 3 to 4 μm to allow resolution by optical microscopy. Although the theoretical resolution of optical microscopy is less than 1 μm, in practice it is extremely difficult to resolve crystals smaller than about 3 to 4 μm in a standard laboratory microscope. In many food products, the crystals are larger than this size and can be readily observed with the optical microscope after careful preparation. The use of polarized light microscopy, which enhances the detection of crystalline material, can help to reveal the presence of smaller (between 1 and 3 μm) crystals. More sophisticated microscopy techniques (ie, modulation contrast, differential interference contrast) for improving contrast of samples also are available.[9]

Laser light microscopy is a relatively new technique for food analysis. Confocal scanning laser microscopy (CSLM), for example, uses laser light to illuminate the first few layers below the surface of the product and provides a three-dimensional view of the microstructure. In CSLM, a dye is excited at a particular wavelength of light, and the fluorescence emission is monitored. By carefully choosing the dye used to stain the sample, certain surfaces can be enhanced, which makes the structural elements more visible. More recently, multiple-photon microscopy techniques have been developed that allow selective visualization of interior structural aspects based on a specific fluorescence emission of certain components of the product.[10] In two-photon microscopy, two photons are required to excite the dye simultaneously; the fluorescence emission of this event can be monitored. This provides greater resolution and deeper penetration into the sample, with less risk of undesired excitations of the dye at levels above and below the confocal plane.

Electron microscopy is the technique of choice when crystals are less than a few microns in size. Either scanning (SEM) or transmission electron microscopy (TEM) can be used, although SEM is by far the most common method because of the relative ease of sample preparation. In SEM, electrons are bounced off the product to give an image of the surface, whereas TEM involves passing an electron beam through a thin film of the product to produce an image of its structure. In SEM, one main disadvantage is that it operates under vacuum; thus products can contain no volatile components such as water. Consequently, preparation of SEM samples involves drying to remove water, and this process may introduce artifacts, or unrepresentative images. More recently, development of an environmental SEM (ESEM) has allowed observation of materials that contain water.[11] Samples now may be studied in their natural state. The ESEM maintains a vacuum for the electron source, but has a pressure gradient that leads to a sample environment that is just below ambient pressure. Thus, the presence of volatiles does not cause visualization problems.

Recently, the development of scanning probe microscopy techniques (scanning tunnelling microscopy or atomic force microscopy) has allowed study of surface characteristics.[12] A probe tip scanning directly at the surface reveals the orientation of the molecules within the material. In crystal studies, these techniques are capable of showing the orientation of molecules within the lattice and can detect the type and nature of imperfections. Although still somewhat in its infancy, these techniques have the potential to uncover, for example, new correlations between crystal structure and growth kinetics.

In all microscopy techniques, sample preparation is critical. Poor sample preparation often can lead to limited or misleading information. Under ideal conditions, microscopy can provide informative, and often beautiful, images of the crystals present in foods. Quantification of these images requires image analysis capabilities, which will be discussed in a later section.

Optical Microscopy

Optical microscopes come in several different varieties.[8,9] They can be operated in either the reflection or the transmission mode. In reflection microscopy, light is reflected off the surface of the sample to provide an image. Transmission microscopy, where light is passed through a thin section of the sample to visualize the internal structure, is used more commonly.

There are several techniques for enhancing light microscopy. They include polarized light, differential contrast, and dark-field microscopy. In each of these techniques, detection of the transmitted light is controlled to enhance visualization of certain structural elements. In foods, the use of polarized light microscopy is common since this technique allows one to distinguish crystals from noncrystalline solids. The crystalline lattice structure rotates polarized light in a specific manner. Two crossed polarizer lenses, which control the vibrational orientation of the light beam, are used to enhance the image of a crystalline material by blocking all light except for the light that is rotated by a crystal. In this way, a crystal appears as a bright region surrounded by a dark background. Images of table sugar taken with polarized light microscopy are shown in Figure 3–4. The object can be

Figure 3–4 Images of (A) Extra-Fine Granulated Sugar and (B) Powdered Sugar (Sucrose Crystals) Taken with Polarized Light Microscopy. Samples dispersed in mineral oil.

further distinguished from the background using more sophisticated contrast techniques. In modulation contrast and interference contrast techniques, the optical gradient at the surface of the object is enhanced to provide additional information about the object. In some cases, further enhancement in contrast of images can be obtained by using ultraviolet (UV) light as the incident light. In this case, it is UV absorption or fluorescence emission of the sample that provides the enhanced image contrast. These techniques are widely used in medical microscopy, but may have special applications in the food industry.

Typical magnifications obtained in optical microscopy are 2 to 200X, with a theoretical lower resolution limit of about 0.2 μm. Lower magnifications (2 to 20X) are used to show larger structures in a product. For example, if there are a few larger crystals contained within a distribution with smaller crystal sizes, low magnification is required to detect these larger crystals. Higher magnifications (40 to 100X) allow observation of crystals down to about 3 to 5 μm in size. To obtain the maximal resolution from optical microscopy, oil immersion lenses are needed. With them, magnifications of 100 to 200X allow observation of crystals as small as 1 μm or so, although care must be taken to preserve the sample if oil immersion is necessary.

Although optical microscopy has many advantages for particle visualization, there are several disadvantages that must be considered. Particles that vary greatly in size (about 150 μm) are difficult to count and size with optical microscopy techniques. This problem is compounded by the limited depth of field in optical microscopy, which can be a problem when imaging a distribution of particles with wide size range. The limited depth of field causes loss of clarity when focusing in on one size range. That is, focusing on large particles causes loss of clarity of small particles, and vice versa. Thus, stratified sampling techniques may be needed to ensure statistical counting of all particles in the size range when a wide CSD exists.

Proper sampling techniques must be used regardless of the type of microscopy. Ideally, to view the internal microstructure of a food product, the material must be prepared in such a way as to keep the structure as intact as possible. However, different experimental techniques may be used depending on the required information. In many cases, a view of the primary crystalline microstructure is required. The crystals in caramel, for example, can be dispersed onto a microscope slide so that the individual crystals can be observed. This does not show how the crystals were oriented within the product, which may be an important aspect of product texture. However, simple dispersion of the crystals allows observation of the size distribution, number, and shape of crystals. If the orientation of the crystals in the product also is important, other techniques (eg, SEM) may be needed to minimize sample disruption prior to visualization.

The sampling technique must not lead to any change in the size or shape of crystals. That is, the crystals must neither dissolve/melt or grow during the time of analysis. This means that the crystalline material must not dissolve nor grow in the selected solvent, and temperature must be controlled during observation.

For example, study of ice crystals in frozen foods using optical microscopy must be accomplished in a controlled temperature environment in order to prevent melting of the ice crystals. This may be done by moving the microscope to a freezer, using a refrigerated cold stage on the microscope, or housing the microscope in a specially designed environment.[13] A refrigerated glove box used to study ice crystals in ice cream is shown schematically in Figure 3–5. The images shown in Figure 1–3 were taken in this unit. A freon compressor unit was installed on a specially built glove box inside of which was housed an optical microscope. The camera attachment for the optical microscope extended outside of the glove box to maintain adequate temperatures. Alternatively, a video camera can be attached to the microscope to enable automated image analysis. Two primary advantages are associated with this system. First, the operator need not work inside a freezer to maintain low temperatures to prevent melting. Second, condensation of water vapor does not interfere with the microscope images.

To prepare a sample for optical microscopy, a small slice of the product is placed on a microscope slide, several drops of an appropriate solvent (such as butanol or kerosene for ice cream) are placed on the sample, a second slide is placed on top of the first, and the two slides are pressed and sheared together to smear the sample into a thin film capable of passing light. In some cases, no dispersant or solvent is needed to provide sufficient separation of crystals.

Typically, preparation of food samples for observation of crystalline microstructure using optical microscopy involves dispersion in an appropriate solvent in order to prepare a thin section for transmission microscopy. In many food products, the crystals are small and numerous, and they are embedded in the matrix of the food product. For example, the sugar crystals in grained caramel cannot be visualized by simply smearing the product between two microscope slides. An organic solvent is required to dissolve and disperse the caramel matrix so that the sugar crystals can be visualized. Choice of organic solvent depends on the nature of the product, but solvents such as kerosene, butanol, and petroleum

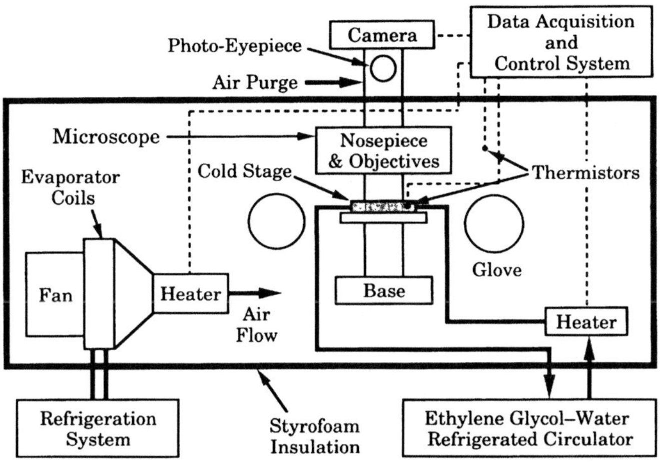

Figure 3–5 Schematic of Refrigerated Glove Box for Studying Ice Crystals in Foods.

ether have been used to study crystals in a variety of foods. Regardless of the solvent chosen, it must be saturated with the solute to ensure that no dissolution occurs during sampling. For example, methanol may be used to disperse sugar crystals in some products. However, sugar is slightly soluble in methanol, and any water in the methanol will dissolve sugar crystals. Thus, the methanol should be saturated with the sugar being studied at the sampling temperature prior to use as solvent. In addition, since solubility is often a function of temperature, both solvent and sample must be maintained at uniform temperature to prevent dissolution or growth during visualization.

Proper choice of solvents, based on refractive index, can sometimes cause certain phases in the product to disappear and others to be enhanced. The principle here is that a refractive difference between the particle and the solution is needed to provide sufficient contrast to see the particle in the light microscope. If a solvent is chosen with the same refractive index as the particle, the particle will disappear in the microscope field. Thus, other structures with different refractive indices can be more clearly distinguished. Stockham and Fochtman[14] provide a table of refractive indices of various solvents useful for microscopic investigation of dust particles. In the food industry, Arbuckle and Cremers[15] discussed the use of different solvents for studying the microstructure of ice cream. They recommended use of acetone (refractive index of 1.357) and ethyl acetate (1.378) for studying fat globules in ice cream since these solvents nearly match the refractive index of ice crystals (1.310). To highlight the ice crystals in the microscope field, a mixture of amyl alcohol and kerosene (1.446) was used to cause the fat globules (1.435–1.456) to disappear.

Generally, it is very difficult to use light microscopy to visualize fat crystals in lipid-based products (ie, chocolate, butter, etc). Fat crystals are typically imbedded in a discrete microstructure (networks), and it is difficult to separate this microstructure without disrupting the interconnections. In some instances, however, fat crystals in foods or during food processing may be observed using optical microscopy, although this is generally limited to relatively dilute suspensions of lipid crystals that can easily be dispersed from one another. Figure 3–6 shows examples of milk fat crystals grown in an agitated suspension and observed under polarized light. The lipid crystals shown in Figure 3–6 were dispersed in mineral oil, although other solvents (other oils) also may be used. The same issues of solubility and temperature control apply in this case as well, and microscopists must always be aware of the stability of their sample during observation. Preliminary tests often are required to verify that a sampling technique is satisfactory for a given system.

A

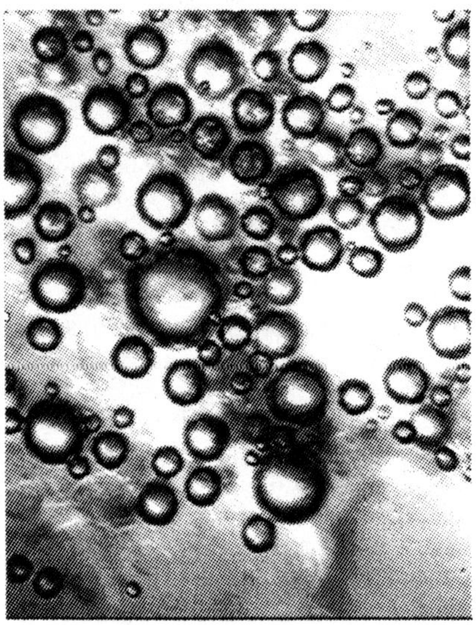

B

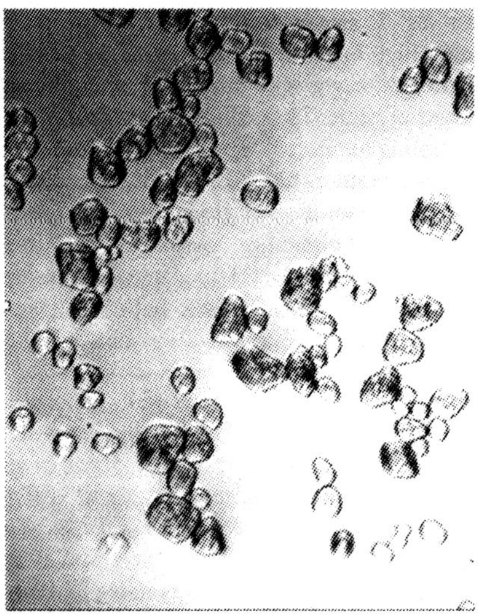

Figure 3–6 Polarized Light Microscope Images of Milk Fat Crystals, Crystallized At 25°C under Agitation (150 rpm) with Different Cooling Rates: (A) Fast Cooling (5.2°C/min) and (B) Slow Cooling (0.2°C/min).

Although microscopy can give useful qualitative information about the crystalline structure, careful preparation of samples may allow quantitative image analysis. However, proper dispersion of crystals is necessary to allow accurate quantification of particle size and shape, particularly when using the modern automated image analysis systems. Crystals must not be in contact with one another, and they must be easily distinguishable from the background. Image analysis techniques will be discussed in more detail in a later section.

Crystallization rates also can be studied with optical microscope techniques. Probably the simplest method is to cool the sample at a constant rate in a controlled temperature cell on a microscope stage. The temperature where nuclei first appear can be characterized as the nucleation temperature. Note that this temperature depends on the rate of cooling, so it is not a fundamental measure of crystallization kinetics. The subsequent growth of nuclei can be characterized by using image analysis techniques. It also is possible to study crystallization under isothermal conditions in a microscope cell. The temperature of the sample must be rapidly cooled to crystallization temperature. The time for first appearance of nuclei is the induction time, and subsequent growth rates of crystals can be evaluated as the change in size with time.

Numerous studies[16–18] have used optical microscopy to study the rates of ice crystal growth and the effects of various components on that growth rate. One often-used technique is to measure the rate of advance of a freezing front. For example, Budiaman and Fennema[16] evaluated the effects of hydrocolloid suspensions on the linear rate of advance of a freezing front. Optical microscopy techniques also have been used to characterize the growth of lipid crystals in the presence of impurities. Smith et al[19] used a similar linear front growth rate to characterize the effects of impurities on growth of trilaurin.

Laser Light Microscopy

CSLM was recently used to evaluate crystalline microstructure in foods.[20,21] In CSLM, fluorescent dyes within the food product are excited by the incident laser light at a certain wavelength and the fluorescence emission spectra of the dye is detected. Appropriate choice of dye allows one to illuminate certain structural elements in a food so that a specific microstructure can be distinguished. Use of laser light excitation and detection of the fluorescence emission allows visualization into the interior of the sample. Thus, CSLM allows observation of internal structure to depths of 30 to 50 μm from the surface, depending on the opacity of the material.

In CSLM, a fluorescing dye must be present to allow visualization of internal microstructure. This dye must be carefully chosen to provide visualization of the desired component. For example, a mixture of Nile blue and Nile red has been found to give the sharpest images of lipid crystals in CSLM in the author's laboratory. However, some experimentation may be necessary to determine the optimal dye (or combination of dyes) for visualization of specific structures. The dye may be added to the sample prior to manufacture or the dye may be diffused into the sample from the surface just prior to visualization. Adding dye to the formulation prior to manufacture results in the best dispersion of the dye at the desired structural interface in the product. However, this is generally limited to experimental investigations. Plus, any effect of the dye on changes in crystalline microstructure (it may act as an impurity and alter crystallization behavior) must be considered. To examine commercial products, it is necessary to apply the dye (dispersed in a suitable solvent or applied as a powder) to the surface of the product. The dye then must diffuse into the matrix and adsorb to the desired structural interface. Diffusion of dye into a product must be done very carefully as changes to structural elements in the product may take place due to the presence of the solvent.

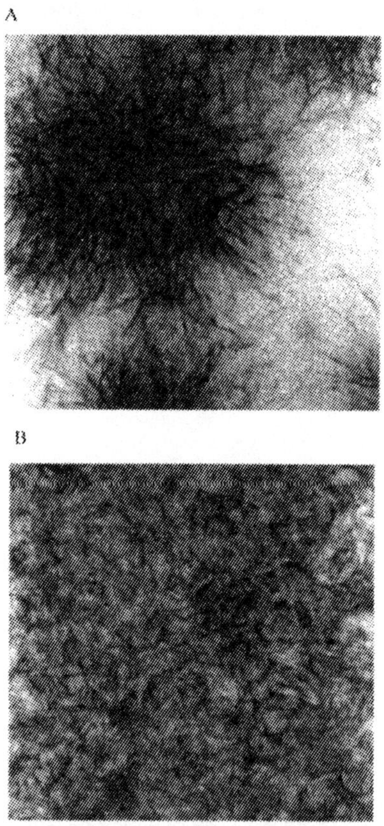

Figure 3–7 Comparison of Crystals Formed during Quiescent Cooling of a Milk Fat Model System (50:50 High-Melting in Low-Melting Milk Fat Fractions). Images taken by confocal scanning laser microscopy for (A) slow cooling at 0.2°C/min and (B) fast cooling at about 3°C/min.

Figure 3–7 shows a CSLM micrograph of milk fat crystals in a sample allowed to cool on a microscope slide. In this case, Nile blue (which contained Nile red at low levels) was incorporated into the liquid fat as it cooled and allowed observation of the lipid microstructure. The effects of cooling rate on lipid crystalline microstructure are apparent. Also, these images clearly show the network structure that can be formed between lipid crystals. This network structure has a significant impact on product texture, and it also makes it extremely difficult to get good quantitative analysis of the true crystalline structure. The use of CSLM now allows quantification of these complex structural aspects without disrupting the sample.

One unique application of CSLM is that three-dimensional images can be obtained. The images taken at different depths into the sample can be put together to provide a composite three-dimensional reconstruction of the microstructure. Modern imaging technology allows this three-dimensional structure to be viewed from all angles, which greatly enhances our understanding of the internal microstructure of materials viewed with the CSLM.

Another recent advance in laser light microscopy is multiple-photon excitation scanning laser microscopy.[10] In this case again, fluorescing elements within the product are excited by a laser light source, but of higher wavelength (eg, 1,050 nm). The fluorescing element may be an auto-fluorescing component of the food or a dye added to enhance resolution. In this technique, two or three photons of higher wavelength (lower energy) are required to induce fluorescence so that the focal plane can be controlled very closely. The fluorescence emission from this plane then can be collected in a photomultiplier tube to provide an image. Depending on the sample, depths of up to 100 μm into the product may be observed. By tailoring the wavelength of the incident light to the fluorescence peak of the dye, either two or three photons are required to excite the dye. This allows added control of the image and provides greater resolution. Marangoni and Hartel[20] recently described this technique for visualization of lipid crystal networks.

Electron Microscopy

For smaller crystals or observation of finer detail within a product, electron microscopy is required.[11] In this case, a beam of electrons impacts the product sample, and very high magnifications (100,000–300,000) with a resolution around 5 nm may be obtained. SEM involves illuminating a sample surface using high-energy electrons. This allows visualization of a three-dimensional structure at a fracture surface of the product. In TEM, electrons are passed through a thin section of the product to visualize the internal structure.

For electron microscopy, preparation techniques that leave images of crystals intact must be developed. This typically involves freezing and drying since most electron microscopes operate under vacuum conditions. Samples for electron microscopy must be dry so that water vapor from the sample does not attenuate the electron beam and reduce visibility. In addition, electrons can build up on nonconducting materials and cause a charging effect when using SEM. Thus, samples for SEM are usually coated with a thin layer of metal (gold or platinum) to provide a conducting surface. Because electron microscopy requires extensive sample preparation, the presence of artifacts, or fictitious images, is a constant concern. Great care must be taken during sample preparation to ensure that representative images are obtained.

In recent years, many food technologists have been using cold-stage SEM (or cryo-SEM). In this case, the sample is frozen rapidly to solidify the structure of the material without introducing detectable ice crystals.[22] The sample is then freeze-fractured to provide a surface for imaging of the internal structure. As long as the sample remains frozen, even within the SEM, there is no need for drying. This technique provides excellent resolution of samples that initially contain high levels of water such as ice cream. Sargent[23] provides a review of the advantages and applications of low-temperature SEM.

An example of the use of SEM for structural analysis in a food product is shown in Figure 3–8, which shows an image obtained by cold-stage SEM of the structural elements in ice cream. Here, the ice crystals can be seen in the perspective of the surrounding matrix elements (air cells, fat globules, etc). This technique allows visualization of the intact structure of the product without any dispersion or dilution. Quantitative analysis of these structural elements also can be performed on these images as long as sufficient numbers of particles can be counted. Thus, the SEM can be used to provide both qualitative and quantitative information about the actual structure of a food product. This is necessary for correlation with physical properties of a food product.

A relatively recent innovation in electron microscopy is the ESEM. In the ESEM, a pressure gradient is used that allows the sample to be maintained at close to atmospheric pressure while maintaining the vacuum needed at the electron source. Thus, sample preparation is much less critical since the sample can contain water.[11] Danilatos[24] summarizes work on ESEM, although most of the applications do not involve foods.

Optical Techniques

Optical particle counters are available for measuring a wide range of crystal sizes, from nanometers to over several millimeters. In these techniques, a light beam is passed through a dispersed sample, and some aspect of the light beam is measured and correlated to the number, size distribution, and shape of the particles. A wide range of systems is available,[2] with measurements generally based on one of the following five mechanisms: (1) obscuration of light as particles pass through the beam; (2) scattering of the light in a forward angle; (3) scattering of light at an angle to the light beam (usually 90°); (4) detecting the phase shift in the light as a particle passes through the beam; or (5) measuring the time taken for a particle to traverse between two light beams. Allen[2] provides a description of many of the commercially available units; Burchard[25] discusses the theories underlying particle size measurement by light-scattering techniques.

Typically, optical systems are used in the batch mode (field scanning), although continuous versions of equipment exist that allow on-line measurement of particle characteristics during processing. In batch operation, the particle sample to be analyzed is dispersed in a carrier fluid that is inert to the particles and that has a significantly different refractive index than the particles to be measured. Light passes through the suspension and the signal is related to the type of scattering. By passing the dispersed slurry through

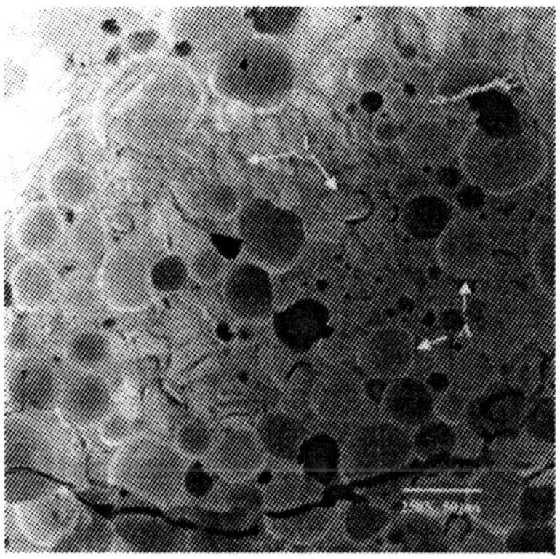

Figure 3–8 Images of Ice Cream Prepared by Freeze-Fracture, Observed Using Cryo-Stage Scanning Electron Microscopy, Showing Air Cells (A) and Ice Crystals (I).

a carefully designed flow cell, the particle characteristics can be measured in a continuous fashion and the information used for process control. The density of the particulate slurry is an important limitation to light-scattering measurements and generally this slurry density must be kept sufficiently low to eliminate multiple scattering events. However, some devices are made to operate even at high (70% solids) slurry densities.

Light-scattering techniques measure particle size distribution by relating the scattered light intensity to particle size through standard scattering theory.[26] Scattered light can be measured at any angle relative to the incident light (see Figure 3–9), although most commonly the forward-scattered light (low-angle light scattering or laser diffraction) is detected. According to scattering theory, the diffraction angle is inversely related to particle size. Thus, the diffraction pattern observed as laser light passes through the particulate sample can be used to quantify a particle size distribution. Modern light-scattering particle analyzers utilize the full, complex, multiple-scattering theories to calculate the particle size distribution from the measured diffraction pattern. They are capable of measuring particle sizes from submicron (0.1 μm) up to several millimeters (2,000 to 3,000 μm), although different techniques are required to cover such a large range in size. Technically, it is necessary to know the optical properties (refractive index and absorption index) of the particles before the size distribution can be calculated. Thus, preliminary measurements may need to be made to obtain the appropriate values for the calculations.

Laser diffraction has many advantages for measuring particle size distribution. These units are simple and easy to use, and they provide a size distribution in minutes. All that is required is a sample that can be adequately dispersed in a solvent or prepared as a dry powder. Also, light-scattering theory applies technically to spherical particles. Thus, an approximation is made for nonspherical particles. Laser diffraction is routinely used, for example, in analysis of particle size in chocolates.[27] Here, an appropriate solvent is used to

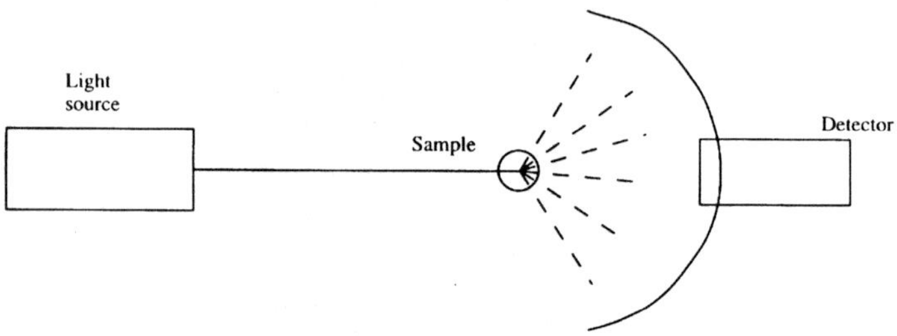

Figure 3–9 General Principle of Light Scattering.

dissolve away the lipid phase of melted chocolate. The slurry is diluted for application to the sample cell in the laser diffraction unit and particle size distribution generated. Jeffery[28] discusses the use of low-angle laser light scattering to determine the size distribution of sugar crystals in fondant. However, in many food systems, separating the particulate material from the food matrix without changing the particle size is not always an easy task. Chawla and deMan[29] used light scattering to measure lipid crystals in shortenings and compared their results against polarized light microscopy and SEM images. Lipid crystals were dispersed in isobutanol for light scattering. Their results suggested that similar sizes were measured from light scattering and microscopy.

Although laser diffraction is probably the most common optical technique, other variations also are available. Light extinction techniques measure the reduction of intensity as light passes directly through a dispersed slurry sample. The reduction in light intensity is related to the projected area and number of particles. The extinction coefficient also is a function of particle size, temperature, and refractive index of the particles. Light blockage techniques essentially measure the shadow of a single particle traversing through a light beam. In this case, the particle slurry flows through a flow cell designed to allow one particle at a time to pass through the beam. The photodiode detector on the other side of the cell measures the intensity of light as particles pass through the cell. The amount of light cut off, or the decrease in intensity of the light beam, is related to the cross-sectional area of the particle. These systems can be used to size particles ranging in size from a few microns up to several millimeters accurately.

Photon correlation spectroscopy (PCS) is another optical technique available for measuring size distributions of very small particles (5 nm to 3 μm). In this case, the frequency spectrum of the intensity changes of the scattered light is detected. These intensity changes occur as light is scattered from moving particles. The particle size distribution is calculated based on the Stokes-Einstein equation of translational diffusion while the particles are moving under Brownian motion. Although PCS has the capability to measure extremely small particles—something often very important in food systems—the system has limitations for use in the food industry due to constraints on sample preparation.

Other techniques for obtaining information about very small particles, in the nanometer range, include small-angle X-ray and small-angle neutron scattering.[30]

Another particle sizing technique utilizes the reflection of a focused beam of laser light to indicate presence and size (chord length) of crystal. In this technique, a probe is placed directly into the system to be characterized. The laser light is focused in the slurry just beyond the window of the probe and the light reflected off of particles is collected in a detector system within the probe. The laser light source moves rapidly around in a circle so that many particles can be counted in a short time as the slurry sample moves past the probe surface. Chord length, or the length of the particle that provides sufficient reflectance of the light, is correlated to the time duration of the reflected pulse and the speed of the laser light source in its circular path. The number of particles is correlated with the number of chord measurements obtained. This detection method provides a measure of the distribution of chord lengths, the size in the measurement direction, for a particle. For a sphere, chord length varies from near zero at the top and bottom to the true diameter at the equator. Irregular-shaped crystals have a different chord length distribution as do particles of different size. Thus, the distribution of chord lengths measured in this technique provides an indirect measure of size distribution. This information can be used directly for comparison purposes, but must be interpreted carefully for the true size distribution.

Optical techniques may be used to study crystallization rates under special circumstances. One of the simplest techniques for studying progress of crystallization is to measure the change in turbidity as light is transmitted through a sample as it crystallizes. An initially transparent, supersaturated solution is placed into a spectrophotometer to measure light transmittance. In some systems, temperature can be controlled by circulating coolant around the spectrophotometer cuvette. Also, some systems allow agitation within the cuvette by a magnetic stirring mechanism. Temperature control and agitation are important to obtain representative crystallization data as is the choice of wavelength for the incident light. A wavelength far away from any absorption peaks of either solvent or solute is necessary for turbidity measurements. Crystallization then can be characterized by bringing the sample to the desired temperature and recording the change in turbidity under isothermal conditions, as shown in Figure 3–10. The time required for the turbidity to change can be characterized as the induction time for onset of massive nucleation since sufficient numbers of particles are necessary to scatter light. The rate of turbidity increase gives a measure of the crystallization rate, although this rate is a combination of nucleation and growth mechanisms since scattering is a function of both number and size of particles. Thus, this turbidity technique does not give a fundamental measure of crystallization rate nor does it measure the true onset of nucleation. However, this technique may be quite useful for rapid and easy comparison of different systems.

Another method to obtain information about crystallization rates using turbidity measurements is to cool the sample through the crystallization zone slowly while constantly monitoring turbidity.[31] In this dynamic turbidity method, the temperature of onset of crystallization is recorded. The slope of the increasing turbidity curve also may be used

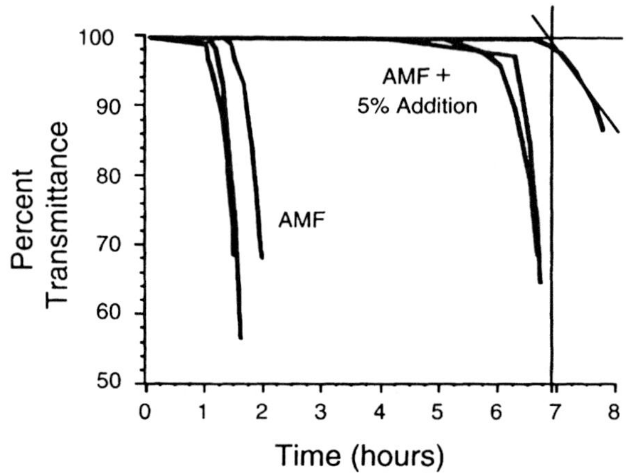

Figure 3–10 Change in Turbidity (% Transmittance) with Time for Crystallization of Anhydrous Milk Fat (AMF) and AMF with Added Emulsifiers.

to characterize a rate of crystallization, but again these are not fundamental measures of crystallization kinetics. The measurements depend on scanning rates.

Although optical techniques are relatively easy and simple to use, proper sampling techniques and careful interpretation of the results are required to obtain meaningful information. In particular, the laser diffraction systems require accurate information on the optical properties of the material being measured. These optical properties often change with temperature. Thus, temperature fluctuations during sampling must be minimized.

Thermal Techniques

Another method to evaluate crystallinity in foods is by using a calorimeter. For example, the presence of crystals in a food may by documented from the melting profile in a calorimeter. Also, crystallization within the food often can be studied by controlled cooling of the sample within the calorimeter. Although many types of calorimeters exist and can be used to study crystallization, the differential scanning calorimeter (DSC) is used most frequently in food analysis. In a DSC, the difference (in either temperature or heat input) between a small sample contained in a DSC pan and a reference pan (usually empty) is used to characterize changes associated with crystal formation or melting. As temperature is increased or decreased in both pans at the same time, melting endotherms and crystallization exotherms associated with phase changes in the material, as well as any reaction or state transition, in the sample pan can be characterized.

The primary advantages of studying crystallization in a DSC are the small sample size required and the high sensitivity. However, because the sample size is so small (1 to 40 mg) and there is no agitation, the results from DSC studies often are difficult to interpret directly in terms of process operating conditions. Thus, DSC is most commonly used (in the melting mode) to verify what type and how much of a certain crystalline material are present in a food. Even this use is somewhat limited since other transitions (ie, gelatinization, moisture loss, chemical reactions, etc) also occur during heating and result in complex thermographs that often are difficult to interpret. Probably the best use of a DSC for studying crystallization is for comparison of effects in relatively simple (or model) systems.

For example, to document the existence of crystals in a grained sugar confection (grained caramel, fudge, etc), a small sample can be placed in a DSC pan and scanned upward in temperature. The melting profile exhibits several transitions, depending on the nature of the sample and the observation conditions. If there is significant water content in the product, sugar dissolves as the temperature increases. This often results in a small peak related to heat of solution as sugar dissolves, although this may not always be observable. If there is a high level of crystalline sugar present in the product, there is a large endothermic transition in the region of 170°C to 190°C, characteristic of sugar (sucrose or lactose) melting. In the case of a caramel that contains both crystalline sucrose and lactose, a complex melting peak in the range of 150°C to 200°C is often obtained. The melting peaks of the sucrose and lactose crystals overlap and are difficult to separate.

The rate of scanning is very important to observation of the desired transitions. In the above example on sugar in caramels, a high scanning rate (10 to 20°C/min) is desired to minimize competition of dissolution of sugar crystals at higher temperatures (since the solubility curve increases with increasing temperature). In other situations, slower (1°C/min) scanning rates are desired to maintain pseudo-equilibrium conditions within the sample. This will be discussed further below in the context of lipid crystallization and melting.

If a sugar glass is heated in a DSC, several transitions can be observed. The first transition is the change in heat capacity associated with the glass transition. That is, the change from glassy to amorphous state can be seen as a change in the baseline (heat capacity). Since this is a low-energy transition, a derivative (first or second) curve is often used to quantify the transition (see Chapter 4 for more details). Upon further heating, the sugar molecules gain sufficient energy to crystallize and an exotherm is observed. The temperature of this exotherm depends on the rate of heating in the DSC and the type of materials being studied. Further heating of the crystallized mass then leads to a sugar melting endotherm at about 180°C, the characteristic melting point of sucrose. These transitions are shown in Figure 3–11. Also shown are the effects of increasing levels of corn syrup on these transitions. In this case, corn syrup increases both the glass transition temperature (although this is difficult to see from this graph) and the crystallization temperature, but reduces the total amount of crystalline material (lower ΔH of melting peak). At corn syrup levels above 50%, crystallization is completely inhibited over the time scale of this experiment.

Analysis of lipid samples in a DSC is quite complex since polymorphic transitions also may occur. When a lipid crystallizes, it typically forms first in a lower-stability

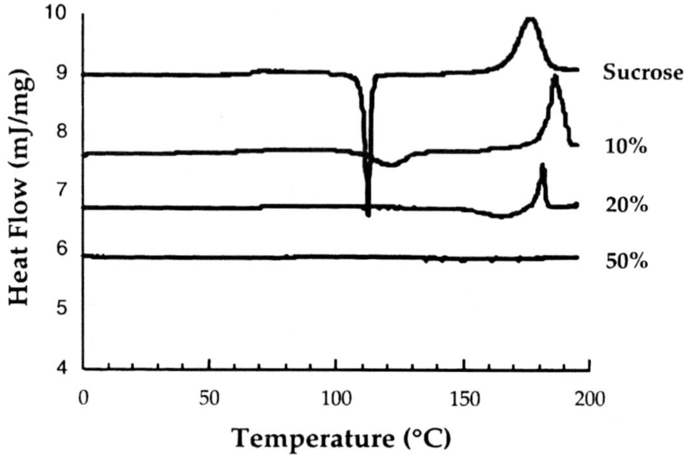

Figure 3–11 Characteristic Differential Scanning Calorimeter Heating Curves for a Sucrose Glass Show the Slight Shift in Baseline for the Glass Transition, a Crystallization Exotherm (Down) and a Melting Endotherm (Up).

polymorph and then transforms to polymorphs with higher stability (see Chapter 2). Both the initial crystallization and the polymorphic transition give rise to energy peaks on the DSC, and these peaks depend greatly on the rate of heating or cooling. In addition, the latent heats of fusion increase with increasing stability of the polymorphic form (from α to β). If a sample is cooled rapidly in the DSC, a relatively small exotherm associated with the α-polymorph may be observed. If this sample is then reheated slowly, the α-polymorph melts (endothermic) and the more stable β′-polymorph forms (exothermic). Upon further heating, the β′-polymorph melts, giving rise to an endothermic melting peak, and the highest stability β-polymorph crystallizes, giving rise to an exothermic peak, as seen in Figure 3–12. Finally, as temperature increases above the melting point, an endothermic β melting peak is observed. Note that verification of polymorph identification must come from an independent measure, such as X-ray diffraction. If the unstable polymorphic form is heated rapidly in the DSC, a completely different pattern may be observed, as seen in Figure 3–12. The shape of the DSC melting profile is dependent on the relative rates of the polymorphic transformations and the rate of heating. If heating is sufficiently rapid, it is possible to bypass crystallization exotherms of more stable polymorphs. Thus, only melting endotherms may be visible.

The exact nature of DSC transitions is often complicated in natural fats by the presence of a variety of glyceride species with diverse melting points. Thus, we often only observe one or two melting endotherms for natural fats. Milk fat, with an extremely

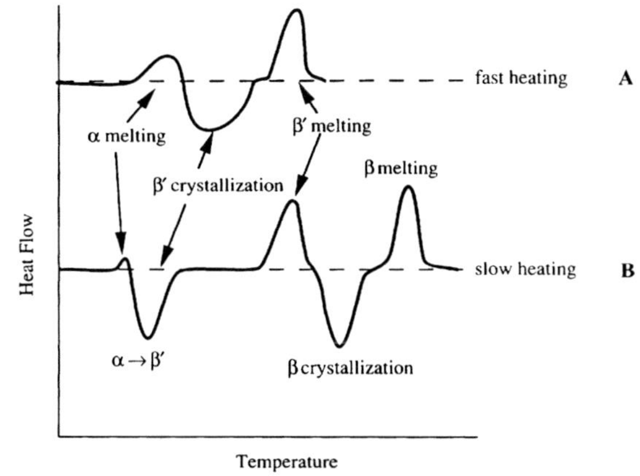

Figure 3–12 Schematic Drawing of Expected Melting Curves for a Lipid in Differential Scanning Calorimeter: (A) α-Polymorph Heated Rapidly to Show Conversion to β′-Polymorph without Formation of β-Polymorph; (B) α-Polymorph Heated Slowly to Show Conversion to β′-Polymorph and Then to β-Polymorph.

wide diversity of molecular species (over 400 different molecular species), has three melting peaks when heated slowly in a DSC (Figure 3–13). These three peaks are broadly characterized as low melting (below about 18°C), middle melting (from about 18°C to 28°C), and high melting (from about 28°C to 40°C); they are generally thought to correspond to the different compositional categories of milk fat.[32] However, Ten Grotenhuis et al[33] suggest that the occurrence of these peaks may, in part, be due to polymorphic transitions from the α to the β′ form. Their conclusions were based on the use of X-ray spectroscopy to identify different crystalline structures associated with the DSC melting peaks. Also shown in Figure 3–13 are the DSC melting curves of both tempered and untempered cocoa butter. Tempering of chocolate, as will be discussed in Chapter 7, involves controlling temperatures so that the stable β-polymorph is formed. A single melting peak is observed. The untempered cocoa butter contains both β′ and β forms. Thus, two melting peaks are observed.

To study crystallization as it occurs in a food or model system, the DSC can be used either in scanning mode or in isothermal mode. In scanning mode, the sample is cooled from melt temperature (above melting point) through the region of interest and transitions are noted. The onset and shape of enthalpic transitions are dependent on the crystallization kinetics of the sample and the rate of scanning. If scanning is very slow, transitions can be quite sharp, with much of the crystallization being completed in a narrow range of temperature. On the other hand, more rapid scanning causes crystallization peaks to be spread out and have less detail. In general, scanning rates are chosen to allow adequate detail for the purpose of the study. Because of the kinetic effects of scanning,

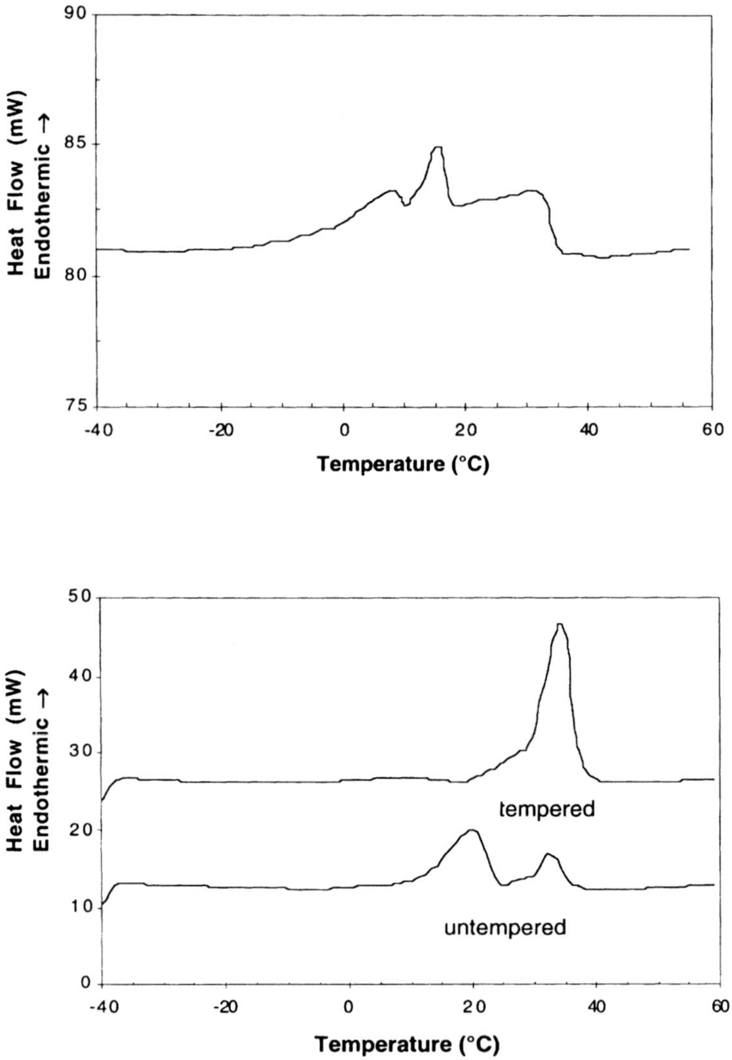

Figure 3–13 Comparison of Differential Scanning Calorimeter Melting Curves for Milk Fat (Top) and Cocoa Butter (Bottom) Both Tempered and Untempered. Scanning rate of 10°C/min for milk fat and 5°C/min for cocoa butter.

the temperature for onset of crystallization is generally dependent on scanning rate. With increasing scanning rates, the system can be subcooled further and so the temperature for onset of crystallization decreases. That is, slow scanning rates generally result in higher temperatures for onset of crystallization. Scanning rate also influences the enthalpy change (ΔH) observed during crystallization. If scanning is slow, there is generally sufficient time for crystallization to be completed and ΔH reaches its maximum value depending on the phase volume of crystalline material. Rapid scanning generally results in lower ΔH observations. Due to these influences of scanning rate on observed crystallization behavior, data from such experiments must be used with caution. Two examples of cooling scans for crystallization of cocoa butter and milk fat are shown in Figure 3–14. Cocoa butter exhibits a relatively sharp crystallization exotherm at about 16°C. This corresponds to formation of either the α- or β′-polymorph, depending on the cooling rate. Confirmation of which polymorph forms requires a second measurement, usually X-ray diffraction, to verify the DSC observation. Cooling of milk fat results in multiple and wider exothermic transitions, compared with cocoa butter (Figure 3–14), although the onset temperature is nearly the same (15°C). The broader nature of the crystallization exotherm is due to the more complex chemical structure of milk fat compared with cocoa butter. In milk fat, high-melting glycerides crystallize first, followed by crystallization of the lower-melting glycerides as the temperature is further reduced.

The DSC also can be used in the isothermal mode to study crystallization. In this case, the sample is cooled quickly and then held at the desired temperature. Once crystallization begins, an exothermic peak is observed, which continues until crystallization is complete (Figure 3–15). The time required for the initial change in enthalpy gives a measure of induction time for onset of nucleation. The energy under this peak is correlated with the total amount of crystalline material formed (based on the total ΔH). However, when these samples are subsequently melted, the melting enthalpy is typically higher than the crystallization enthalpy. This probably occurs due either to incomplete crystallization or to the release of latent heat during the later stages of crystallization that is so slow that it is not registered by the DSC. Nevertheless, comparisons in crystallization rates between systems can be made in this way. For example, Metin and Hartel[34] used this technique to quantify the effects of milk fat fractions on crystallization of cocoa butter and Kawamura[35] studied crystallization of palm oil.

More recently, the capabiiities of the DSC have been expanded to provide additional information about the nature of phase transitions. In modulated or dynamic DSC, temperature scans are accompanied by temperature fluctuations above and below the mean temperature profile. That is, over the normal temperature increase or decrease, an oscillating temperature profile is superimposed (Figure 3–16). Thus, the material experiences thermal fluctuations above and below the mean temperature with the goal that transitions that are reversible and irreversible can be distinguished. For example, when a sample melts due to a temperature increase, it may or may not recrystallize once the temperature decreases during the thermal cycle, depending on the kinetics of crystallization and the nature of the transition. If a metastable form melts, it is unlikely that the same form will recrystallize during the cooling cycle. However, if a stable form melts, recooling will likely

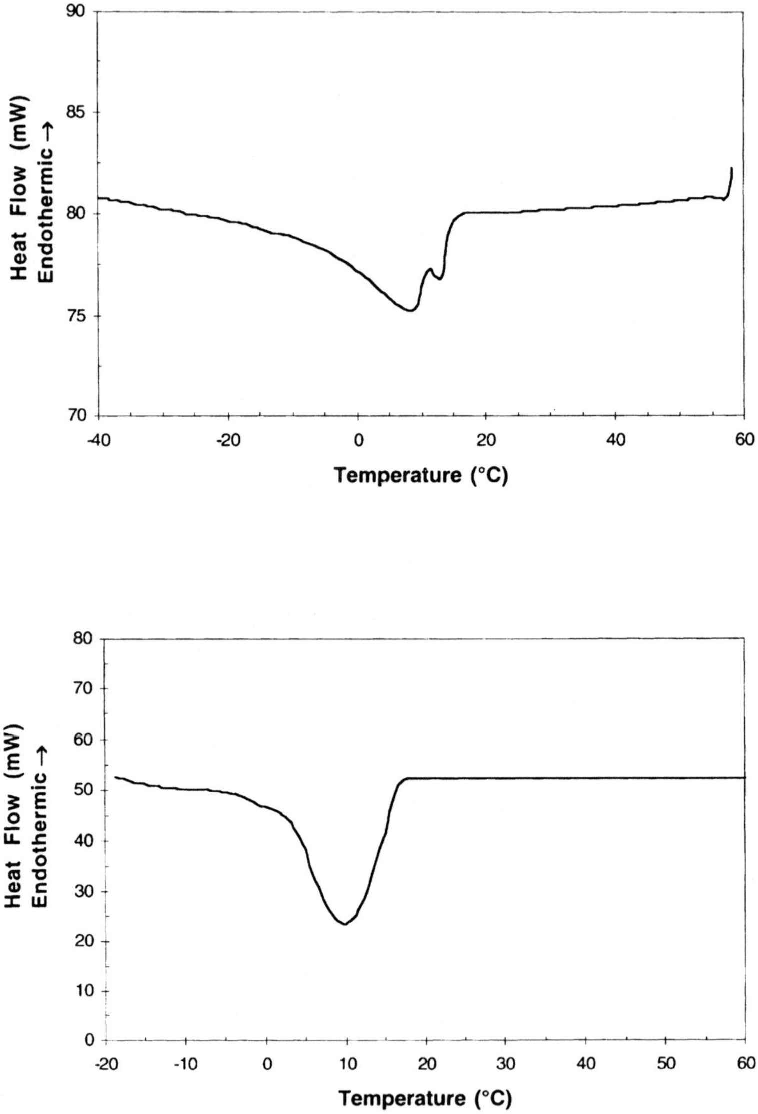

Figure 3–14 Comparison of Differential Scanning Calorimeter Crystallization (Cooling) Curves for Milk Fat (Top) and Cocoa Butter (Bottom). Scanning rate of 10°C/min.

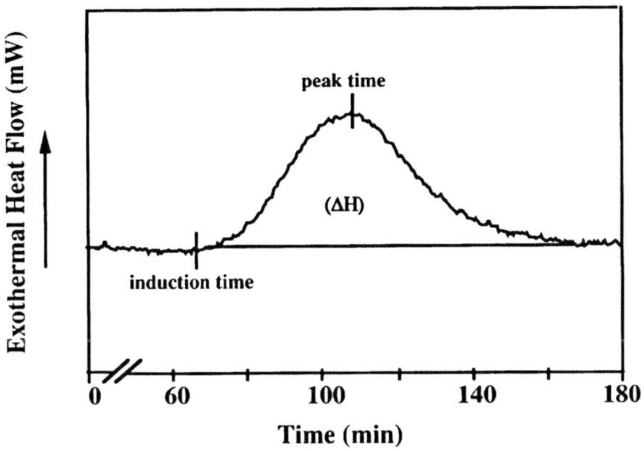

Figure 3–15 Isothermal Crystallization of Cocoa Butter in a Differential Scanning Calorimeter.

result in formation of the same phase, although the kinetics of formation and the rate of heat transfer (cycling rate) determine the nature of the recrystallized phase. Based on these principles, modulated DSC has been used to separate reversible and irreversible thermal events and to improve resolution for transitions that occur over a similar temperature range. For example, this technique has been used to study glass transitions[36,37] and phase transitions (crystallization and polymorphism) in lipids.[38,39]

Another thermal technique for studying phase transitions in foods is the reaction calorimeter. In this device, each of the diverse energy terms (heat removal through the walls of the vessel, heat input due to agitation, exothermic or endothermic energies of transitions, etc) are taken into account to allow more accurate representation of crystallization processes. The primary advantage of this type of analysis over the DSC is that materials can be crystallized while being agitated. Despite the potential advantages of this technique, it has not been used very extensively in the food industry.

Spectroscopy

Several types of spectroscopy techniques are used to characterize the crystalline nature in foods. Foremost of these for crystal analysis is probably X-ray diffraction spectroscopy. However, other spectroscopic techniques also have been used, including nuclear magnetic resonance (NMR), Raman, and infrared (IR) spectroscopy. Typically spectroscopy is used to determine the presence of crystalline material in a product and to identify which crystalline structures (eg, different polymorphs) are present. However, spectroscopy techniques also can be used to determine the onset of crystallization and to follow the rate of crystallization in foods and model systems.

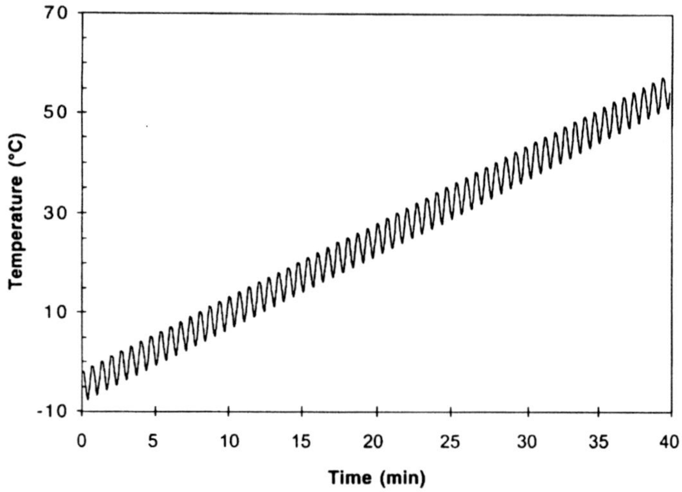

Figure 3–16 Typical Temperature-Time Profile for Modulated or Dynamic Scan in a Differential Scanning Calorimeter.

X-ray Spectroscopy

The ordering of molecules in the crystal lattice gives a unique scattering profile when X-rays are passed through the material. The spectrum of scattered X-rays (intensity of scattered X-rays with angle of observation) allows delineation of the crystalline lattice structure.[40] Physicists have long used X-ray diffraction spectra to characterize and identify molecular structures in the solid state. Recently, advances in computing have allowed physicists to predict X-ray scattering profiles based on conformational arrangement of a molecule in a lattice structure. In contrast to ordered solids, molecules in the liquid state cause only a weak scattering of X-rays and give a very diffuse X-ray spectrum. Glasses also give a diffuse X-ray spectrum. For example, the X-ray spectra of crystalline sucrose is compared with that of the glassy state in Figure 3–17.

X-ray diffraction spectroscopy to analyze the molecular orientation within the lattice is best done using a single crystal. By shining X-rays at a specific feature (face) of a crystal, the orientation of the molecules in the lattice can be determined. However, often it is difficult to grow crystals of sufficiently large size and purity for X-ray spectroscopy. In fact, proteins often are grown in space in an attempt to grow large, undisturbed crystals for structure analysis. In addition, it is sometimes difficult to stabilize a lipid polymorph in an unstable state sufficiently long to obtain a single crystal X-ray spectrum. Thus, powder X-ray spectroscopy often is used, where the scattering spectrum of a crystalline powder of the material is analyzed. This is probably the most common technique used in

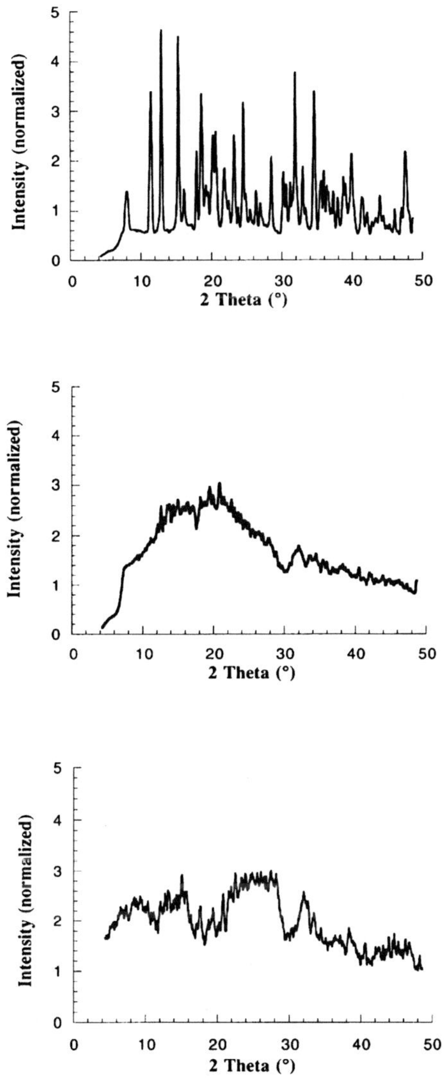

Figure 3–17 Comparison of X-ray Spectra for Crystalline (Top) and Freeze-Dried (Middle) Sucrose. Bottom figure is a 50:50 mixture of freeze-dried sucrose and corn syrup.

the food industry since it allows us to evaluate crystals contained within a food structure. For example, powder X-ray spectroscopy can be used on solid chocolate. The resulting spectrum is a combination of scattering from both the crystalline sucrose and cocoa butter present in the solidified chocolate. This complexity makes it difficult to obtain details about either crystalline structure (sugar or fat).

The presence of crystals in a food is easily distinguished by X-ray spectroscopy. For example, a grained caramel (containing sugar crystals) would easily be distinguished from an ungrained caramel (without sugar crystals) using X-ray spectroscopy. The ungrained caramel would give a diffuse scattering spectrum, whereas the grained caramel would give ordered intensity peaks according to the type and amount of crystals present. Even the size of crystals can be distinguished, to some extent, from X-ray scattering data under ideal conditions, based on the width of the spectral peaks. However, in most foods, the influence of other components limits the resolution of X-ray results. Low levels of crystallinity may be difficult to distinguish from the underlying baseline spectrum of the material. For example, low levels of crystalline sugar in a sugar glass are not detectable until about 5 to 10% of the material is crystallized. Once crystalline peaks are observed, the mass of crystalline material can be estimated from the area under the peaks. However, this must be done with caution since, in many instruments, the incident light intensity may vary slightly.

X-ray spectroscopy also may be used to characterize changes in crystallinity over time in a food. For example, Jouppila and Roos[41,42] studied crystallization of lactose from spray-dried milk powders. As the powders picked up moisture from the air during storage, they converted from the glassy to amorphous state and eventually crystallized. The onset of crystallization was quantified as the time necessary to observe a crystalline spectrum by X-ray spectroscopy. Furthermore, different crystalline forms of lactose, as distinguished by the differences in X-ray spectra, were found to form at different conditions. By quantifying the intensity of the X-ray diffraction peaks, Jouppila and Roos[41,42] were able to correlate the increase in percent crystallinity of this powder sample with the relative humidity of the air in which the powder was stored. In this way, X-ray spectroscopy data can be used to generate kinetic information about rates of crystallization.

X-ray spectroscopy is commonly used to differentiate between lipid polymorphs.[43] In fact, the basic X-ray spectra are used to define the different polymorphs. An α-polymorph (hexagonal crystalline structure) is characterized by a strong short-spacing peak near 4.15 Å, whereas the β′-polymorph (orthorhombic forms) has two short-spacing peaks near 3.80 Å and 4.20 Å (and a doublet in the 720 cm^{-1} region of the infrared absorption spectrum). The β-polymorph (triclinic structure) often is characterized as satisfying neither of the above criteria.[40] The β-polymorph generally has a relatively strong line at 4.6 Å with several weak lines at about 3.8 Å and 5.4 Å. The shape of the X-ray spectra obtained from lipid polymorphs is dependent on the arrangement of the molecules into an ordered lattice structure. From the shape of these spectra, the solid-state physicist can determine the molecular packing within the lattice and the exact orientation of the molecules within the unit cells. Specifically for lipids, the X-ray spectra allow determination of the spacings between molecules in the lattice (see Chapter 2). The long spacings represent the total length of the alkyl chains in the long direction of the molecule, whereas the short spacings represent the distance between adjacent molecules in the ordered lattice.[40]

For example, X-ray spectra have been used to differentiate the two different β-polymorphs in chocolate, as related to the formation of bloom. Figure 3–18 shows the change in the short spacings of the X-ray spectra as the content of the β-polymorph increased in the mixture of pure cocoa butter in the two polymorphic forms. Such changes were used to study the formation of bloom in chocolate during different storage times and with different lipid composition in the chocolate.[44]

The crystallinity of starch also has been documented by X-ray diffraction studies. Zobel[45] summarized the different forms (allomorphs) of starch that give the different X-ray patterns observed in nature. The different patterns (A,B, C, and V) are due either to different molecular structures of the starch or to different complexes that form with the starch molecule. Other references to the crystalline nature of starches as studied by X-ray diffraction can be found in Clark[46] and Buleon et al.[47]

In most X-ray units, obtaining a sample spectra could take anywhere from a few minutes to an hour to achieve sufficient intensity to characterize a sample accurately. Thus, X-ray spectroscopy could not be used for real-time measurement of crystallization events unless a single angle was used. Recently, a higher intensity X-ray source, based

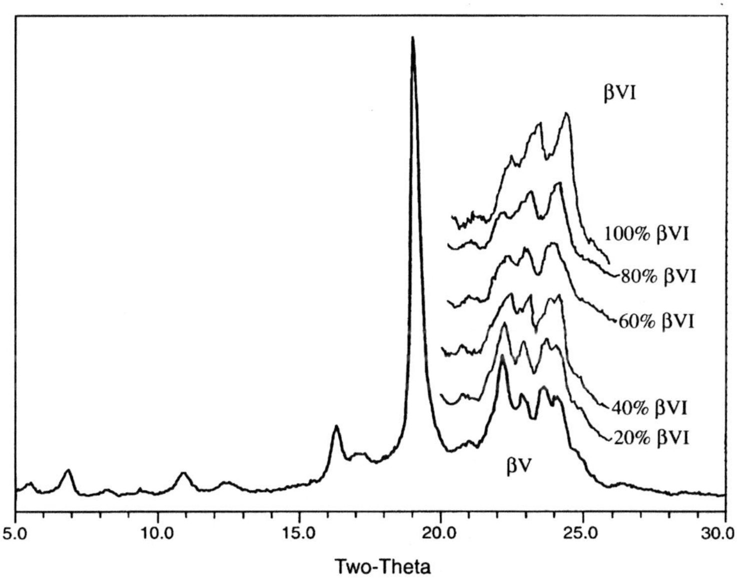

Figure 3–18 Change in X-ray Spectra with Increasing Concentration of the β VI (Most Stable) Form of the β-Polymorph in Cocoa Butter.

on synchrotron radiation, became available for X-ray spectroscopy that allows a full spectra to be obtained in seconds. Several research groups have used this technique to study polymorphism and crystallization of lipids in real time. Minato et al[48] studied the phase behavior of a binary mixture of triacylglycerols. Ueno et al[49] studied liquid crystal formation and polymorphic crystallization of sn-1,3-distearoyl-2-oleoyl glycerol (SOS). Van Malssen et al[50,51] studied crystallization and polymorphism of cocoa butter in a real-time X-ray powder diffractometer.

Nuclear Magnetic Resonance

The absorption of radio-frequency electromagnetic radiation by molecules while in a magnetic field can be used to characterize many aspects of the molecules (type, structure, etc).[52] In this case, the nuclei of the atoms are responsible for absorbing the energy. A related technique, electron spin resonance (ESR) spectroscopy, is similar to NMR but an unpaired electron (free radical) provides the signal.[53] By applying a burst (pulse) of radio-frequency radiation, the molecules are excited into a higher energy spin state. The relaxation, or release of nonradiative energy, upon cessation of the radio-frequency burst as the nuclei revert back to their original spin state is then used to characterize the molecular state. The molecules that are primarily studied in NMR are ^{1}H and ^{13}C (^{19}F and ^{31}P also can be used) because of their normal spin states. Sutcliffe[54] describes the basic principles and applications of NMR and ESR spectroscopy.

The relaxation spectrum from pulsed NMR is generally characterized using two relaxation times (T_1 and T_2). The primary decay of the elevated energy state due to the pulse of radio-frequency radiation is characterized by T_1, often called the spin-lattice relaxation. T_1 provides a measure of the average lifetime of the nuclei in a higher energy state. It depends on the mobility (specifically, rotational correlation times) of the molecules in the matrix under study. For example, in solids and viscous liquids, molecular mobility is low and T_1 is large. Increasing temperature or melting into less viscous liquids causes a reduction in T_1 as the free volume of the molecules expands and the molecules are better able to move around. The average lifetime in the excited state decreases as the molecules become more mobile. The secondary or longer time decay, characterized by a spin-spin relaxation (T_2), involves interactions between neighboring molecules that influence each other's spin states. In crystalline solids and viscous liquids, both T_1 and T_2 are very low, indicating the strong interactions between molecules in these states of limited mobility. Based on these principles of molecular response, NMR can be used to quantify the relative mobility of molecules in a sample.[55] Gidley et al[56] show the difference in high-resolution, solid-state ^{13}C NMR spectra for crystalline and amorphous sucrose. Thus, NMR is an extremely useful tool for studying crystallization and glass transition under certain circumstances.

One widely used application of NMR in food crystallization is to measure the solid fat content (SFC) of a lipid sample. In these instruments, the lipid sample is placed in the holder at the desired temperature, a pulse of magnetic field is applied, and the decay of spin states is used to characterize the percentage of liquid versus solid (crystalline) fat. In the simplest and most easily used versions of these units, the sample is placed into the sample holder and the SFC is displayed directly. Typically, the signals measured at two times after a magnetic field pulse are used to calculate SFC.[57] For example, melting profiles, as characterized by the SFC change with increasing temperature for several different fats, are shown in Figure 3–19. These data were obtained by tempering the fat (to promote maximum amount of crystal phase volume), allowing equilibration at each temperature, and measuring SFC in the pulsed NMR.

Arishima et al[58] used ^{13}C cross-polarization and magic-angle spinning NMR to characterize the polymorphic forms of several triacylglycerols. The distinct molecular motions of the molecules in the different polymorphic crystalline forms were used to identify each polymorph. Identification and further verification of polymorphic forms were performed by X-ray diffraction and DSC.

NMR also can be used to study isothermal crystallization rates of lipid materials in a very simple fashion. A lipid sample is placed in the NMR holder, quickly brought to the desired temperature, and allowed to crystallize. Alternatively, samples from a separate crystallizer can be quickly loaded into an NMR tube and measured for SFC at a single temperature. Temperatures must be carefully controlled to ensure that no melting or further

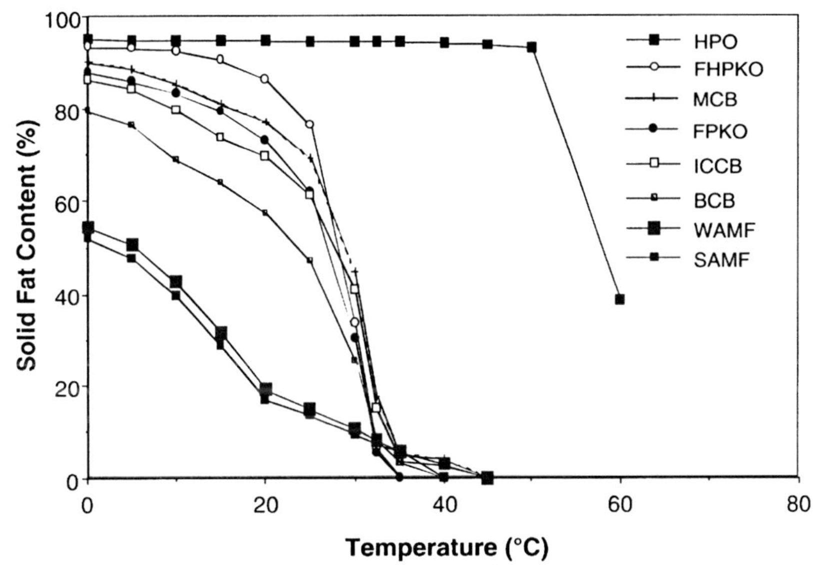

Figure 3–19 Solid Fat Content Melting Profiles Obtained by Pulsed Nuclear Magnetic Resonance for Numerous Fats. HPO, hardened palm oil; FHPKO, fractionated, hydrogenated palm kernel oil; MCB, Malaysian cocoa butter; FPKO, fractionated palm kernel oil; ICCB, Ivory Coast cocoa butter; BCB, Brazilian cocoa butter; WAMF, winter anhydrous milk fat; SAMF, summer anhydrous milk fat.

crystallization can occur as samples are transferred. The rate of crystallization is measured by characterizing SFC in the unit as time progresses. Figure 3–20 shows a typical trace of SFC with time during crystallization of a lipid, in this case a high-melting component of milk fat in a low-melting component.[59] From such data information on induction time required for onset of crystallization can be obtained and, from the slope of the line, a measure of the rate of crystallization can be calculated. However, these data do not provide fundamental information that can be used to differentiate kinetic rates of nucleation and growth.

NMR and ESR also have been used to characterize glass transitions in foods.[60,61] Lloyd et al[61] used a discontinuity in the spin-lattice relaxation time (T_1) at different temperatures to determine the glass transition temperature of a lactose powder. These results correlated well with data for glass transition obtained from DSC and physical property analysis. Hemminga[60] observed a discontinuity in the rotation correlation time for a spin probe in a sugar glass as temperature was increased through the glass transition zone. The temperature where this discontinuity occurred was taken as the glass transition temperature.

Magnetic resonance imaging (MRI) techniques now are available that can be used to characterize phase transitions in foods.[62,63] A relaxation time map from a food product in a magnetic field gradient can be used to document phase changes, as in freezing or thawing of fruits and vegetables or solidifying of lipid foods. In MRI, ice is invisible due to its short relaxation time. Thus, a product that is freezing has a very distinct transition from bright

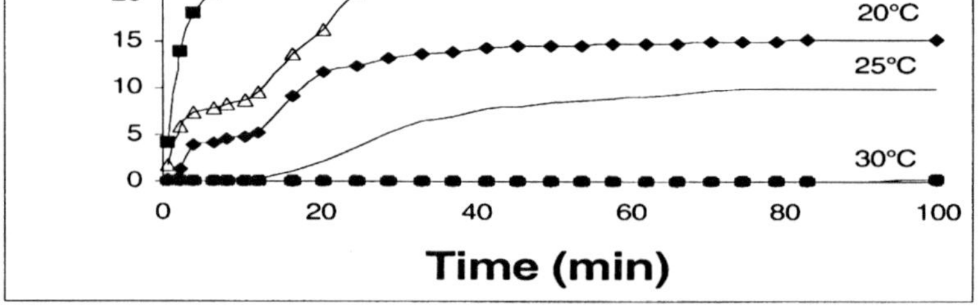

Figure 3–20 Change in Solid Fat Content (SFC) with Time (Obtained by Pulsed Nuclear Magnetic Resonance) for Anhydrous Milk Fat Crystallized at Different Temperatures.

to dark as the freezing front traverses the product. However, the resolution of even the best MRI systems is still insufficient for detailed resolution of phase transitions.

Raman and Infrared Spectroscopy

The vibrational interactions between adjacent atoms of molecules (either intra- or intermolecular) can be distinguished by using either Raman or IR spectroscopy. From these techniques, information on the molecular interactions within a crystal, glass, or fluid can be obtained.[64] Such information includes molecular structures (conformations, functional groups, etc), aggregation states (molecular packing, etc), and intermolecular forces (between neighboring molecules of a lattice).[65]

In Raman spectroscopy, the molecules are excited to a higher vibrational energy state by application of a laser light source; the radiation emitted as the molecule decays to a lower energy state is studied. The emitted radiation is shifted in wavelength from the incident radiation (the Stokes lines). The characteristic wave number shifts due to different molecular vibrations depend on the nature of the molecules and their interaction with their neighbors. Thus, differences in Raman spectra can be used to distinguish between the crystalline and liquid states,[66] as well as between different polymorphic crystalline forms.[65] For example, Raman spectroscopy has been used to provide evidence for clustering of molecules in a supersaturated liquid.[67]

IR spectroscopy, another technique that characterizes vibrational mobility of molecules, also can be used to study molecular structure. In this case, the complex absorption spectrum of the molecules to IR radiation (either mid-IR or near-IR) is characterized to provide information about the vibrational energies of the atomic interactions of the molecules. The vibrational spectra obtained by IR techniques are quite sensitive to changes in molecular conformation and intermolecular interactions. Recent advances in IR technology, such as Fourier-transform (FT) IR and its method of application (diffuse reflectance, attenuated total reflection, etc), make IR technology useful for characterizing molecular orientation and interactions within supersaturated solutions and crystals.

The use of an IR microscope allows analysis of the IR spectra from a domain size less than 100 μm^2 in size and this allows distinction of different phases or states within a food.[68] Wilson et al[69] used a mid-IR technique to follow changes in amylopectin during staling of bread. A line width reduction of one of the peaks in the IR spectrum was related to a decrease in the molecular mobility of amylopectin molecules.

One important application of Raman and IR spectroscopy techniques to the study of crystallization has been the identification of lipid polymorphs.[65] For example, Yano[70] evaluated the molecular structures of the different polymorphs for several triacylglycerols by using IR spectroscopy. Minato et al[71] used FT-IR to study microscopic structures and conformations of binary mixtures of triacylglycerols.

Other Techniques

There are numerous other techniques that have been used to study the crystalline nature of foods. A few of these are briefly discussed here.

Ultrasonic Velocity

Since crystalline material attenuates ultrasonic waves differently than the liquid phase, it is possible to use ultrasonic velocity measurements to monitor crystallization processes.[72] In this technique, a pulse of ultrasonic energy at low power (below cavitation threshold) is applied to the sample and the echo is recorded. When crystallization of a sample occurs, there is a significant change in ultrasonic velocity, as measured by the time observed between pulse and echo. This difference can be used to quantify the onset of freezing or melting. For example, Povey[72] shows the use of the pulse-echo ultrasonic velocity technique to study crystallization of lipids in emulsions. Also, Hodate et al[73] used ultrasonic velocity measurements to characterize crystallization rates of palm oil emulsions.

One drawback to this technique is that the scattered ultrasonic wave is influenced strongly by particle size. That is, as particle size changes, the ultrasonic velocity recorded by the pulse-echo technique also changes. Thus, this technique is somewhat limited to onset of crystallization and melting. In addition, ultrasonic energy is known to increase temperature and pressure within a sample, factors that could alter the crystallization behavior of a system. The use of low-energy ultrasonic waves minimizes this effect; however, one should use caution when interpreting data from this technique. As always, confirmation of results should be provided by using two or more techniques.

Dielectric Measurements

As a glassy material is heated and undergoes a transition to a more fluid-like state, the dielectric constant increases due to the increase in molecular mobility.[74] The dielectric constant can be measured by applying an alternating electric field to the food as temperature is scanned. A response similar to measurements of mechanical properties is obtained. The change in dielectric constant as temperature is increased has been used to study, for example, the glass transition region of amorphous sugars.[75]

Rheologic Measurements

An indirect measure of the crystalline nature of foods often can be obtained from rheological measurements. When a fluid crystallizes, the dispersed solid phase causes an increase in viscosity of the solution,[76] which is related to the amount of crystalline phase and the nature of the crystals (ie, size, distribution, shape, etc). Under carefully controlled conditions, this viscosity increase can be used to characterize the production of crystals in a process. For example, Loisel et al[77] recently measured the torque on a stirring motor as an indirect measure of fluid viscosity during crystallization of cocoa butter in chocolate in a specially designed crystallizer. This torque was used to approximate the amount of crystalline cocoa butter formed during tempering. Note that both solution and dispersed phase affect the viscosity measurement in a slurry. When particulates are present in a fluid, the apparent viscosity generally increases with an increase in dispersed phase volume. However, if the particles interact (aggregation, etc), viscosity is no longer a simple function of particle concentration. Thus, rheological measures of crystallization must be used with caution.

Crystallization of sucrose in high-concentration systems was studied by evaluating the changing torque on a stirrer.[78] As crystallization occurred, two competing phenomena influenced the apparent viscosity (and torque on the stirrer). As the supersaturation was relieved during massive crystallization, the viscosity of the continuous phase decreased. At the same time, the apparent viscosity increased due to the formation of the dispersed crystalline phase. These competing effects resulted in first a sharp decrease in torque (and apparent viscosity), due to the decrease in solution viscosity, followed by a dramatic increase due to the presence of many small crystals. At the point of highest stirrer torque, there were many small crystals present, and the mass had a "plastic" nature. Following this brief period of nucleation, the torque on the stirrer decreased to nearly its original value as the solution desupersaturated. This study again documents that correlation of viscosity changes to quantify crystallization must be done with care.

Computer Modeling

Recent advances in computer technology have allowed substantial improvements in computer modeling of crystalline structure based on intermolecular interactions.[79] These programs incorporate the molecular interactions between atoms, molecules, and side chains of molecules to allow prediction of the lattice structure. By minimizing the lattice energy between adjacent molecules, the most likely lattice structures and their subsequent X-ray scattering spectra can be predicted. Combining computer modeling with experimental X-ray diffraction measurements has allowed substantial improvements in our understanding and prediction of how molecules organize together in the crystal lattice.[80] These programs also allow prediction of polymorphic behavior, although their use in the food industry is still in its infancy.

Combined Techniques

The above techniques can be used to provide useful information about crystallinity in foods and crystallization rates. However, application of a single technique to study crystallization kinetics in commercially important systems leads to potential misunderstandings since each technique has limitations. In general, it is necessary to apply several techniques to ensure a complete picture of crystallization. Sato et al[81] studied the polymorphic behavior of mixtures of fatty acids by combined X-ray diffraction, DSC, optical microscopy, and Raman scattering. Lammers et al[82] measured the particle size distribution of milk powders with several different techniques (sieving, sedimentation, electrozone sensing, light scattering, and photomicroscopy) and found that each technique gave different results. Thus, it is important to understand fully the differences between techniques.

For example, phase behavior and kinetics of crystallization can be studied with systems that utilize several different measurements. A reaction calorimeter can be used to provide agitation and temperature control while monitoring thermal events during crystallization. The material is cooled according to a temperature profile of interest, and the onset and progress of crystallization are followed by monitoring the enthalpic changes associated with phase transition. The thermal measurements can be backed up, for example, by use of turbidity measurement, a light-scattering particle size measurement, and a submersible microscope probe for visualization of crystalline product. These techniques provide different information about the particulate material and can be used to better interpret the calorimeter data. In this system, the effects of various operating conditions and product formulations can be studied to provide useful information to better control commercial operations.

CRYSTAL SIZE

For some simple shapes (spherical or cubic particle), the definition of crystal size is straightforward since the diameter or edge length is a unique measure of dimension. However, most crystals do not form as spheres or cubes. As discussed in Chapter 2, crystal shape can range from plate-like to needle-like with an assortment of shapes in between. Definition of a single crystal size becomes more difficult as shape changes. Furthermore, each of the size measurement techniques described above measures a certain characteristic of the crystal (weight, area, etc) and associates a size with that characteristic. However, the size that determines sedimentation rate is not the same dimension of importance in light scattering or for electrozone sensing. Thus, the choice of crystal size depends on the measurement system and the particular geometry of the crystal. A disk-shaped crystal sized by image analysis gives a different result and requires a different typical dimension than a needle-shaped crystal measured by Coulter Counter. If two techniques are used to measure the size of the same particles, the results generally will be quite different since the measurement dimension and technique will be different.[82] Caution must be used when interpreting such data. Table 3–1 presents definitions of some commonly used sizes.

In sieve analysis, the characteristic size is related to the narrowest dimension of the crystal. As the sieves are shaken and the particles orient themselves at the sieve openings, it is this narrowest dimension that determines whether or not the particle will pass through the sieve mesh. For example, sieving spherical particles will give different results than sieving needle-shaped crystals.

Electrozone sensing (Coulter Counter) provides an equivalent spherical diameter since the particle displacement or volume is measured. From the measured volume of the particle, an equivalent spherical diameter is calculated from the equation for volume of a sphere. Again, spherical particles give different results than needle-shaped crystals, and shape of the particle is important to interpretation of the results from electrozone sensing. It often is necessary to use microscopy to obtain a shape factor for proper interpretation of size data from the electrozone sensing technique.

Optical techniques provide different measures of size depending on the technique used. For example, laser diffraction techniques provide a measure of the volume of the particles as light is diffracted. However, light obscuration techniques provide a measure of the projected area of the particles.

Image Analysis

Image analysis of photomicrographs (whether optical or electron) results in a projected area, and the equivalent circular diameter (or perimeter diameter) often is used.

Table 3–1 Definitions of Commonly Used Particle Size Measures

Name	*Definition*
Volume	Diameter of a sphere having the same volume as the particle
Surface	Diameter of a sphere having the same surface area as the particle
Surface/volume	Diameter of a sphere having the same external surface to volume ratio as the particle
Drag	Diameter of a sphere having the same resistance to motion as the particle (moving at the same velocity in a fluid of the same viscosity)
Stokes	Free-falling diameter in the laminar flow regimen (from Stokes' law, Equation 3.1)
Projected area	Diameter of a circle having the same projected area as the particle
Perimeter	Diameter of a circle having the same perimeter as the projected outline of the particle
Sieve	Width of the minimum square aperture through which the particle will pass
Longest chord	Length of the longest aspect of the particle
Shortest chord	Length of the shortest aspect of the particle
Feret diameter	Distance between pairs of parallel tangents to the projected outline of the particle in some fixed direction
Martin diameter	Chord length parallel to some fixed direction that bisects the area of the particle

Image analysis also allows observation of more details of the crystal habit and thus allows more detailed analyses of crystal size. Some characteristic dimensions that may be obtained through image analysis include:

- Martin's diameter: Length of a line that bisects the particle in a fixed direction
- Feret's diameter: Distance between two tangents on opposite sides of the particle, parallel to a fixed direction
- Longest dimension: Maximum length in any direction
- Minimum dimension: Minimum length in any direction

These dimensions are shown in Figure 3–21. Image analysis also allows one to characterize the shape factor of crystals. Quantifying the ratio of projected area to perimeter, for example, gives a measure of the circularity of the particle.

Choice of particle dimension by image analysis depends on the nature of the crystals being studied and perhaps on how that size will be related to some attribute or functionality of the food. In some cases, the longest dimension may be the most important since a crystal may be detected in the mouth according to this characteristic. However, the choice of characteristic size often is not that critical as long as one understands that each measurement technique has its own inherent biases and uses the size data appropriately.

Crystal images using microscopy can be sized either manually or automatically. Manual sizing requires either tracing each crystal perimeter, matching a crystal with a circle of equivalent diameter, or measuring some other standard dimension of the crystal. Semi-automated systems involve manual crystal sizing coupled with automatic data entry into a computer. For example, some aspect of crystal size (perimeter, chord length, etc) can be manually characterized from either a photograph or a direct image from a microscope

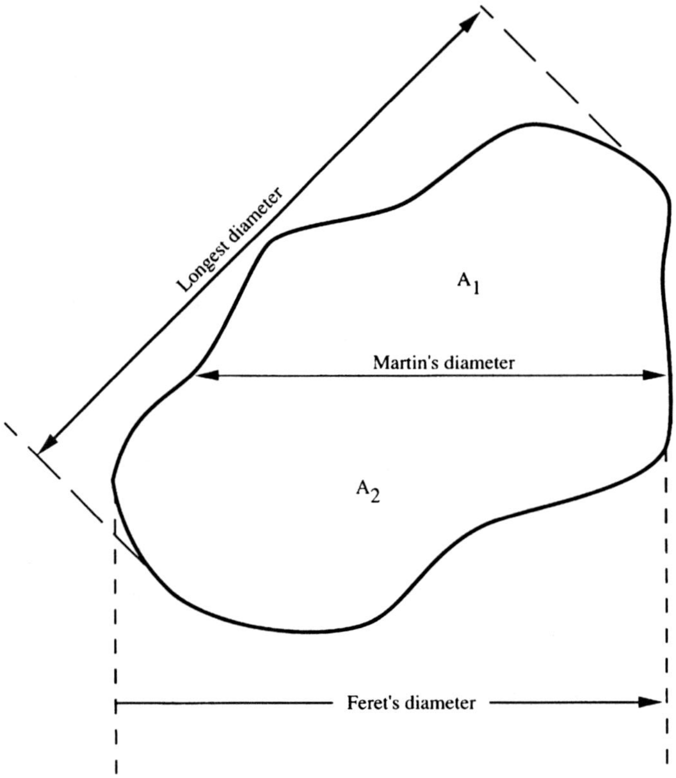

Figure 3–21 Characteristic Dimensions of Particles from Image Analysis (Note That $A_1 = A_2$).

on a digitizing board connected to a computer that logs each data point. However, caution is needed when developing the guidelines used to define crystal size (eg, whether adjacent particles are considered separate or fused). A set of guidelines must be chosen so that the operator consistently analyzes a distribution of particles in the same way. Operator-to-operator differences also should be minimized by proper choice of these guidelines. Although some subjectiveness may arise in the choice of guidelines, consistent application of the guidelines will allow comparison of differences in size distributions.

Automated systems are available that measure particle size and number by using differences in contrast between particles and the background. These systems generally analyze each pixel of an image on a gray scale (typically 256 shades of gray from white to black) and then allow the operator to set a threshold value that gives good distinction between particle and background. The particle size is then measured according to the areas of different contrast. Automated image analysis systems can produce characteristic size distributions in a matter of minutes. However, automated systems generally suffer from an inability to distinguish particles that are overlapping. It is relatively easy for the eye to distinguish whether two adjacent particles are just in contact or whether they have actually started to bridge to form a single particle. In many food applications, it is very difficult to obtain a microscope image of the particles where each individual particle is completely separated from its neighbors. Thus, the operator must either edit these adjacent particles manually, according to some preset guidelines (eg, if less than 25% of the area of two particles is in contact, consider the particles to be two separate entities), or program the computer to look for these interactions automatically. The human eye is still generally better than the computer at distinguishing entities in contact, and some manual editing generally is required during automated image analysis. Nevertheless, commercially available software and computers are getting significantly better and less expensive.

Although sample collection and preparation are important for any sizing technique, the use of image analysis from micrographs requires special attention throughout the process. Collection of samples for analysis must be done so that an accurate representation of the particle structure is obtained. For analysis of particles within foods (such as caramel, ice cream, etc), this requires the operator to obtain a representative sample from the product and then prepare it in such a way that the particles do not change with time. For example, samples of frozen foods or ice cream must be obtained so that ice crystals do not melt or grow during sampling. If a solvent is used to disperse the particles, the solvent must be inert to the particles being measured. In addition, the sampling technique itself might determine the aspect of the particles being observed and introduce biases into the results. The smearing technique described above for ice cream causes the ice crystals to orient in their most stable dimension on the microscope slide. That is, they sit with the largest surfaces in contact with the slide, so a somewhat biased aspect (the largest side) of the ice crystal is viewed in this way. In contrast, preparation of ice cream samples for SEM allows visualization of ice crystals in their random, three-dimensional arrangement in the actual product. The choice of sampling technique often is based on convenience and the desired information.

Once a sample has been prepared for microscopy, images must be taken to provide a representative count of particles in all size ranges. If particle size covers a wide range, from submicron to over 100 μm, for example, it may be necessary to use different magnifications to ensure statistical counting of all particle sizes. The fields of view for image analysis also must be chosen to ensure a representative particle size distribution is measured. After a cursory glance at the entire sample, the operator should randomly choose a field of view for the initial analysis. Images then are taken as the sample is traversed in both directions (x and y directions on the microscope stage). Sufficient numbers of images must be taken so that a significant number of particles can be counted. In general, it is necessary to take several different fields of view so that a count of several hundred particles can be obtained. To be safe, statistical analyses should be done to ensure that sufficient numbers of particles have been counted and that the same distribution can be measured each time the analysis is done on the same sample.

Shape Factors

Shape factors often are used to convert between characteristic dimensions. For example, volume displacement measurements, such as the size determined by Coulter Counter, allow calculation of an equivalent spherical diameter. A shape factor, k_v, relates the volume measure, V, and the associated diameter, L, as

$$V = k_v L^3 \tag{3.2}$$

Since the Coulter Counter correlates an equivalent spherical diameter, the shape factor is defined by the volume equation for a sphere and $k_v = \pi/6$. When projected areas, A, are converted to size, an area shape factor must be used as defined by

$$A = k_a L^2 \tag{3.3}$$

Shape factors, both area and volume, depend on the crystal habit and the method of size measurement. Shape factors determined from sieving irregular particles generally fall between 0.5 and 0.7.

Another parameter often used to characterize shape of particles is called sphericity. *Sphericity* is the ratio of the surface area of a sphere with the same volume as the particle to the actual surface area of the particle. A sphere would have a sphericity of 1, and an irregular-shaped particle would have a sphericity of less than 1. A corresponding parameter for two-dimensional images is called circularity. This is defined as

$$\text{Circularity} = \frac{(\text{Perimeter of particle})}{4\pi\,(\text{Projected area of particle})} \tag{3.4}$$

A change in shape of particles over time can be characterized by their change in sphericity or circularity, measures that are often found in automated image analysis systems. Aguilera and Stanley[6] provide a review of the application of image analysis to food microstructure measurements.

SIZE DISTRIBUTIONS

The CSD can be represented in several ways, depending on how the information will be used. The raw size distribution data obtained often are the number of crystals,

N, contained within a range of size (ΔL), area, volume, or mass. Alternatively, this size distribution may be given as the area, volume, or mass of particles contained within an interval of area, volume, or mass. These data can be plotted according to the individual parameter (size, area, mass, etc) in each size increment (differential) or as a continuous sum of the parameter with increasing size (cumulative). These distributions may be plotted as number versus size, area versus size, or volume (mass) versus size.

Note, the choice of number- or volume-based analysis often comes down to the end use of the data. Use of volume- (or mass-) based distributions gives more weight to the larger particles, since one large crystal contains more mass than hundreds of small crystals. Conversely, use of number-based distributions gives more weight to the small particles. A few large crystals may have a significant sensory impact; thus, a weight distribution provides the best information. Conversely, smaller particles with high surface area may have more impact on viscoelastic properties (viscosity and yield values, for example); in this case, a number (or population) distribution gives the desired information. Beckett[27] suggests that the surface area distribution also is important for particles in chocolate, since each sugar particle needs to be coated with cocoa butter; thus, the surface area of particles is the main influence on chocolate viscosity.

Differential distributions relate the number of particles (N_i) in each size increment (L_i) to size (N_i versus L_i). It is more common, though, to plot the fraction of the total number of particles in each size increment (N_i/N_t versus L_i), as shown in Figure 3–22. Normalizing the size distribution by dividing the number count in each increment by the total number of

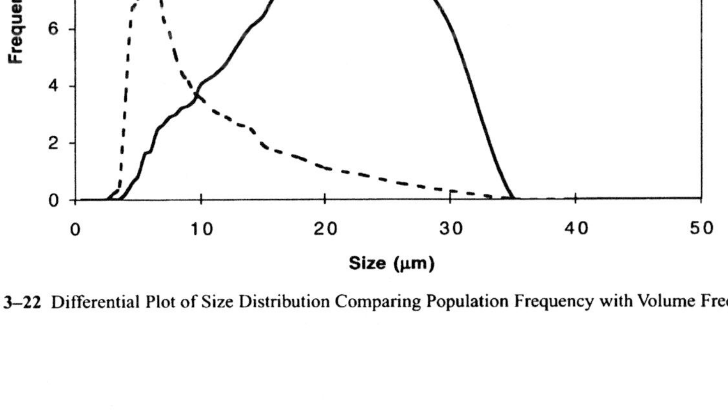

Figure 3–22 Differential Plot of Size Distribution Comparing Population Frequency with Volume Frequency.

particles (N_t) counted allows different distributions to be compared. For differential plots, the number frequency is plotted at the center point of each size increment. Differential size distributions also may be plotted in histogram form, where bars are used to denote the magnitude of the particle count in each channel. Differential distributions also may be presented to show either the area or volume (mass) distribution of particles, depending on the size measurement technique. Figure 3–22 compares how the distribution of the same particles shifts to larger size when plotted as either a population or volume measurement.

When there are several distributions being compared, it often gets too confusing to compare more than a few differential distributions on one graph. In this case, it is best to plot cumulative distributions, where the total number of crystals below a certain size (ΣN_i) are plotted in each size increment (L_i). In this case, the summed value of all crystals counted below a certain size range is plotted at the starting size of each size increment (see Figure 3–23). Again, it is best to plot the cumulative fraction of particles so that the y axis extends from 0 (meaning no crystals) to 1 (meaning all the crystals counted). The primary advantage of these plots is that it is quite easy to determine the number of particles that are less than a certain size. For example, a threshold detection size may be set for quality concerns in a product. If most of the particles remain below this size, then the product is acceptable. A cumulative distribution plot easily allows one to visualize when the particles are too large to meet this requirement. Cumulative plots also may be put on an area or mass (or volume) basis, depending on the measurement technique and the desired information.

One additional method of plotting CSDs is the density distribution. Here, the particle count is normalized by dividing by the size increment ($N_t/\Delta L_t$). When increments of varying size are used, it brings the data closer together for graphical purposes. In many cases, it is a good idea to use continuously increasing size increments for size measurements. That is, at smaller crystal sizes, smaller size increments are used because there are many crystals. At larger sizes, however, there are often fewer crystals. Thus, larger size increments are used to obtain statistically significant numbers of counts in the larger size ranges. This distribution has not been applied to food analysis.

DISTRIBUTION STATISTICS

Sometimes size distribution data are fit to distribution models that can be characterized by a few simple parameters. This allows mathematical manipulation of the CSD for modeling and control purposes, for example. To define these distributions, the following terminology is used.[83] The normalized density function $f(L)$ is

$$\int_{L_{min}}^{L_{max}} f(L)dL = 1 \tag{3.5}$$

The fraction of the distribution found in the size range from L to $L + dL$ is dF [$= f(L)\, dL$]. The cumulative function $F(L)$ is given by

A

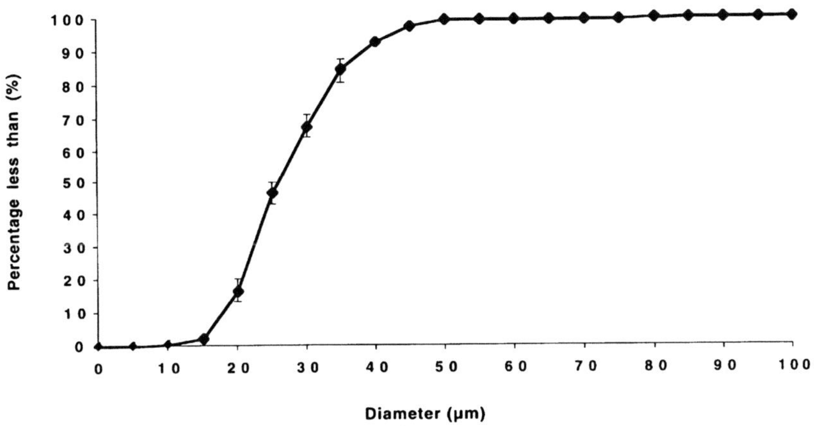

B

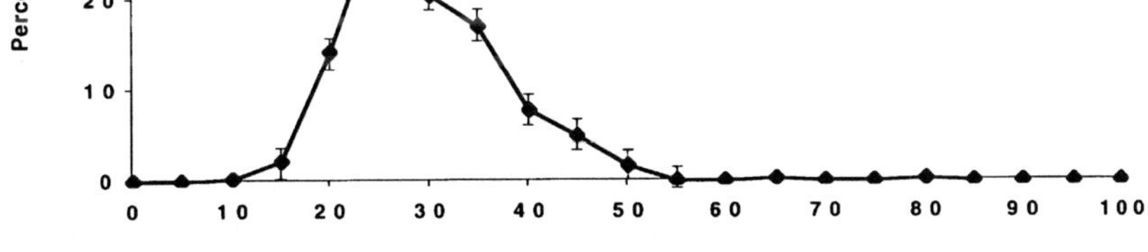

Figure 3–23 Size Distributions of Ice Crystals in Ice Cream: (A) Cumulative Distribution and (B) Differential Distribution.

$$F(L) = \int_{L_{min}}^{L} f(L)dL \tag{3.6}$$

The mean of the distribution is given as

$$\bar{L} = \int_{L_{min}}^{L_{max}} Lf(L)dL \tag{3.7}$$

The variance, σ^2, is given as

$$\sigma^2 = \int_{L_{min}}^{L_{max}} (L-\bar{L})^2 f(L)dL \tag{3.8}$$

The coefficient of variation of the distribution, CV, is given as

$$CV = \frac{\sigma}{\bar{L}} \tag{3.9}$$

The range of sizes, L_{min} to L_{max}, is the range of the independent variable over which the distribution is defined and normalized. For any experimental distribution, these values may be calculated using the above equations.[83]

Another mathematical approach to characterizing size distributions is through the moments of the size distribution. These are defined as

$$m_j = \int_{L_{max}}^{L_{max}} L^j f(L)dL \tag{3.10}$$

where j can range from 0 to infinity. Generally, only the first few moments (m_0 to m_4) of a distribution are calculated. These moments can be used to characterize the physical properties of the crystalline material represented by the size distributions.[83] The zeroeth moment is related to the total number of particles present, whereas the third moment is related to the total volume or mass of crystals. The first and second moments are related to the total length and total surface area of the crystal population, respectively. Moments often are used for further analysis of crystallization systems. For example, when a mass balance on the total amount of crystalline material present is needed, the third moment is used. In a suspension of crystals in solution, the total suspension density (kg solids/m^3 solids-free solution) can be given as:[83]

$$M_T = \rho k_v m_3 \tag{3.11}$$

where M_T is the suspension density, ρ is crystal density, and k_v is volume shape factor.

The arithmetic mean defined above is only one method for defining an average size of the distribution. Other definitions are sometimes used depending on the nature of the distribution and the intended use. The geometric mean is calculated as

$$\log(\bar{L}_g) = \int_{L_{min}}^{L_{max}} \log(L) f(L) dL \tag{3.12}$$

The surface mean diameter is

$$\bar{L}_s = \left(\int_{L_{min}}^{L_{max}} L^2 f(L) dL \right)^{1/2} = \bar{L}_{2,0} = \left(\frac{m_2}{m_0} \right)^{1/2} \tag{3.13}$$

The volume mean diameter is

$$\bar{L}_v = \left(\int_{L_{min}}^{L_{max}} L^3 f(L) dL \right)^{1/3} = \bar{L}_{3,0} = \left(\frac{m_3}{m_0} \right)^{1/3} \tag{3.14}$$

The volume-surface mean (or Sauter mean) is

$$\bar{L}_{vs} = \frac{\int_{L_{min}}^{L_{max}} L^3 f(L) dL}{\int_{L_{min}}^{L_{max}} L^2 f(L) dL} = \bar{L}_{3,2} = \left(\frac{m_3}{m_2} \right) \tag{3.15}$$

The weight mean diameter (DeBroucker mean) is

$$\bar{L}_w = \frac{\int_{L_{min}}^{L_{max}} L^4 f(L) dL}{\int_{L_{min}}^{L_{max}} L^3 f(L) dL} = \bar{L}_{4,3} = \left(\frac{m_4}{m_3}\right) \tag{3.16}$$

The choice of the appropriate mean diameter to use to represent size distribution data depends on its intended use. Sometimes the number distribution of crystal size is most pertinent and $L_{1,0}$ is used. This may be because the number of small crystals has a significant impact on product quality or process parameters. In chocolate, it is the small particles that most influence the amount of lecithin that must be added to cause the desired drop in viscosity, and a population-based mean size is appropriate. However, at other times, the mass-based distribution is most pertinent, and $L_{4,3}$ is often used. Sensory smoothness may best be related to particle size by using $L_{4,3}$ since it is often the few larger crystals present that most influence sensory coarseness.

The above statistical expressions for characterizing size distributions are applicable regardless of the shape or nature of the distribution. That is, no assumptions are necessary about the nature of the distribution in order to calculate mean size. However, oftentimes we fit the raw CSD data to some known, well-characterized distribution.

EMPIRICAL DISTRIBUTIONS

Several standard distributions are often used to characterize CSD data.[83-85] They include the normal, log-normal, and gamma distributions, which are defined in Table 3–2. Many other distributions have been used to describe particulate distributions, but these three can be used to fit a wide range of size distribution data. The normal distribution, sometimes called the Gaussian distribution, has particles evenly distributed both above and below some mean size. Thus, the normal distribution is symmetrical about the mean size, and the spread of sizes can be characterized by the variance. However, in many natural systems, there is a skewness in the distribution such that there is a sharper decrease in small-sized crystals and a more gradual decrease in larger-sized crystals. This skewness to larger sizes often can be accounted for by using a logarithmic size axis. Thus, a log-normal distribution has the same form as a normal distribution, but with the linear size axis replaced with the logarithm of the size. When plotted on a graph with an x axis of log L, a log-normal distribution should appear to be normally distributed about the mean size.

Sometimes natural distributions do not fit either the normal or log-normal distributions. In other fields, a variety of other distributions often are used.[2] However, a general form of a distribution that can be fitted to most experimental data is the gamma distribution. The two parameters obtained from this distribution, a and b, do not have the same physical significance as the parameters obtained from the normal distributions (mean size

Table 3–2 Characteristics of Size Distributions

Property	*Normal*	*Log-Normal*	*Gamma*
Distribution, f(L)	$\frac{1}{\sqrt{2\pi}\sigma}\exp\left\{-\frac{(L-\bar{L})^2}{2\sigma^2}\right\}$	$\frac{1}{\sqrt{2\pi}\log(\sigma')}\exp\left\{-\frac{\log^2(L/\bar{L}')}{2\log^2(\sigma')}\right\}$	$\frac{L^a\exp\left\{-\frac{aL}{b}\right\}}{\Gamma(a+1)(b/a)^{a+1}}$
Size range	$-\infty$ to ∞	0 to ∞	0 to ∞
Size parameter	$\bar{L}$	$\bar{L}'$	b
Width parameter	$\sigma = \frac{(L_{0.84} - L_{0.16})}{2}$	$\sigma' = \frac{L_{0.84}}{L_{0.16}}$	a
Mean size	$\bar{L} = \bar{L}_{1,0}$	$\bar{L}'\left[\exp\left\{\frac{1}{2}\log^2(\sigma')\right\}\right]$	$\frac{b(a+1)}{a}$
Coefficient of variation	$\frac{\sigma}{\bar{L}}$	$\left[\exp\left\{\log^2(\sigma')\right\}-1\right]^{1/2}$	$\frac{1}{(a+1)^{1/2}}$

and variance). However, the gamma distribution can be used to generate a mathematical expression of size distribution for modelling purposes.

Fitting experimental data to these distributions is typically done using a computer curve-fitting package to minimize differences between individual points and a predicted distribution. For the normal and log-normal distributions, probability plots can be made that demonstrate how well experimental data fit one of these distributions. In these graphs, the particle number distribution is plotted on a probability axis versus the size of the increment, either on a linear (normal) or logarithmic (log-normal) scale. If the experimental data fit the distribution, the resulting line will be straight.

In many cases, particle size distributions fall somewhere between normal and log-normal in shape. That is, the distribution is skewed somewhat to larger sizes, but it is not sufficiently skewed to fit the log-normal distribution. For example, the distribution of sizes of ice crystals in ice cream typically fall somewhere between normal and log-normal. However, in one study[86] the ice crystal size distribution in ice cream fit the log-normal distribution quite well. The differential distribution for these data is shown in Figure 3–24A; the same data are represented on log-normal probability coordinates in Figure 3–24B. Clearly, this distribution fits the log-normal distribution quite well. In this case, the size distribution can be represented simply by two parameters, the log-normal mean size and the variance of sizes.

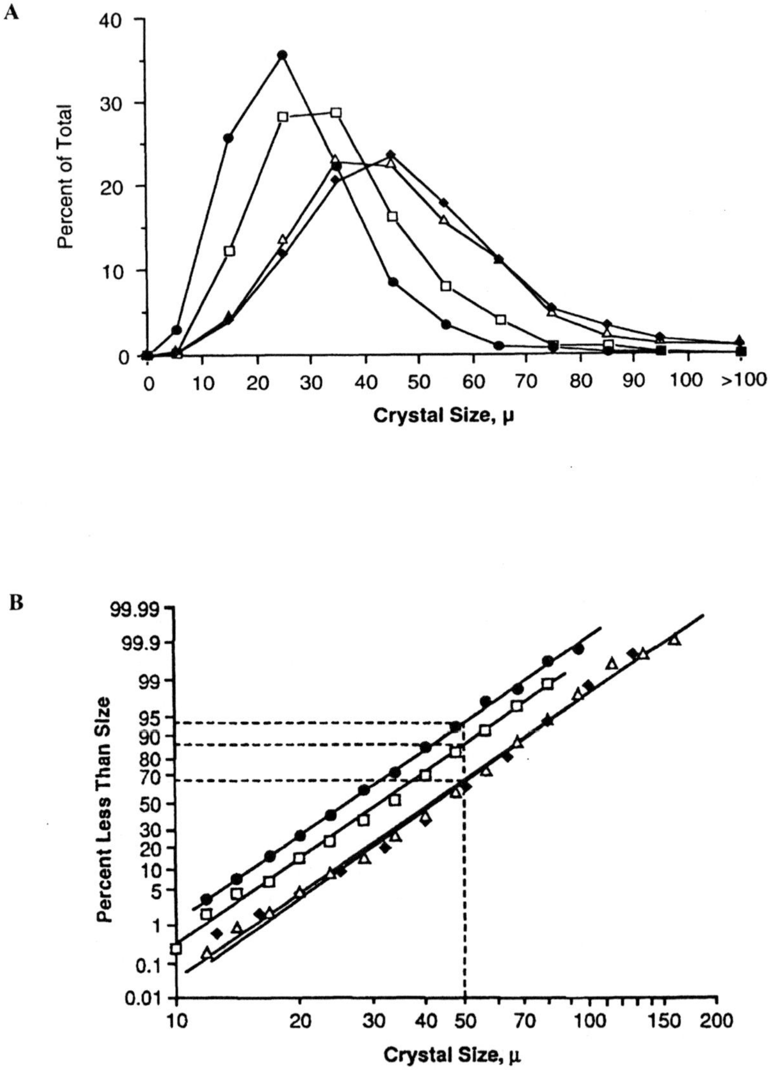

Figure 3–24 Size Distributions of Ice Crystals in Ice Cream at Different Stages of Manufacture and Storage: (A) Differential Distribution; (B) Log-Normal Probability Plot. Ice cream at draw from freezer (•), after hardening (❑), after 7 weeks of storage at –20°C (Δ) and after 14 weeks storage at –20°C (◆).

SUMMARY

Characterizing the nature of crystals in a food product is an important step in quantifying the physical and organoleptic properties. The crystalline characteristics, along with other structural elements, impact the mechanical/rheological properties as well as shelf life of a product, as seen by the interrelationships in Figure 1–1. However, obtaining a true measure of the crystalline characteristics often is not a simple task. Many different techniques are available, and the food technologist must understand the advantages and limitations of each technique to obtain the desired information. In addition, quantification of the crystalline structure through size distribution data requires careful analysis.

REFERENCES

1. Patience D, Hartel RW, Illingworth D. Crystallization and pressure filtration of anhydrous milk fat: mixing effects. *J Am Oil Chem Soc.* 1999;76(5): 585–594.
2. Allen T. *Particle Size Measurement.* 5th ed. New York: Chapman and Hall; 1997:1.
3. Siebert PC. Simple sedimentation methods. In: Stockham JD, Fochtman EG, eds. *Particle Size Analysis.* Ann Arbor, MI: Ann Arbor Science; 1979:45–55.
4. Hartel RW. *A Kinetic Study of the Nucleation and Growth of Sucrose Crystals in a Continuous Cooling Crystallizer.* Fort Collins, CO: Colorado State University; 1980. PhD dissertation.
5. Stanley DW, Aguilera JM, Baker KW, Jackman RL. Structure/property relationships of foods as affected by processing and storage. In: Rao MA, Hartel RW, eds. *Phase/State Transitions in Foods* New York: Marcel Dekker; 1998:1–56.
6. Aguilera JM, Stanley DW. *Microstructural Principles of Food Processing and Engineering.* 2nd ed. Gaithersburg, MD: Aspen Publishers; 1999.
7. Flint O. *Food Microscopy.* Oxford: Bios Scientific; 1994.
8. Rochow TG, Tucker PA. *Introduction to Microscopy by Means of Light, Electrons, X-rays or Acoustics.* 2nd ed. New York: Plenum Press; 1994.
9. Helmsley DA, ed. *Applied Polymer Light Microscopy.* London: Elsevier Applied Science; 1989.
10. Xu C, Zipfel W, Shear JB, Williams RM, Webb WW. Multiphoton fluorescence excitation: new spectral windows for biological linear microscopy. *Proc Natl Acad Sci.* 1996;93:10763–10768.
11. Heertje I, Paques M. Advances in electron microscopy. In: Dickinson E, ed. *New Physico-Chemical Techniques for the Characterization of Complex Food Systems.* Glasgow: Blackie; 1995:1–52.
12. Miles MJ, McMaster TJ. Scanning probe microscopy in food-related systems. In: Dickinson E, ed. *New Physico-Chemical Techniques for the Characterization of Complex Food Systems.* Glasgow: Blackie; 1995:69–85.
13. Donhowe D, Hartel RW, Bradley RC Jr. Ice crystallization processes during manufacture and storage of ice cream. *J Dairy Sci.* 1991;74(10): 3334–3344.
14. Stockham JD, Fochtman EG, eds. *Particle Size Analysis.* Ann Arbor, MI: Ann Arbor Science; 1977.
15. Arbuckle WS, Cremers LF. Fat-smoothness research. *Ice Cream Field.* October 1954:98, 102, 155.
16. Budiaman ER, Fennema OR. Linear rate of water crystallization as influenced by temperature of hydrocolloid suspensions. *J Dairy Sci.* 1987;70(2):534–546.
17. McClellan, MR, Morris GJ, Grout BWW, Hughes K. Light microscopy of foodstuffs during freezing and thawing. In: Bald WB, ed. *Food Freezing: Today and Tomorrow.* London: Springer-Verlag; 1991:171–185.
18. McKenzie AP, Rapatz GL. Transition from polyhedral to spheroidal ice crystals and vice versa, in partly frozen aqueous suspensions. *Biodynamica.* 1968;10(214):219–227.
19. Smith PR, Cebula DJ, Povey MJW. The effect of lauric-based molecules on trilaurin crystallization. *J Am Oil Chem Soc.* 1994;71(12):1367–1372.
20. Marangoni AG, Hartel RW. Visualization and structural analysis of fat crystal networks. *Food Technol.* 1998;52(9): 46–51.

21. Brooker BE. Imaging food systems by confocal laser scanning microscopy. In: Dickinson E, ed. *New Physico-Chemical Techniques for the Characterization of Complex Food Systems*. Glasgow: Blackie; 1995:53–68.

22. Caldwell KB, Goff HD, Stanley DW. A low-temperature scanning electron microscopy study of ice cream. I. Techniques and general microstructure. *Food Structure*. 1992;11:1–9.

23. Sargent JA. Low temperature scanning electron microscopy: advantages and applications. *Scanning Microscopy* 1988;2(2):835–849.

24. Danilatos GD. Bibliography of environmental scanning electron microscopy. *Microscopy Res Tech*. 1993;25:529–534.

25. Burchard, W. Light scattering. In: Ross-Murphy SB, ed. *Physical Techniques for the Study of Food Biopolymers*. Glasgow: Blackie; 1994:151–214.

26. Hayes W, Loudon R. *Scattering of Light by Crystals*. New York: John Wiley & Sons; 1978.

27. Beckett, ST. Control of particle size reduction during chocolate grinding. In: *Proceedings of the Pennsylvania Manufacturing Confectioners' Association(PMCA) Production Conference*. Center Valley, PA: PMCA: 1994:136–143.

28. Jeffery MS. Particle size: the technology of fondant. In: *Proceedings of the Pennsylvania Manufacturing Confectioners' Association(PMCA) Production Conference*. Center Valley, PA: PMCA; 1987:23–25.

29. Chawla P, deMan JM. Measurement of the size distribution of fat crystals using a laser particle counter. *J Am Oil Chem Soc*. 1990;67(5):329–332.

30. Eastoe J. Small-angle neutron scattering and neutron reflection. In: Dickinson E, ed. *New Physico-Chemical Techniques for the Characterization of Complex Food Systems*. Glasgow: Blackie, 1995.268–295.

31. Toro-Vazquez JF, Gallegos-Infante A. Viscosity and its relationship to crystallization in a binary system of saturated triacylglycerols and sesame seed oil *J Am Oil Chem Soc*. 1996;73(10):1237–1246.

32. Timms RE. The phase behaviour and polymorphism of milk fat, milk fat fractions and fully hardened milk fat. *Aust J Dairy Technol*. 1980;35:47–53.

33. Ten Grotenhuis E, van Aken GA, van Malssen KF, Schenk H. Polymorphism of milk fat studied by differential scanning calorimetry and real-time X-ray powder diffraction. *J Am Oil Chem Soc*. 1999;76(9):1031–1039.

34. Metin S, Hartel RW. Crystallization kinetics of blends of cocoa butter and milk fat or milk fat fractions. *J Thermal Anal*. 1996;47(5):1527–1544.

35. Kawamura K. The DSC thermal analysis of crystallization behavior in palm oil. *J Am Oil Chem Soc*. 1979;56:753–758.

36. Izzard MJ, Ablett S, Lillford PJ, Hill VL, Groves IF. A modulated differential scanning calorimeter study: glass transitions occurring in sucrose solutions. *J Thermal Anal*. 1996;47:1407–1418.

37. Bell LN, Touma DE. Glass transition temperatures determined using a temperature-cycling differential scanning calorimeter. *J Food Sci*. 1996;61(4):807–810, 812.

38. Singh SK, Jalali AF, Alden M. Modulated temperature differential scanning calorimetry for examination of tristearin polymorphism: 1. Effect of operational parameters. *J Am Oil Chem Soc*. 1999;76(4):499–505.

39. Singh SK, Jalali AF, Alden M. Modulated temperature differential scanning calorimetry for examination of tristearin polymorphism: 2. Isothermal crystallization of metastable forms. *J Am Oil Chem Soc*. 1999;76(4):507–510.

40. Hernqvist L. Crystal structures of fats and fatty acids. In: Garti N, Sato K, eds. *Crystallization and Polymorphism of Fats and Fatty Acids*. New York: Marcel Dekker; 1988: 97–138. Surfactant Science Series, Vol. 31.

41. Jouppila K, Roos YH. Water sorption and time-dependent phenomena of milk powders. *J Dairy Sci*. 1994;77: 1798–1808.

42. Jouppila K, Roos YH. Glass transitions and crystallization in milk powders. *J Dairy Sci*. 1994;77: 2907–2915.

43. deMan JM. X-ray diffraction spectroscopy in the study of fat polymorphism. *Food Res Int*. 1992;25:471–476.

44. Bricknell J, Hartel RW. Relation of fat bloom in chocolate to polymorphic transition of cocoa butter. *J Am Oil Chem Soc*. 1998;75(11):1609–1616.

45. Zobel HF. Starch crystal transformations and their industrial importance. *Starch*. 1988;40:1–7.

46. Clark AH. X-ray scattering and diffraction. In: Ross-Murphy SB, ed. *Physical Techniques for the Study of Food Biopolymers*. Glasgow: Blackie; 1994:65–150.

47. Buleon A, Le Bail P, Colonna P, Bizot H. Phase and polymorphic transitions of starches at low and intermediate water contents. In: Reid DS, ed. *The Properties of Water in Foods. ISOPOW 6*. London: Blackie; 1998:161–178.

48. Minato A, Ueno S, Yano J, et al. Synchrotron radiation X-ray diffraction study on phase behavior of PPP-POP binary mixtures. *J Am Oil Chem Soc*. 1996;73(11):1567–1572.

49. Ueno S, Minato A, Seto H, Amemiya Y, Sato K. Synchrotron radiation X-ray diffraction study of liquid crystal formation and polymorphic crystallization of SOS (sn-1,3-distearoyl-2-oleoyl glycerol). *J Phys Chem B*. 1997;101:6847–6854.
50. Van Malssen K, Peschar R, Schenk H. Real-time X-ray powder diffraction investigations on cocoa butter. I. Temperature-dependent crystallization behavior. *J Am Oil Chem Soc*. 1996;73(10):1209–1215.
51. Van Malssen K, Peschar R, Schenk H. Real-time X-ray powder diffraction investigations on cocoa butter. II. The relationship between melting behavior and composition of β–cocoa butter. *J Am Oil Chem Soc*. 1996;73(10):1217–1223.
52. Belton PS, Delgadillo I, Gil AM, Webb GA, eds. *Magnetic Resonance in Food Science*. London: Royal Society of Chemistry; 1995.
53. Gray R. Electron spin resonance spectroscopy for detection of irradiated foods. In: Dickinson E, ed. *New Physico-Chemical Techniques for the Characterization of Complex Food Systems*. Glasgow: Blackie; 1995:86–108.
54. Sutcliffe LH. Basic principles and applications of magnetic resonance. In: Belton PS, Delgadillo I, Gil AM, Webb GA, eds. *Magnetic Resonance in Food Science*. London: Royal Society of Chemistry; 1995:3–17.
55. Ablett S, Darke AH, Izzard MJ, Lillford PJ. Studies of the glass transition in malto-oligomers. In: Blanshard JMV, Lillford PJ, eds. *The Glassy State in Foods*. Loughborough: Nottingham University Press; 1993:189–206.
56. Gidley MJ, McArthur AJ, Darke AH, Ablett S. High resolution nuclear magnetic resonance spectroscopy of solid and semi-solid food components. In: Dickinson E, ed. *New Physico-Chemical Techniques for the Characterization of Complex Food Systems*. Glasgow: Blackie; 1995:295–318.
57. Van den Enden JC, Haighton AJ, Van Putte K, Vermaas LF, Waddington D. A method for the determination of the solid phase content of fats using pulse nuclear magnetic resonance. *Fette Seifen Anstrichmittel*. 1978;80:180.
58. Arishima T, Sugimoto K, Kiwata R, Mori H, Sato K. ^{13}C cross-polarization and magic-angle spinning nuclear magnetic resonance of polymorphic forms of three triacylglycerols. *J Am Oil Chem Soc* 1996;73(10):1231–1236.
59. Herrera ML, Hartel RW. Effect of processing conditions on physical properties of a milk fat model system. Microstructure. *J Am Oil Chem Soc*. 2000; 77(11):1197–1204.
60. Hemminga MA. Analysis of molecular motions in food products. In: Belton PS, Delgadillo I, Gil AM, Webb GA, eds. *Magnetic Resonance in Food Science*. London: Royal Society of Chemistry; 1995:43–53.
61. Lloyd RJ, Chen XD, Hargreaves JB. Glass transition and caking of spray-dried lactose. *Int J Food Sci Technol*. 1996;31:305–311.
62. Hills BP. Magnetic resonance imaging in food science. In: Dickinson E, ed. *New Physico-Chemical Techniques for the Characterization of Complex Food Systems*. Glasgow: Blackie; 1995:317–341.
63. McCarthy, MJ. *Magnetic Resonance Imaging in Foods*. London: Chapman and Hall; 1994.
64. Banwell CN, McCash EM. *Fundamentals of Molecular Spectroscopy*. 4th ed. London: McGraw-Hill; 1994.
65. Kobayashi M. Vibrational spectroscopic aspects of polymorphism and phase transition of fats and fatty acids. In: Garti N, Sato K, eds. *Crystallization and Polymorphism of Fats and Fatty Acids*. New York: Marcel Dekker; 1988:139–187. Surfactant Science Series, Vol. 31.
66. Taylor LS, Zografi G. The quantitative analysis of crystallinity using FT-Raman spectroscopy. *Pharm Res*. 1998;15(5): 755–761.
67. McMahon PM, Berglund KA, Larson MA. Raman spectroscopic studies of the structure of supersaturated KNO_3 solutions. In: Jancic JJ, deJong EJ, eds. *Industrial Crystallization '84*. London: Elsevier; 1984:229–239.
68. Wilson RH. Recent developments in infrared spectroscopy and microscopy. In: Dickinson E, ed. *New Physico-Chemical Techniques for the Characterization of Complex Food Systems*. Glasgow: Blackie; 1995:175–195.
69. Wilson RH, Goodfellow BJ, Belton PS, Osbourne BG, Oliver G, Russell PL. A comparison of fourier transform mid-infrared spectroscopy and near infrared spectrascopy with differential scanning calorimetry for the study of staling of bread. *J Science Food Agriculture*. 1991;54:471–483.
70. Yano J. *Vibrational Spectroscopic Study on Structures and Polymorphic Transformations of Triacylglycerols*. Osaka, Japan: Osaka University; 1998. PhD dissertation.
71. Minato A, Yano J, Ueno S, Smith K, Sato K. FT-IR study on microscopic structures and conformations of POP-PPO and POP-OPO molecular compounds. *Chem Phys Lipids*. 1997;88:63–71.
72. Povey MJW. Ultrasound studies of shelf-life and crystallization. In: Dickinson E, ed. *New Physico-Chemical Techniques for the Characterization of Complex Food Systems*. Glasgow: Blackie; 1995:196–213.
73. Hodate Y, Ueno S, Yano J, et al. Ultrasonic velocity measurement of crystallization rates of palm oil in oil-water emulsions. *Colloids Surfaces A*. 1997;128:217–224.

74. Roos YH. *Phase Transitions in Foods*. San Diego: Academic Press; 1995.
75. Noel TR, Ring SG, Whittam MA. Dielectric relaxations of small carbohydrate molecules in the liquid and glassy states, *J Phys Chem* 1992;96:5662–5667.
76. Ross-Murphy SB. Rheology of biopolymer solutions and gels. In: Dickinson E, ed. *New Physico-Chemical Techniques for the Characterization of Complex Food Systems*. Glasgow: Blackie; 1995:139–156.
77. Loisel C, Lecq G, Keller G, Ollivon M. Dynamic crystallization of dark chocolate as affected by temperature and lipid additives. *J Food Sci*. 1998;63(1):73–79.
78. Shastry AV. *Batch Crystallization of Sucrose from High Concentration Syrups*. Madison, WI: University of Wisconsin-Madison; 1991. Thesis.
79. Docherty R. The application of computational chemistry to the study of molecular materials. In: Myerson AS, Green DA, Meenan P, eds. *Crystal Growth of Organic Materials*. Washington, DC: American Chemical Society; 1996:2–14. ACS Conference Proceedings Series.
80. Fagan PC, Hammond, RB, Roberts KJ, Docherty R, Edmondson M. An ab initio approach to crystal structure determination using high resolution powder diffraction and computational chemistry techniques. In: Myerson AS, Green DA, Meenan P, eds. *Crystal Growth of Organic Materials*. Washington, DC: American Chemical Society; 1996:22–27. ACS Conference Proceedings Series.
81. Sato K, Yano J, Kawada I, Kawano M, Kaneko F, Suzuki M. Polymorphic behavior of gondoic acid and phase behavior of its binary mixtures with asclepic acid and oleic acid. *J Am Oil Chem Soc*. 1997;74(9):1153–1158.
82. Lammers WL, van der Stege HJ, Walstra P. Determination of the particle size distribution of non-agglomerated milk powders. *Neth Milk Dairy J*. 1987;41:147–160.
83. Randolph AD, Larson, MA. *Theory of Particulate Processes*. 2nd ed. New York: Academic Press; 1988.
84. Irani RR, Callis, CF. *Particle Size: Measurement, Interpretation and Application*. New York: Wiley; 1963.
85. Tavare NS. *Industrial Crystallization*. New York: Plenum Press; 1995.
86. Donhowe DP. *Recrystallization in Ice Cream and Frozen Desserts*. Madison, WI: University of Wisconsin-Madison; 1993. PhD dissertation.

Chapter 4

Solution Characteristics and Glass Transition

Since crystallization involves a liquid to solid transition, it is important to understand the characteristics of the liquid state. Conditions in the liquid state have an enormous impact on the amount and type of crystals formed as well as on the rate at which crystallization proceeds. Since the kinetics of crystallization control the nature of the crystalline microstructure (crystal size distribution [CSD], polymorph, shape, etc), an understanding of the nature of the liquid phase is required.

In many systems (sugars, salts, organic acids, etc), crystals are formed from solutions where the solute crystallizes based on its solubility in a solvent. For the most part, the important solvent in foods is water. There are other systems, however, where crystallization occurs from the melt, or the molten state of the crystallizing species. Water by itself is an example of a melt system since ice forms when liquid water is cooled below the freezing point. In simple solution and melt systems, the driving force for crystallization is based on well-understood principles of phase behavior. In solutions, the crystallizing species must have a concentration greater than some solubility concentration (C_s), which represents the equilibrium condition between molecules in liquid and solid phases. In melt systems, that equilibrium is found at the melting (or freezing) point (T_m). Melt systems must be cooled below T_m in order for crystallization to occur.

In lipid systems, the driving force may be based on either solution or melt systems. Pure triacylglycerols in the molten state are cooled below their melting point to initiate crystallization, whereas triacylglycerols crystallized from an organic solvent are treated as a solution system, just as sugar in water. However, crystallization of natural fats, which contain a wide range of triacylglycerol species, may be thought of as a combination of solution and melt systems. In actuality, the components of natural fats with higher melting points dissolve to a certain extent in the more liquid components (lower melting) of the material. For example, milk fat contains a small percentage of highly saturated triacylglycerols and an abundance of triacylglycerols with short-chain and unsaturated fatty acids (lower melting point). These more liquid components act as a solvent for the high-melting components. Thus, when natural fats such as milk fat are cooled slightly below their highest melting point, a solution of high-melting triacylglycerols in liquid fat is obtained. This may be the case during fractionation of fats. However, when fats are cooled substantially below the highest melting point, there is so much crystalline material formed that the system may be treated as a melt crystallization system.

PROPERTIES OF FLUIDS

Liquid Structure

The nature of the molecular organization prior to crystallization controls the formation of nuclei and their subsequent growth. Thus, some understanding of this molecular organization is necessary to control crystallization for the desired characteristics.

Before crystallization can occur, sufficient driving force (supersaturation or supercooling) for nucleation must be developed. Solute concentration and temperature are the primary determinants of supersaturation in solution systems, although other factors (impurities) may be involved as well. In melt systems, decreasing temperature is the means to generate supercooling. An understanding of liquid behavior, especially near the nucleation point, is critical to controlling crystallization processes. However, our knowledge of molecular interactions, especially as they lead to nuclei formation, is still incomplete.

Consider the addition of sucrose molecules to water. At dilute concentrations of sucrose, there is a hydration layer containing several (estimated to be up to about six) water molecules surrounding each sucrose molecule. These water molecules are not tightly bonded to the sucrose molecule, but rather undergo continuous detachment and then another one undergoes reattachment, as with many molecular processes. Thus, at any one time, there may be an average of about six water molecules, but they are never the same ones for very long.

Thus, there is a continuous migration of water molecules to and from the sucrose molecule. As the sucrose content is increased, there is more competition between water molecules for the hydration sites around the sucrose. In addition, the viscosity of the solution increases. At sufficiently high levels of sucrose, the hydration sites of each sucrose molecule may not be filled, and the average hydration level decreases. As concentration increases, the sucrose molecules begin to interact among themselves as the number of water molecules is limited. Note that there are still plenty of water molecules even at very high sugar concentrations. A 67% (weight/weight [w/w]) saturated solution of sucrose in water contains approximately 9.3 times more water molecules than sucrose molecules. Even at a concentration of 80% (w/w), there are about 4.7 times more water molecules than sucrose molecules. However, the availability of water molecules to interact with sucrose molecules decreases sufficiently such that sucrose-sucrose interactions can take place. Eventually, these sucrose-sucrose interactions result in the formation of clusters of molecules that eventually lead to nuclei formation.

Richardson et al[1] used ^{17}O nuclear magnetic resonance (NMR) measurements to probe concentrated sugar solutions and correlated molecular relaxation rate with increasing solution concentration. Relaxation rate increased linearly with concentrations up to about 40% (w/w) concentration, which was attributed to the effects of viscosity increase on relaxation rate. Above 40% concentration, relaxation rate increased exponentially with increasing concentration, which suggested that increased hydrogen bonding and/or sucrose-sucrose interactions were occurring. Mathlouthi[2] suggests that the transition from more sucrose-water association to sucrose-sucrose association occurs at about 40%, based on data from numerous techniques (viscosity, NMR, and Raman spectroscopy). However, no discontinuities or slope changes are observed in either viscosity or spectroscopy data[1] as concentration exceeds the solubility concentration (67%). Once crystallization had occurred, relaxation rate decreased to the value obtained for saturated sugar solution regardless of how much crystal phase volume had been formed.[1] These measurements provide useful information about the nature of molecular interactions in concentration solutions; however, the exact nature of these interactions is difficult to define from spectroscopic experiments.

It has been postulated that formation of subcritical embryos in solution must precede formation of a stable nucleus. This makes intuitive sense, since somehow the molecules in solution must form together into a unit that can develop a crystal lattice structure. There is even some experimental evidence for these clusters, although the link between these clusters and nuclei formation is inconclusive. Concentrated solutions of sucrose[3] or citric acid[4] held in a long vertical column developed density gradients when allowed to sit quiescently under certain conditions. Density gradients only were found in supersaturated solutions (although not at all conditions) and were not observed in undersaturated solutions. For sucrose, these gradients were attributed to flotation of clusters with lower density than the surrounding solution,[3] as might occur if sugar molecules were surrounded by more water molecules than found in the bulk. It often is postulated that these clusters must be the precursors to nuclei.

Other spectroscopic techniques have been used to study solution structure. By using Raman spectroscopy, McMahon et al[5] suggested that clusters (dimerization) were found in highly concentrated solutions of potassium nitrate (KNO_3). Another study[6] using Raman spectroscopy on ammonium phosphate showed that the molecules in the liquid state were organized as a dimer with a certain orientation. However, this orientation was different from that found in the crystal lattice, suggesting that the solution-phase dimer must have to dissociate or rearrange before the molecules could incorporate into a crystal lattice. This liquid-phase association may actually hinder the formation of crystal nuclei as molecular bonds must be broken before they can be reformed into the desired lattice orientation. Studies[7,8] using fluorescence spectroscopy also document the existence of such structures in supersaturated solutions.

Despite studies that seem to suggest some structure in supersaturated solutions, there is very little direct evidence that clusters formed in solution lead to nuclei.[9] Solute-solute and solute-solvent interactions undoubtedly play an important role in nucleation. However, our understanding of the specific nature of these interactions important to crystallization is still limited. This remains an active area of research.

In melt systems, such as water and simple fats, similar self-organization into subunits is suspected. In triacylglycerols (and other fats), it has been proposed that the molecules associate into a lamellar form.[10,11] In this case, the triglyceride molecules of natural fats organize with their fatty acids lined up with their adjacent neighbors, as shown schemati-

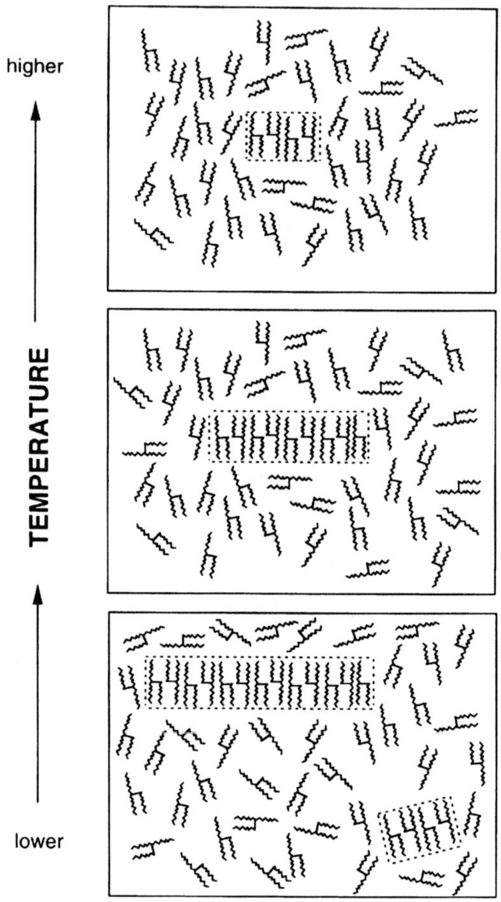

Figure 4–1 Representation (Highly Schematic) of Lamellar Ordering of Triacylglycerols in the Liquid State. Bent chains represent fluid triacylglycerols.

cally in Figure 4–1. The size of these structures is thought to increase as temperature decreases. When temperature is lowered below the melting point, crystallization occurs as the alkyl chains orient and freeze.[12] This process is demonstrated in Figure 4–2.

It has been widely speculated that liquid water molecules cluster together in three-dimensional arrays due to hydrogen bonding.[13] These binding forces (electrostatic forces, electron delocalization energy, etc) are responsible for some of the physical properties of

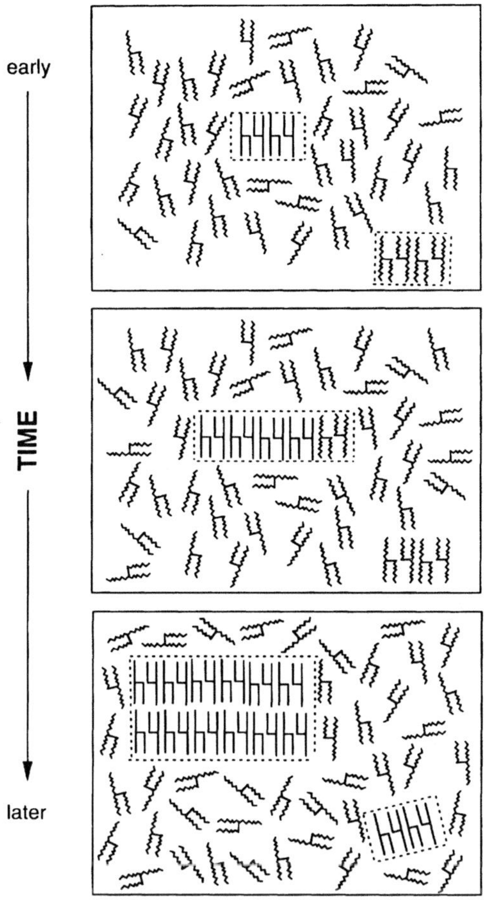

Figure 4–2 Proposed Mechanism (Highly Schematic) for Nucleation of Triacylglycerols. Straight chains indicate crystallized triacylglycerols, whereas bent chains represent fluid triacylglycerols.

water (high heat capacity, high melting point and heat of fusion, etc).[14] These clusters in liquid water do not have any long-range order, but are dynamic entities with small size (a few molecular diameters) and a rapid turnover rate (formation and decay). Kennedy[15] suggests that the lifetime of a cluster of water molecules is on the order of 10 ps, as determined by neutron-scattering experiments. Whether these clusters of water molecules lead to ice formation upon cooling is unclear.

Macroscopic Properties

Several macroscopic properties of solutions and melts change dramatically as concentration increases and/or temperature decreases.[16] Figure 4–3 schematically shows the sharp increase in viscosity and decrease in molecular diffusivity (a measure of mobility) as concentration increases or temperature decreases in a solution system such as sucrose or citric acid. No apparent discontinuities are evident in these curves that indicate the solubility concentration has been exceeded. Since mass transfer plays an important role in crystallization, these solution properties must be important in understanding and controlling nucleation and growth processes. In viscous systems, in particular, molecular mobility (diffusivity) may be sufficiently slow that kinetic constraints control crystallization processes. This will be discussed in greater detail in the section on the glassy state.

Mutarotation

A solution property of importance in crystallization of certain species is mutarotation. Many molecules, such as reducing sugars and organic acids, take more than one form

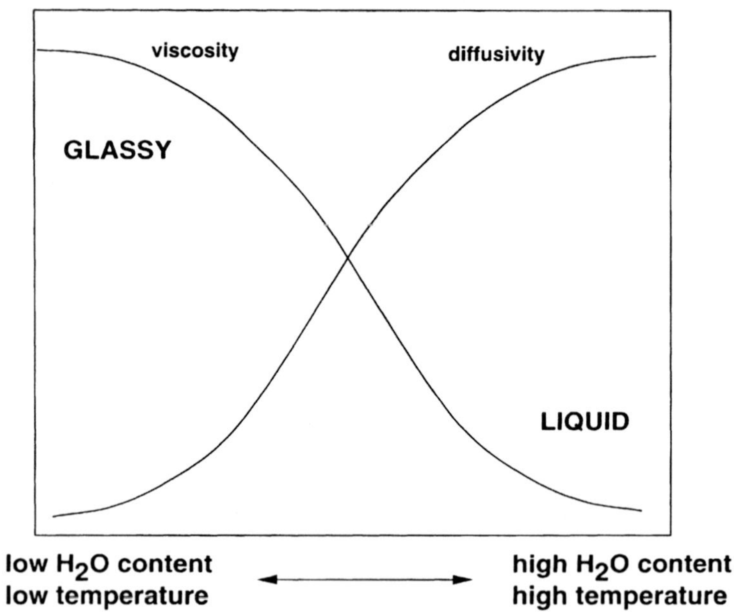

Figure 4–3 Viscosity and Diffusivity Changes with Concentration and Temperature.

in solution.[17,18] The transformations between α and β anomers in solution or between the various ring-shaped tautomers may impact crystallization. For example, lactose and glucose may take either the α or β form in solution, depending on the direction of the reducing hydroxyl (OH) group ("up" or "down"). An equilibrium reaction occurs between these two forms (Figure 4–4) and, at any given solution condition (temperature, pH, etc), there is an equilibrium concentration of both forms. In lactose solutions at room temperature and normal pH, there is about two thirds β-lactose and about one third α-lactose.[19] A similar ratio was found for glucose.[18] This ratio is a function of concentration, temperature, and pH. The ratio shifts to lower α-lactose content as lactose concentration increases and temperature increases.[20] Interestingly, it is the α-lactose that crystallizes under these conditions even though it is present at the lower concentration in solutions with temperatures less than 93.5°C. This is related to the relative solubility of the two forms, which will be discussed in a later section. More complex mutarotation behavior is found for fructose,[18] since substantial amounts of the furanose form are found in solution.

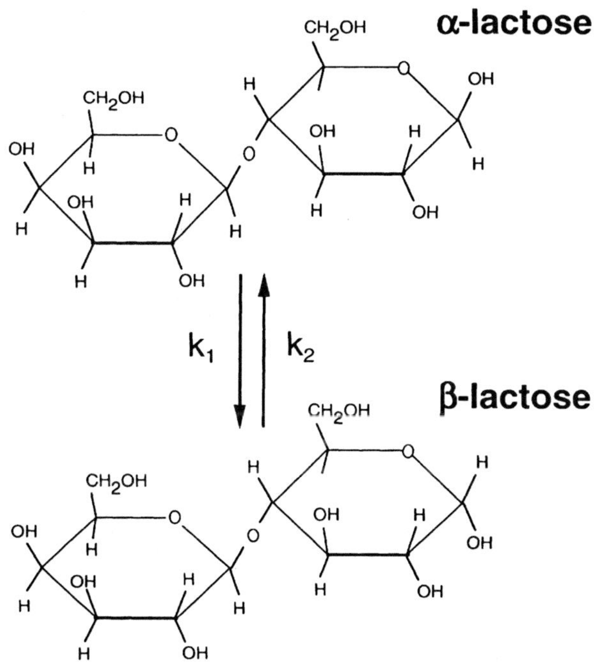

Figure 4–4 Mutarotation of Lactose.

Mutarotation rate is a complex function of solution conditions, such as temperature, concentration, pH, and presence of impurities. For lactose, the mutarotation rate is essentially constant for pH values between 2 and 7, and dramatically higher rates are found at pH values below 2 and above 7.[21] The mutarotation rate decreases as temperature is decreased, with an activation energy of 16 to 17 kcal/mol.[20] Salts have been found to increase mutarotation rate of lactose,[20] whereas sucrose causes a decrease in mutarotation rate.[22]

When α-lactose crystallizes from solution at temperatures less than 93.5°C, the equilibrium is shifted such that there is an excess of β-lactose in solution and mutarotation must occur to regain equilibrium between species in solution. This two-step process for crystallization is shown schematically in Figure 4–5. Although the mutarotation reaction is generally rapid, there are conditions when mutarotation rates can limit the rate of crystallization. In particular, growth of a slurry of lactose crystals with high surface area, particularly at low temperatures, may result in mutarotation limitation of growth. It has been suggested that this situation could occur during lactose refining.

PHASE DIAGRAMS

The equilibrium conditions (pressure, temperature, and concentration) between solid and liquid are shown on a phase diagram. These relationships between equilibrium states

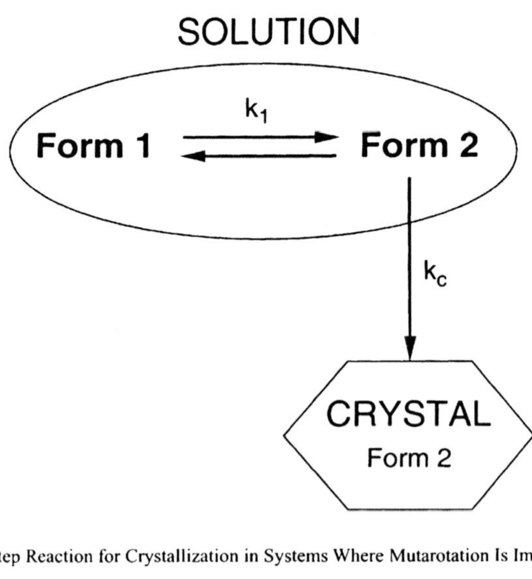

Figure 4–5 Two-Step Reaction for Crystallization in Systems Where Mutarotation Is Important.

of a material are critical in the control of crystallization. The phase diagram provides a map of the conditions where different phases can be found if equilibrium conditions are attained. From the phase diagram, we can determine the total amount of material that may be crystallized at any given condition. However, it is important to point out that foods are usually not at equilibrium, and the phase diagram does not necessarily tell us what will happen in complex food systems. Kinetic constraints may have significant impact on the crystallization processes and may lead to formation of incomplete crystal phase volume or different phases than predicted by the phase diagram. Furthermore, in complex food products containing many ingredients, it is often difficult to generate a phase diagram. Thus, phase diagrams may be difficult to obtain in real foods and, when available, only provide a guideline for crystallization in foods. Nevertheless, delineation of the phase behavior in a food needs to be understood as well as possible to allow proper control of crystallization during processing and storage.

A phase diagram for a pure component is often given in terms of pressure and temperature, although a third axis to represent volume also may be used.[14,23] The transitions of water between vapor, liquid, and crystal states, due to pressure and temperature changes, are detailed on the phase diagram, as seen in Figure 4–6. Within the region of crystal formation, there are up to 10 different types (polymorphs) of ice crystal that form,[14,24] depending on temperature and pressure during ice formation. However, only one form is stable at normal pressures and temperatures.

The Gibbs phase rule relates the number of components (C) and phases (P) with the degrees of freedom (F) allowed at equilibrium:

$$P + F = C + 2 \tag{4.1}$$

For a one-component system, like water, with two phases present (ice and liquid water), there is only one degree of freedom. That means that only one variable—pressure or temperature—may be specified independently; the others are set as governed by the thermodynamic rules.

For a binary system ($C = 2$), the phase diagram indicates the equilibrium relationships between solute and solvent. Since the solvent of primary importance in foods is water, we are most interested in binary phase diagrams with water. Numerous types of phase behavior are observed for binary mixtures, depending on the nature of interactions between the two molecules.[25–28] Mullin[25] suggests that three types of phase diagrams (solid solution, eutectic, and compound formation) are most important. Phase diagrams of binary mixtures of triacylglycerols fall into one of four categories,[26,28] as shown in Figure 4–7. Phase behavior of various lipids are discussed in Sato[29] and Lutton.[30]

In the first case, the two components may form a simple solid solution (Figure 4–7A). The concentration of solid and liquid phases at equilibrium can be easily determined at any condition. Some mixtures of pure triacylglycerols, particularly triacylglycerols with similar chain length and degree of unsaturation (ie, SSS/ESS and POS/SOS, where S is

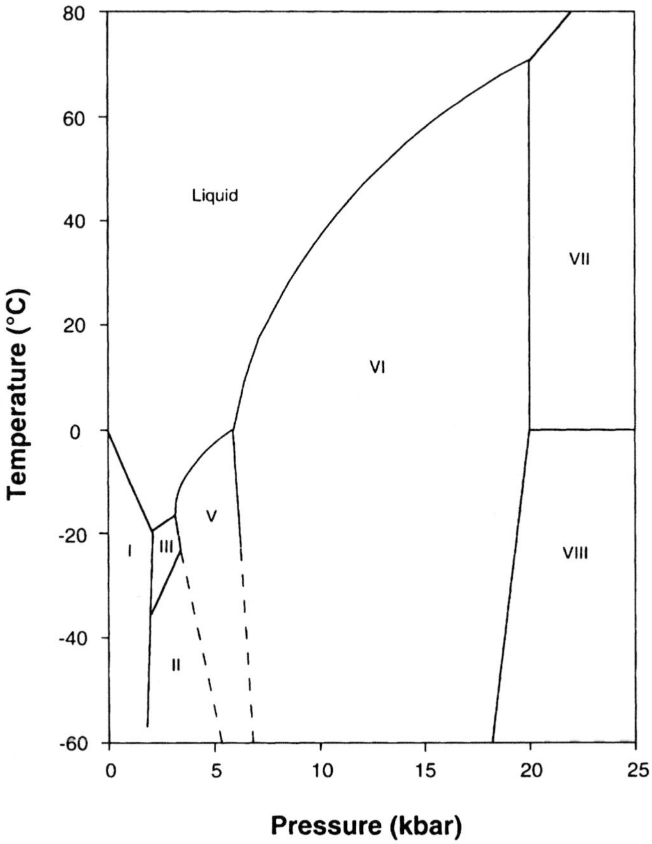

Figure 4–6 Phase Diagram of Water.

stearic acid, E is elaidic acid, P is palmitic acid, and O is oleic acid), fit into this category.[28] In this case, the solidus and liquidus lines are always separated, except at the end-points, and there is a region of two-phase mixture with well-determined concentrations. There is no true separation of components in phases (pure A or pure B). That is, when the system solidifies, the solid phase is a mixture of the two components at the ratio governed by the position on the phase diagram. This, in part, leads to limited fractionation efficiencies for natural fats as similar triacylglycerols crystallize together in solid solutions.

In eutectic binary systems (Figure 4–7B), the two molecules interact in such a way that the intersection of the equilibrium curves intersects at a temperature below the

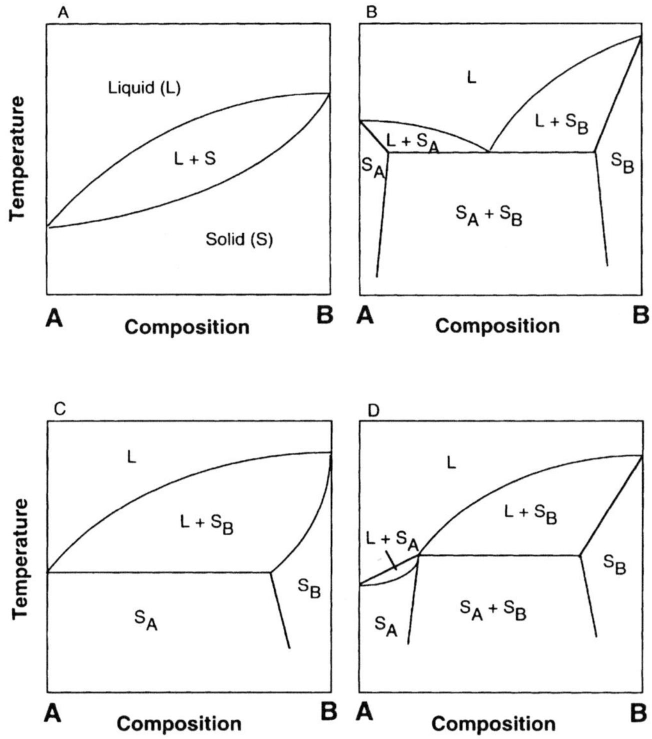

Figure 4–7 Phase Behavior in Binary Systems: (A) Monotectic, Continuous Solid Solution; (B) Eutectic; (C) Monotectic, Partial Solution; and (D) Peritectic.

melting point of either pure component. Binary mixtures of salts and sugars with water often show eutectic behavior. For example, the phase diagram of sucrose shows eutectic behavior,[31,32] as shown in Figure 4–8.[31,32] In a true eutectic system, a solid solution forms when temperature is below the eutectic temperature. The concentration of the eutectic solid is defined by the phase diagram at the intersection of the solvent and solute curves. In sugars, however, a eutectic solid is not formed due to the slow kinetics of sugar crystallization. Thus, ice continues to freeze independently of the sucrose at temperatures below the eutectic point. Many salts of importance in foods form eutectics with water. Fennema et al[14] provide a table of eutectic temperatures for various salts. These salts may indeed solidify as a eutectic solid during freezing of foods. Some binary mixtures of triacylglycerols form eutectics, where the melting point of the mixture falls below the melting point of either of the pure components. The triacylglycerol pairs PPP/SSS,

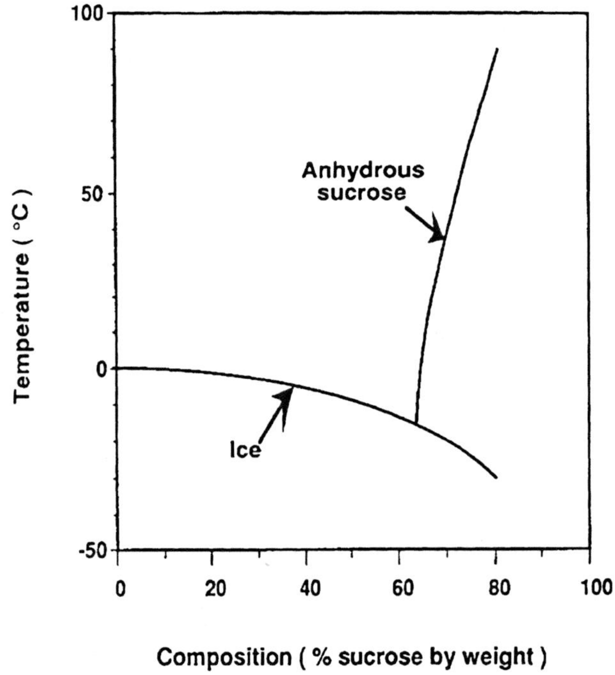

Figure 4–8 Phase Diagram for the Sucrose-Water System.

EEE/SOS, POS/PSO, PPP/LLL and PPP/SOS (where P is palmitic acid, S is stearic acid, E is elaidic acid, O is oleic acid, and L is linoleic acid) fall into this category.[26,28]

Triacylglycerols with a wide difference in melting point often exhibit monotectic behavior (Figure 4–7C), where there is a large two-phase liquid region. The triglyceride pairs SSS/OOO, SSS/LLL, PPP/POP, and SSS/SOS have been shown to exhibit monotectic behavior.[26,28] Peritectic behavior (Figure 4–7D) is observed in lipid systems where one of the triacylglycerols contains two unsaturated fatty acids. For example, SOS/SOO and POP/POO exhibit peritectic phase behavior.

For ideal solutions, the solubility of solute in solvent can be obtained from the Hildebrand equation,[33] which can be written as

$$\ln(X) = \frac{\Delta H_m}{R}\left(\frac{1}{T_m} - \frac{1}{T}\right) \tag{4.2}$$

where X is mole fraction of solute at any temperature, T; ΔH_m is molar heat of fusion; and T_m is melting point of solute. Equation 4.2 is based on a thermodynamic equilibrium between solute molecules in crystalline form and liquid form. Timms[33] uses this equation to show the solubility of tristearin (SSS) in several solvents (triolein, OOO; trilinolein, LLL; and paraffin oil) as well as tripalmitin (PPP) in triolein. He points out that solubility according to Equation 4.2 is only dependent on the solute and not on the solvent, since ideal solutions were assumed. In reality, many systems (especially lipid systems) do form solid solutions, and thus Equation 4.2 cannot be applied. In general, Equation 4.2 can be used only when solute and solvent do not interact to form solid solutions. Kloek et al[34] used Equation 4.2 to describe the solubility of fully hydrogenated palm oil in sunflower oil.

Some binary mixtures show compound behavior, where several different crystal forms (eg, hydrates, polymorphs) may form at different conditions. For example, the phase diagram of D-glucose in water (Figure 4–9) documents the conditions at which each of the crystalline forms is in equilibrium. As seen in Figure 4–9, glucose may crystallize in one of several forms (hydrates or anomers) depending on the temperature and concentration of solution. Given sufficient time for equilibrium conditions to be attained, the phases and crystal forms shown in Figure 4–9 will occur at those conditions. Fructose (Figure 4–10) exhibits similar behavior. Mixtures of triacylglycerols also commonly show compound behavior due to polymorphic effects. Also, mixtures of SOS (sn-1,3-distearoyl-sn-2-oleoylglycerol) and OSO (sn-1,3-dioleoyl-sn-2-stearoylglycerol) have been shown to form a compound crystal at mixtures above 50% of OSO in SOS.[35] Compound formation also is observed between SSO/SOS and POP/OPO.[28] The differences in molecular structure of these triglycerides are such that they can form mixed crystals of the two species. This is a common attribute of complex natural lipids.

Wesdorp[36] developed a technique to calculate the phase behavior of triacylglycerol mixtures based on thermodynamic principles. To define the equilibrium condition between the two phases, the chemical potentials of each phase were equated. Interaction parameters between triacylglycerol species were obtained by using a differential scanning calorimeter (DSC) technique. The interaction parameters were used to calculate excess free energies as two triacylglycerols were mixed. Based on these results, the phase behavior of various triacylglycerol mixtures in any polymorphic form was obtained. Solid fat content melting profiles of multicomponent mixtures of triacylglycerols also could be calculated. Wesdorp[36] also provides a review of experimental phase diagrams for mixtures of pure triglycerides and shows excellent fit between theory and experiment in most cases.Roussett et al[37] used the same technique as Wesdorp to study the phase behavior of mixtures of the different polymorphs of SOS and POP (where S is stearic acid, O is oleic acid, and P is palmitic acid).

In Figures 4–8 through 4–10, phase equilibria of both solute (sucrose, glucose, or fructose) and solvent (water) are shown. The solubility curve represents the conditions of temperature and concentration where solute molecules in solution are in equilibrium (chemical potential of the two phases are the same) with those in the crystalline form. Practically speaking, the solubility line represents the concentration, at a given temperature, of a saturated solution. The line representing the phase boundary between dilute sugar

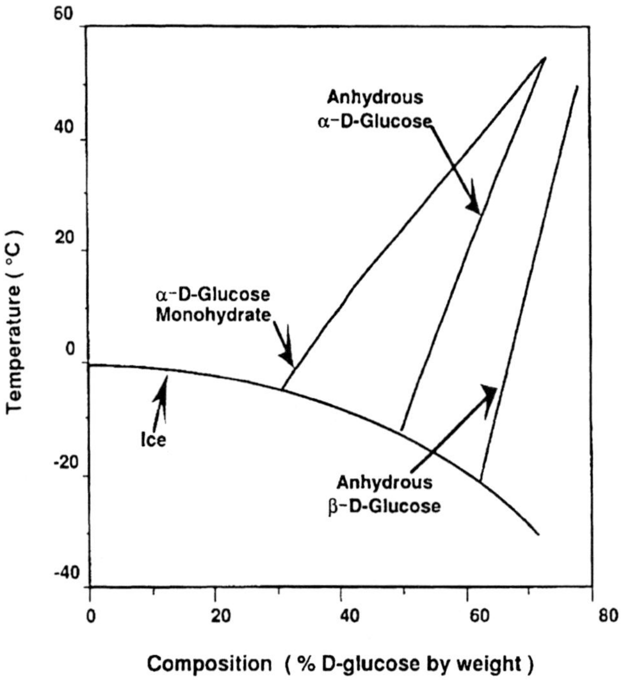

Figure 4–9 Phase Diagram for the Glucose-Water System.

solution and ice is often called the freezing curve or freezing point depression curve. For low-molecular-weight salts and sugars, the freezing point of the solution decreases as concentration increases, according to the colligative properties of the solutes. This occurs in many foods during freezing, where the freezing point is continually reduced as ice crystallizes and the remaining solution is concentrated.

Measurement of the equilibrium curves for solvent (freezing point depression curve) and solute (solubility curve) is required to construct a phase diagram. The freezing point of a mixture can be measured in a variety of ways.[14] One simple method is to freeze the solution slowly, induce nucleation of ice at some temperature slightly below the suspected freezing point, and observe the temperature changes that occur as freezing progresses. Due to the release of latent heat during freezing, the temperature in the sample chamber increases slightly and reaches a maximum value. As freezing continues, the temperature begins to decrease as the solution becomes more concentrated. The temperature at the plateau is used to determine the melting point for that initial concentration of solute

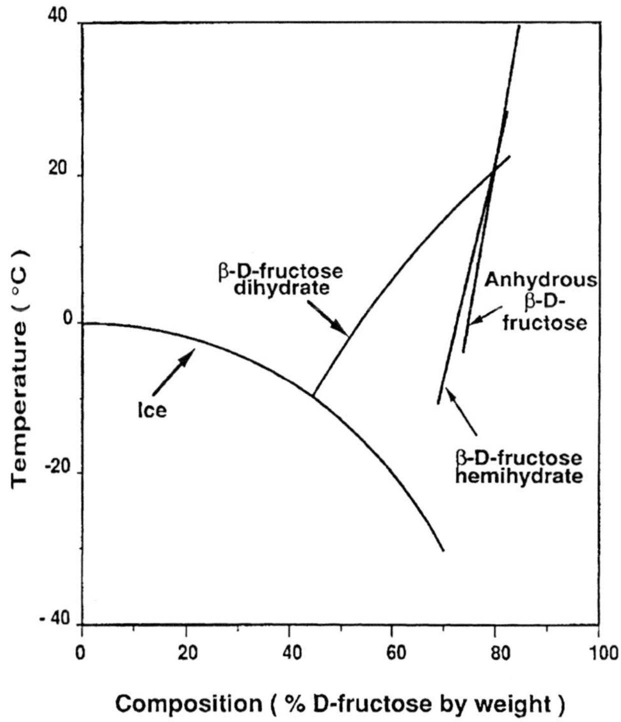

Figure 4–10 Phase Diagram for the Fructose-Water System.

in solution. Corrections due to supercooling of the solution, either by use of standards or by an extrapolation curve for different supercoolings, are necessary to find the true freezing point. In many units, the corrections are already calculated during calibration against a standard and the freezing point temperature is read directly off the unit display. Freezing point depression curves for several sugars are shown in Figure 4–11. At elevated concentrations of solute (over about 40 to 50%), the temperature method becomes less accurate as the kinetics of ice crystallization slow dramatically. Thus, it is difficult to measure freezing point temperature accurately in systems with high concentration and viscosity.

The freezing point temperature, based on the colligative effect of small molecules, can be calculated from thermodynamic principles[14,38,39]:

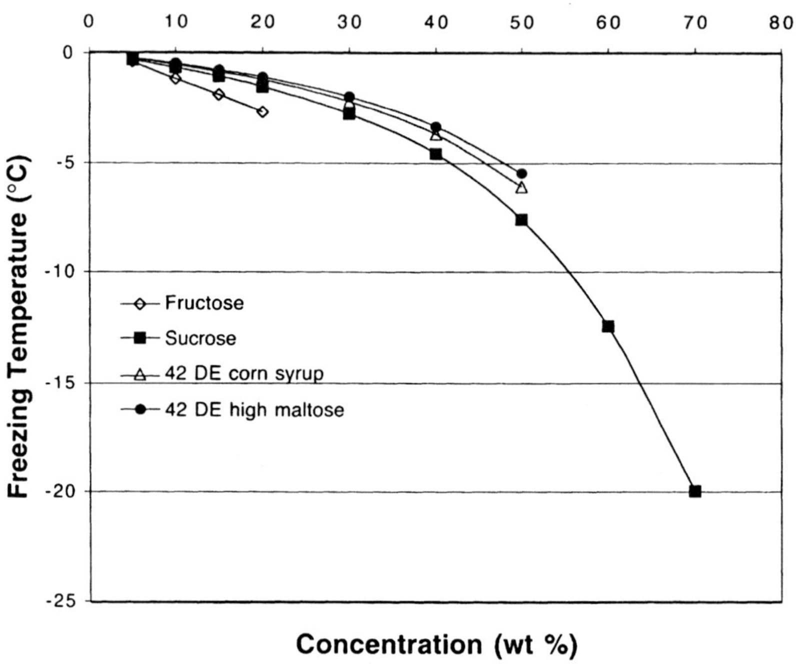

Figure 4–11 Freezing Point Depression Curves for Fructose, Sucrose, 42 DE (Dextrose Equivalent) Corn Syrup, and 42 DE High Maltose Corn Syrup.

$$\frac{\Delta H}{R}\left[\frac{1}{T_0}-\frac{1}{T}\right]=\ln(X_w) \tag{4.3}$$

where ΔH is molar heat of fusion (kJ/mol), R is the ideal gas constant, T_0 is freezing point of pure water, T is freezing point of water in solution of solute (A) with water content expressed as mole fraction X_w. Equation 4.3 was derived by setting chemical potentials of the two phases (water and ice) equal and utilizing Raoult's law, which relates vapor pressure to concentration. Several assumptions are made in this derivation,[14] with the three most notable assumptions being (1) ideal solutions, (2) constant value of ΔH over the freezing point range, and (3) dilute solutions. Note that Equation 4.3 is basically the same as Equation 4.2, and it was derived in essentially the same way.

Heldman and Singh[38] further simplify Equation 4.3, again only for dilute solutions, as

$$\Delta T_f = \frac{RT_0^2 MW_w m}{1000\Delta H_f}$$

(4.4)

where ΔT_f, is the freezing point depression (T_0–T), MW_w is the molecular weight of solvent (water), m is molality of solute (mol/kg of solvent), and ΔH_f is the mass heat of fusion (kJ/kg).

The application of these equations to real food systems gives only an estimate of freezing point depression. The combined effects of each solute in a complex food lead to the actual freezing point depression. In ice cream, for example, the sugars (sucrose, lactose, and corn syrup) interact with the salts present from the milk to alter freezing point.[40] To use Equation 4.3 in food systems, the mole fraction of water can be obtained from

$$X_w = \frac{W_w / MW_w}{(W_w / MW_w) + (W_s / MW_s)}$$

(4.5)

where W_w is weight fraction of water, W_s is weight fraction of solids ($1 - W_w$), and MW_s is an effective molecular weight of solids in solution. However, this method also is limited by the ability to express complex food products by some effective molecular weight. Schwartzberg[41] also developed a semi-empirical method for estimating freezing point and the amount of unfrozen water in foods with mixtures of solutes.

To generate the freezing point depression curve for an ice cream mix, typically a single measurement is made on the mix (35 to 42% total solids); the rest is calculated according to a standard equation based on composition.[42] This empirical approach has been developed over the years to provide a reasonable approximation of the true freezing point depression curve. There also is some evidence that stabilizers (gums, etc) added to ice cream mix may affect freezing point depression, although not through the colligative effect developed above since they have high molecular weight. Blond[43] discusses the unexpected freezing points obtained from certain hydrocolloids at low concentration. Carboxymethylcellulose, pectin, polyvinyl alcohol, and gelatin were studied at low concentrations (less than 10%), and their freezing point temperatures were measured. Freezing point depressions of greater than 0.4°C were observed in these solutions even though the equations developed above would predict no freezing point depression based on large molecular weight. In particular, a relationship between aging of the gelatin structure and freezing point depression was noticed. These anomalous freezing point temperatures for high-molecular-weight materials often have been attributed to slow growth of ice crystals (which affects the measurement of T_f), although capillarity in a gel system also has been suggested as the cause.[43]

In binary systems that show eutectic behavior, a temperature and a concentration are reached where the phase equilibrium lines for solute and water intersect. At this point, called the eutectic point, further crystallization typically results in a solid solution containing both solute and water. Eutectic points can be found in the literature for a wide range of solutes in water, including salts and sugars.[14] However, in the case of most sugars, a eutectic compound does not usually form. Due to kinetic constraints—sugar crystallization occurs very slowly at these low temperatures—the solution usually cools below the eutectic point, the solute continues to become supersaturated, and ice continues to form without the formation of sugar crystals or a solid solution. In salt solutions, however, eutectics are often observed, and eutectic compounds may be formed during freezing of foods.

The equilibrium condition between solid and liquid phases of solute represents the solubility concentration of solute. At this thermodynamic equilibrium, the chemical potentials of the solute molecules in both liquid and solid phases are equal. The solubility, or saturation, concentration is the maximum amount of solute that can be solubilized at a given temperature. Think of adding sugar to water to make a dissolved sucrose solution. At low temperatures (as in iced tea), only a small amount of sucrose can be dissolved; some sugar crystals may remain on the bottom of the glass. In hot tea, however, substantially more sucrose can be dissolved; typically no sugar crystals remain on the bottom of the cup. Thus, the solubility concentration for sucrose, as well as for many other substances, increases with increasing temperature, in a "normal" fashion. Solubility curves for a number of sugars and sugar alcohols are shown in Figure 4–12 and for some salts and organic acids in Figure 4–13.

The examples shown in Figures 4–12 and 4–13 all have solubility increasing with temperature in the "normal" fashion. However, there are some salt systems (calcium sulfate [$CaSO_4$], lithium carbonate [Li_2CO_3], lithium sulfate [Li_2SO_4], sodium sulfate [Na_2SO_4], manganese sulfate [$MnSO_4$]) that show an inverse solubility behavior in water.[25] That is, they become less soluble as the temperature increases. This may be important where water must be treated prior to being heated into steam so that certain salts do not crystallize from solution and form a scale layer on the heating pipes. Hard water, containing high concentrations of dissolved minerals such as calcium and magnesium, forms a scale layer on the heating surface as these minerals crystallize. These minerals that form "temporary" hardness in water potentially cause problems in boiler systems unless the water is adequately treated beforehand.

To measure solubility concentrations, it is necessary to approach an equilibrium between solid and liquid phases in solution. Mullin[25] discusses the variety of techniques used for measurement of solubility. Probably the most common method of measuring solubility concentrations involves placing an excess of crystalline material in a certain amount of water and allowing the full extent of dissolution. Often this requires maintaining constant temperature and agitation for several days to a week. The concentration in the solution phase is then measured to obtain the solubility concentration for that temperature. The process is repeated at different temperatures to obtain the solubility curve. One should also measure the equilibrium condition from the other direction, with solute crystallizing out of supersaturated solution until equilibrium is attained. If true equilibrium has been attained in these experiments, both measurements should give identical results. However,

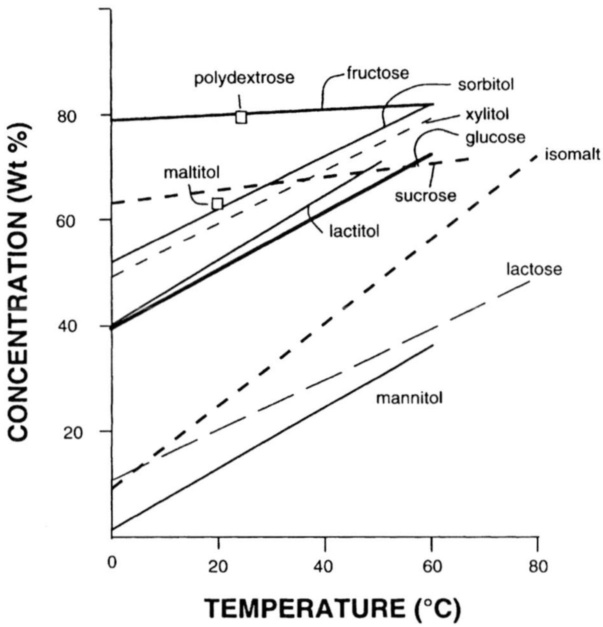

Figure 4–12 Solubility Curves for Various Sugars and Sugar Alcohols.

---

this is not always the case. Kinetic constraints (slow dissolution or crystallization) at low supersaturations or undersaturations often limit the accuracy of these measurements.

Phase diagrams are most useful to predict the maximum amount of crystalline material (equilibrium phase volume) that can be formed in a system at certain conditions. However, this equilibrium phase volume may not necessarily exist in a food since crystallization may be constrained by kinetic factors. Sugar crystallization in a hard candy is one example where some small crystals may be dispersed through a system of such high concentration (and viscosity) that crystallization cannot be completed. Even though the system is highly supersaturated in sucrose, crystallization does not occur because of the limited mobility of the sucrose molecules. Similar results may be found for ice crystallization in a rapidly frozen food. There may not be sufficient time for the maximum amount of ice to form before viscosity limitations prevent further ice formation. In fact, the principle of cryo-preparation for SEM (see Chapter 3) is based on this effect. The effects of viscosity (or mobility) limitations on crystallization will be discussed in the section on glass formation.

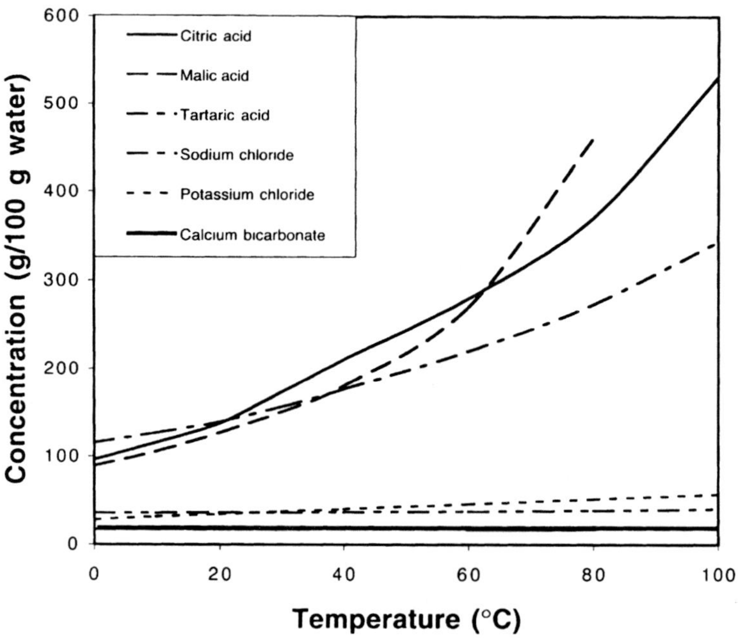

Figure 4–13 Solubility Curves for Several Salts and Organic Acids.

Multicomponent Phase Diagrams

For mixtures of three components, ternary phase diagrams (often called triangle diagrams) can be constructed.[25,44] In ternary phase diagrams, the pure components lie at the corners of the triangle and points along the edge lines represent binary mixtures. Ternary mixtures of the components are found within the triangle itself. Ternary phase diagrams have not been used extensively in the food industry, although several examples can be found. For instance, ternary phase diagrams for various mixtures of salts in water can be found in Mullin[25] and Tavare.[44] Ternary phase diagrams of mixtures of sucrose-water with either fructose, glucose, or sodium chloride (NaCl) are shown in Bubnek and Kadlec.[45] A ternary phase diagram for sucrose-water-impurity, treating all nonsucrose impurities in impure sugar solution (for refining) as a single compound, also is shown in Bubnek and Kadlec.[45] Timms[26] discusses an approach for predicting the phase behavior of multicomponent fat mixtures. Marangoni and Lencki[46] show ternary phase diagrams for mixtures of the three different components of milk fat (high-melting, middle-melting, and low-melting components).

Effect of Particle Size on Phase Equilibrium

The solubility curves shown in phase diagrams apply to the equilibrium relationship between liquid molecules and a solid boundary of infinite extent. Very small crystals, with a high radius of curvature, exhibit elevated solubility due to interfacial tension effects. This elevated solubility is known as either the Ostwald-Freundlich, Gibbs-Thomson, or Gibbs-Kelvin effect[25] and can be approximated by the following expression (Kelvin equation):

$$\ln\left(\frac{C_r}{C_s}\right)=\frac{2MW\sigma}{\rho RTr} \tag{4.6}$$

where C_r is solubility concentration for a particle of radius, r; C_s is solubility concentration for a plane crystal surface; MW is molecular weight of solute; σ is interfacial tension of the crystal surface; R is the gas constant; T is absolute temperature; and ρ is crystal density. Smaller particles exhibit higher solubilities, although only extremely small crystals—less than 1 to 10 µm, depending on the system (interfacial tension)—exhibit the Ostwald effect to any measurable extent. A similar equation applies for melt systems,[47] where the melting point, $T(r)$, depends on crystal size, r.[48,49]

$$T_\infty - T(r)=\frac{2\sigma T_\infty}{\rho \Delta H_f r} \tag{4.7}$$

where T_∞ is the melting point of ice, σ is the interfacial tension between solution and ice crystal, ρ is crystal density, and ΔH_f is latent heat of fusion. Smaller ice crystals have lower melting points, making them less stable. For example, a 40-µm diameter ice crystal in a sucrose-water solution has a subcooling of 0.0024°C compared to an infinitely large crystal, whereas a 1-µm crystal has a subcooling of 0.046°C.[50] These differences in melting point, although very small, become important during storage of frozen foods. These small driving forces can result in significant changes in ice crystal size distribution over time (recrystallization or ripening), which can lead to loss of product quality. Similar problems are observed during storage of fondant, which contains many small sugar crystals. Further discussion of this phenomenon can be found in Chapter 8.

CALCULATING CRYSTALLINE YIELD

One of the most important aspects of the phase diagram is that it allows calculation of the amount of crystalline material (or maximum yield) to expect from a given system. In order to calculate yield, the initial condition is compared with the final equilibrium condition based on the phase diagram. In simple binary systems, the initial and solubility concentrations are used to calculate yield based on a mass balance. In the simplest case, the product is considered to be composed of a pure crystalline phase; the solution phase contains water and dissolved solute at the solubility concentration. The mass balance equations for a binary system are given as

$$F = P + Y \tag{4.8}$$

and

$$C_i F = C_s P + Y \tag{4.9}$$

where F is total mass of starting solution, Y is the mass of crystalline material (or yield), P is mass of solution in the product, C_i is the initial concentration (in weight percent), and C_s is the solubility concentration. These equations can be used to calculate the amount of crystalline product, Y, and the amount of solution phase in the product. Equation 4.9 assumes that crystallization has been completed so that the concentration in the solution of the final product is the solubility concentration and the maximum crystalline phase volume (maximum yield) has been obtained. Also assumed is that there is complete separation between liquid and crystal phases.

As an example of how these equations can be used, we can calculate Y and P for 100 kg of a sucrose solution with 80% total solids that crystallizes to phase equilibrium at a temperature of 25°C. At this temperature, the saturation concentration (C_s) is 67.5%.[45] Solving Equations 4.8 and 4.9 together simultaneously gives a maximum yield (Y) of 38.5 kg of crystals and a solution phase of 61.5 kg that has concentration of 67.5% total solids.

In refining operations, this yield calculation gives the theoretical amount of crystalline material that can be separated. In many cases, however, the actual yield in the separator is higher due to liquid entrainment within the crystalline phase. In sugar refining, the amount of mother liquor (saturated solution) adhering to the crystals may be about 10% of the crystal weight. In more viscous systems and those where there is a high degree of affinity between crystal and fluid, the amount of liquid entrainment may be substantially higher. For example, in fractionation of fats, the amount of liquid oil entrained within the filter cake may be as high as 70% of the weight of crystalline material.

In some cases (ie, lactose, glucose, etc), waters of hydration crystallize within the lattice structure, and these also must be accounted for in the mass balance. In this case, the molar ratio of water in the lattice (eg, 1 mol water/mol lactose to form the monohydrate) is converted to a mass basis for the crystalline phase; this concentration is utilized in Equation 4.9 to modify the yield (Y) term. Since this concentration is a known parameter (based on waters of hydration), the calculation follows in the same manner as outlined above.

In order to use these equations for systems with impurities, it is necessary to know the solubility concentration in the presence of the impurity. For example, assume we have 10% corn syrup solids in the initial solution (and now 70% sucrose) in the previous example. The solubility concentration of sucrose in the presence of corn syrup solids must be found from existing data,[51] as shown in Table 4–1. If there is 10% corn syrup solids on a total solution basis, then there are about 27 g of corn syrup solids per 100 g water (10% of initial solution) and the solubility of sucrose is about 139 g sucrose per 100 g water (52.2% of saturated solution). In this case, for 100 kg of initial syrup, there will be 37.2 kg of crystalline solid (compared with 38.5 kg in the case of pure sucrose). The solution phase will have a corn syrup solids concentration of 15.9%, as found from a mass balance on corn syrup solids. In combination with the sucrose concentration of 52.2%, the total solids concentration of the solution phase will be 68.1% (as compared with the 67.5% for the pure sucrose). As more corn syrup is added, the crystalline yield decreases and the solution concentration increases.

For more complex food systems, as in most food products, it is much more difficult to perform this calculation. Typically, phase diagrams for such complex foods are not available. For example, the equilibrium condition for sucrose or lactose in a caramel formulation has not been determined.

SUPERSATURATED SYSTEMS

Supersaturation Driving Force

In order to have a driving force for crystallization, the actual concentration in solution must exceed the equilibrium saturation (or solubility) concentration or the temperature of a melt system must be lowered below the melting point. Thermodynamically, there must be a difference in chemical potential between the concentrated solution and the equilibrium condition. The larger the chemical potential difference, the greater the driving force for

Table 4–1 Solubility of Sucrose in 42 Dextrose Equivalent (DE) Corn Syrup at 30°C

Corn Syrup Solids (g/100 g water)	*Sucrose Solubility (g/100 g water)*	*Corn Syrup Content* (weight %)*
0	215	0.0
10	180	3.4
20	157	7.2
30	131	11.5
40	100	16.7
50	70	22.7

* Percentage of corn syrup solids in total solution (g/100 g solution) at saturation.

crystallization. As crystallization occurs in a closed system, this chemical potential driving force decreases until the equilibrium condition is attained and crystallization ceases. The thermodynamic driving force, or chemical affinity, ϕ, is defined by the difference in chemical potential between phases.[25]

$$\mu_s - \mu_l = -\Delta\mu = \phi \tag{4.10}$$

Here, μ_s and μ_l represent chemical potential of solid and liquid phases, respectively. Chemical potential is related to activity by

$$\mu = \mu^0 + RT \ln (a) \tag{4.11}$$

where μ^0 is a reference state chemical potential and a is the activity.

Solution Systems

For ideal solutions, which is typically not the case as related to crystallization in foods, activity is directly related to concentration; and Equations 4.10 and 4.11 can be combined to give a definition of chemical affinity in terms of solution concentration.

$$\phi = RT \ln\left(\frac{a}{a_{eq}}\right) = RT \ln\left(\frac{\gamma X}{\gamma_s X_s}\right) \approx RT \ln\left(\frac{C}{C_s}\right) \tag{4.12}$$

where, a_{eq} is activity at the equilibrium condition, γ is activity coefficient, γ_s is activity coefficient of the equilibrium solution, X is the mole fraction, X_s is the mole fraction at equilibrium, C is concentration, and C_s is solubility or saturation concentration. For small supersaturations ($C/C_s < 1.1$), a Taylor series expansion of the natural logarithm of supersaturation ratio ($S = C/C_s$) often is used to further simplify Equation 4.12.

$$\phi = RT(S - 1) = RT(s) \tag{4.13}$$

where

$$s = S - 1 = \frac{C - C_s}{C_s} \tag{4.14}$$

Typically, s or S is used as an approximation to supersaturation driving force, particularly at low supersaturations, since concentrations can be easily measured. Another expression often used to indicate degree of supersaturation is the concentration difference between actual solution and saturation conditions.

$$\Delta C = C - C_s \tag{4.15}$$

For high supersaturations ($C/C_s > 1.1$), it is more accurate to use

$$S' = \ln\left(\frac{C}{C_s}\right) \tag{4.16}$$

since the approximation of the Taylor series expansion does not hold. The parameter, S', is closer to the true chemical affinity given in Equation 4.12.

Note that the choice to use concentration, instead of activity, to represent driving force is not thermodynamically correct in all but the simplest situations (not typically found in foods). Nevertheless, concentration is used primarily due to ease of measurement. Some activity data are available for sucrose solutions,[52,53] although crystallization of sucrose has rarely been expressed in these terms.

The units used to express concentration affect the magnitude of the supersaturation value reported. Thus, any value of S should be accompanied by the corresponding units of C used to calculate S. For example, for a sucrose solution of 80% total solids at 20°C, the supersaturation ratio, S, is 1.1443 when concentration is given in g sucrose/kg water but only 1.0434 when concentrations are calculated as g sucrose/kg of solution.[54] The best choice of units depends on the particular system and on the expected results. Mullin and Sohnel[55] discussed the approximation of using concentration measurements instead of activity to define supersaturation. They concluded that the choice of concentration units for use in calculating supersaturation that best matched the true chemical affinity was dependent on the crystallizing system. The simplest units for concentration measurement (eg, weight percent for sugar solutions) may be used when simply comparing crystallization rates at different conditions, whereas units that minimize errors compared with thermodynamic activities are best suited for mechanistic studies of crystallization processes.

Generating Supersaturation

There are essentially three ways to generate supersaturation in solution systems, as shown schematically in Figure 4–14. The first is to heat the solvent prior to dissolving the solute so that high-concentration solutions can be made. Supersaturation is generated by cooling the highly concentrated solution below the temperature at which saturation occurs for that concentration, following line AB in Figure 4–14. The degree of supersaturation is related to the extent by which the solution is subcooled. Making rock candy at home is

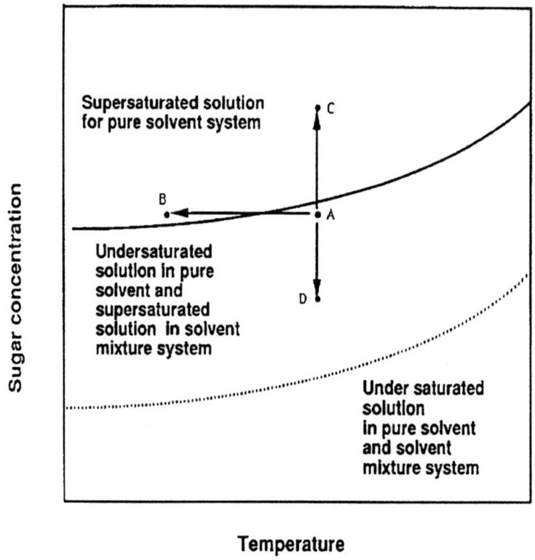

Figure 4–14 Generating Supersaturation in Solution Systems.

done in this way. Water in a pot is heated on the stove, and sugar is allowed to dissolve completely. After sufficient sugar has been dissolved, the pot is removed from the stove and allowed to cool. Sugar crystals then can form on a stick or string after sufficient time at room temperature.

The second method to generate supersaturation is to evaporate solvent from a solution until the saturation concentration has been exceeded, following line AC in Figure 4–14. The extent of supersaturation is related to the difference in concentration between the solution and the solubility concentration at that temperature. When generating supersaturation by evaporation, oftentimes there is a cooling effect as well due to evaporative cooling. Refined sucrose is produced in a vacuum evaporation pan where the rate of evaporation is used to control supersaturation and thus the rate of crystallization.

The third technique for generating supersaturation involves adding a second solvent in which the solute is insoluble. For example, adding alcohol to aqueous sugar solutions causes the sugar to crystallize out since sugar is insoluble in alcohol. The resulting solution, while diluted slightly due to solvent addition, has a net supersaturation due to the change in the solubility line (shown as the dashed curve in Figure 4–14). This technique also is called salting-out, based on the early practice of precipitating proteins from aqueous solution through the addition of salts. Examples of this method of generating supersaturation in food processing are scarce, although some liqueurs, for example, are based on this principle.

Melt Systems

For melt systems, such as water, pure triacylglycerols, and some organic chemicals, the driving force for crystallization is related to the difference between actual temperature and melting or freezing point. The chemical affinity is often written in this case as

$$\phi = \Delta H_f \frac{(T_f - T)}{T_f} \tag{4.17}$$

where ΔH_f is latent heat of formation, T_f is freezing (or melting) temperature, and T is actual temperature of solution.

Mixed Systems

For complex triacylglycerols or their mixtures, as found in natural fats and complex foods, supersaturation may be caused by a combination of solubility and melting point. Higher-melting triacylglycerols are soluble to some extent in more liquid (lower melting point) oils that act as a solvent.

In complex lipid systems with a wide range of triacylglycerols (eg, butter), the complex interactions between triacylglycerols and varying melting points make fundamental crystallization studies extremely difficult. For example, the high-melting components of milk fat (triacylglycerols with primarily long-chain, saturated fatty acids) are soluble to a certain extent in the more liquid components of milk fat (triacylglycerols with primarily short-chain, unsaturated fatty acids). An effective solubility of these high-melting components in the liquid component of milk fat has recently been measured (Liang and Shi, unpublished data, 1987) and is shown in Figure 4–15. The solubility of these high-melting triacylglycerols from milk fat in canola oil also is shown in Figure 4–15. The high-melting triacylglycerols of milk fat are less soluble in canola oil than in the low-melting milk fat components, which is most likely due to the differences in molecular structure (chain length and degree of unsaturation) between the triacylglycerols.

In cases where high-melting components are solubilized into low-melting lipid components, crystallization of lipids can be treated as a solution system. However, this approximation only applies at temperatures close to the melting point (solubility curve) of the fat. At lower temperatures, the intermediate- and low-melting triacylglycerols begin to crystallize, and the system exhibits a combined melt and solution phase behavior. Definition of a supersaturation driving force when both solution and melt behavior are important is difficult, and no fundamental definition of a driving force has been proposed for such a system.

Metastability

Although it is necessary to attain supersaturation before crystallization can occur, not every supersaturated solution will crystallize during the time available for observation.

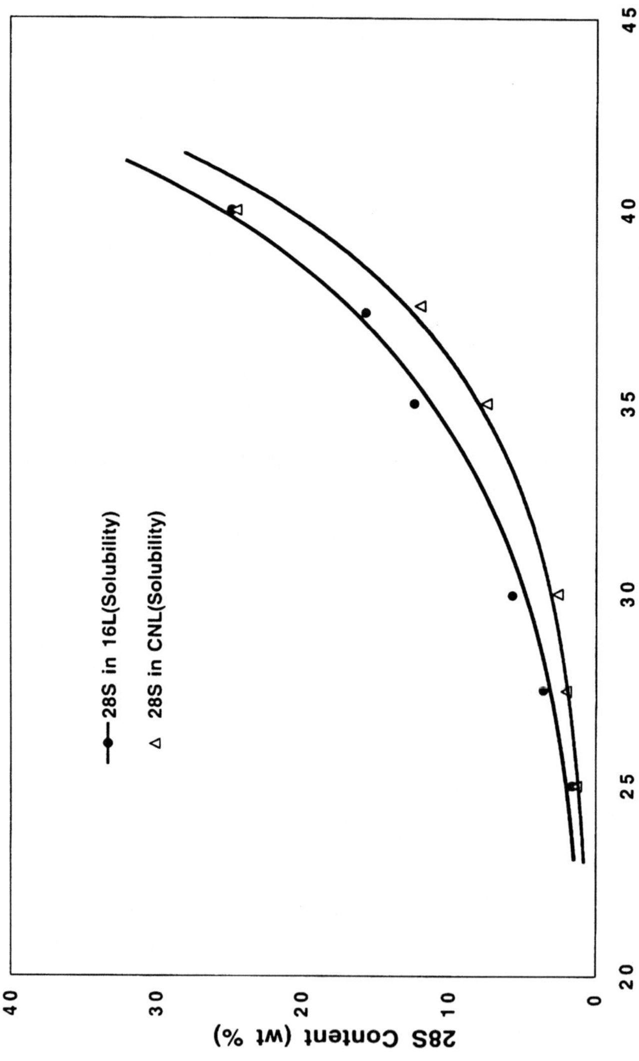

Figure 4–15 Solubility of High-Melting Components (28S) in Low-Melting Components (16L) of Milk Fat or Canola Oil (CNL).

There are many instances where materials form metastable solutions that are supersaturated, but do not crystallize for extended periods of time. Thermodynamically, there is a driving force for crystallization once a supersaturation has been developed. However, other constraints may inhibit formation of nuclei. In many systems, the energy barrier that must be overcome to form a stable nucleus is large enough that low levels of supersaturation do not provide sufficient impetus for crystallization in a reasonable time period. Once crystals do form, however, they grow at any level of supersaturation and continue growing until equilibrium has been attained. Thermodynamically, there is always a driving force for crystallization once the saturation or equilibrium condition has been surpassed (in the metastable zone), but the time scale required for crystal formation to occur may be too long to be observed on a practical scale. These systems are said to be metastable.[44]

Previously, some researchers, particularly those studying sugar refining, defined a metastable zone or region of supersaturation where nucleation does not occur within the time scale of their process. However, this definition of the metastable supersaturation zone (where nucleation does not occur in a certain time) depends on the time scale employed for measuring it and the nature of the crystallizing system. For example, supersaturated sucrose solutions at $S = 1.3$ (based on concentrations in g sucrose/100 g solution) crystallize within a few minutes, at $S = 1.2$ it may take 20 to 30 hours, and at $S = 1.1$ it takes 150 to 200 hours.[56] Lower supersaturations require even longer times before crystallization occurs, well beyond the scope of commercial refining operations. Thus, sucrose solutions at $S < 1.2$ are effectively metastable for the time scale required for crystallization under conditions of importance in sugar refining.[57] However, given enough time at this condition, a solution will eventually crystallize. In systems where the time scale may be equivalent to the shelf life of the food (as in an ungrained caramel, for example), the metastable zone width is essentially nonexistent since the time scale of importance may be over a year.

The extent of this effective metastable zone depends on a number of parameters. The type of system plays an important role since the activation energy for nucleation depends on the molecular, atomic, or ionic structure in solution. Metastability is often demonstrated on "supersolubility" diagrams,[58] as shown for sucrose in Figure 4–16. Similar plots for other sugars, including lactose, glucose, and fructose, also have been published.[54,58] Singh[59] shows similar behavior, although he calls it a supersaturation curve, for crystallization of partially hydrogenated soybean oil. The supersolubility diagram partitions solutions into several zones. In the region below the solubility line, at low concentration and high temperature, the solution is undersaturated. Crystals put in this solution dissolve as the saturation concentration is approached. In the region just above the solubility curve, the solution is supersaturated but metastable, where nucleation will not occur over "normal" time spans. At lower supersaturations within the metastable zone, closer to the solubility curve, longer times are required for nuclei to form. However, any crystals present will grow at this condition, and the supersaturation decreases until equilibrium has been reached. At sufficiently high supersaturations, above the supersolubility line, nucleation occurs spontaneously. This region is sometimes called the labile zone, where crystallization occurs rapidly at these high supersaturations. The supersolubility

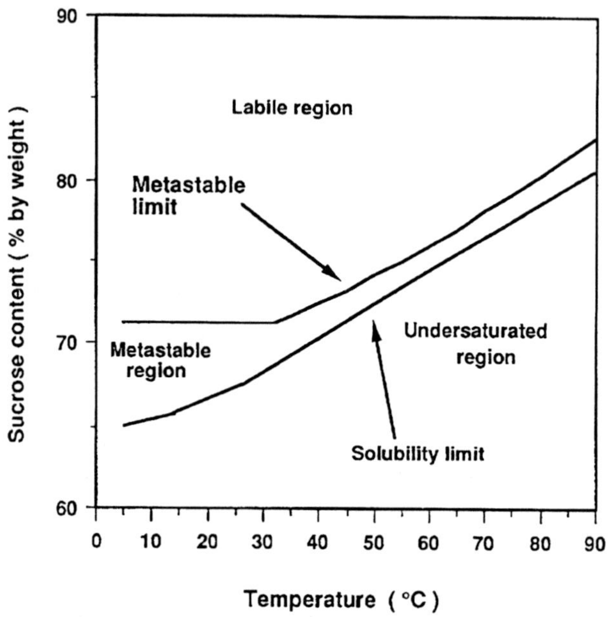

Figure 4–16 Metastable Zone Width for Sucrose.

boundary is not a well-defined curve, but rather an approximate range of conditions that defines relative stability of the solution and immediate, spontaneous nucleation.

Many people define another zone—the forced nucleation or "false grain" zone—within the upper region of the metastable zone. Typically, for refining of sucrose, this falls between supersaturation ratios of $S = 1.2$ to 1.3.[57] In this zone, nucleation can be forced by agitation in the presence of existing crystals through the mechanism of secondary nucleation (to be discussed in Chapter 5). Secondary nucleation occurs due to the presence of these existing crystals. If these crystals were not present, the solution would be metastable and nucleation would not occur. In sucrose refining, crystallizers operate at the upper limit of the metastable zone (about $S = 1.2$), where existing crystals grow but no new ones nucleate. This is done in order to grow a population of seed crystals up to the desired product size in the minimum amount of time (at the highest possible supersaturation). If, however, the solution goes into the "false grain" zone ($S > 1.2$), nucleation is likely to occur and a crop of small crystals develops. The size distribution in the crystallizer then contains the larger product-sized crystals plus the fine crystals formed due to false grain formation. This crystal size distribution is not easily separated in a centrifuge, and the process is not very efficient.

Again, the lines marking the boundaries of these regions are not clearly defined and depend to some extent on other conditions in the crystallizer. Agitation, in particular, affects these boundaries, with higher agitation levels reducing the supersaturations that separate these regions. A solution under high agitation is generally more prone to nucleation than a quiescent solution. Another primary factor is the presence of insoluble foreign impurities, like dust particles. These tend to catalyze the formation of nuclei and also reduce supersaturation levels of the above regions. Soluble impurities or additives, on the other hand, may extend these boundaries of the metastable zone by inhibiting nucleation. In order to control and predict crystallization, it is critical to have a good understanding of these boundaries. Through years of experience, these are fairly well known for established processes such as sucrose refining, but they are not defined for many practical crystallization processes in the food industry.

Another point to note from Figure 4–16 is that the extent of the metastable zone may not be uniform over the range of temperatures from 0 to 90°C. Between 30 and 90°C, the width of the metastable zone for sucrose is fairly uniform, whereas at lower temperatures the zone becomes wider. Note that below about 30°C, the metastable zone limit has a fairly constant concentration despite the decreasing solubility concentration. This occurs, in part, due to increasing viscosity at these lower temperatures, which plays an inhibiting role on nucleation. However, each system is different and may result in different behavior.

Again, it must be stressed that this metastable zone width is only an artifact of the processing conditions and the time required for nuclei formation in some systems. Once a system is supersaturated, crystallization will eventually occur. However, the time scale for nucleation may be very long under certain circumstances, and an understanding of this effective metastable zone width may be used to certain advantage to control processing conditions.

Effects of Impurities/Additives on Phase Equilibrium

Many additives (intentionally added) and impurities (unintentionally added) impact the phase equilibrium of crystallizing species. This is quite important in most crystallizing systems, particularly foods, where product composition is varied for reasons related to texture and flavor. Most foods are complex combinations of proteins, starches, sugars, salts, and fats, either in aqueous or lipid phases. The solubility or melting point of a food component may be affected by many of the other ingredients. For example, salts and other sugars have a significant effect on sucrose solubility in water, and this may have a dramatic impact on sucrose crystallization.

Sugars

Most substances, including fructose and glucose, decrease the solubility of sucrose, although some salts (potassium chloride [KCl] and NaCl most notably) actually increase the solubility of sucrose. Typically, other solutes that compete for the water as a solvent will decrease solubility, although this effect is thought to be related to the water-structuring capabilities of the other component. Sometimes solubility increases on addition of an impurity or additive such as when a soluble complex forms between the sugar and the salt. This occurs in lactose, for example, where calcium (Ca) ions cause formation of a soluble lactose complex.[60]

Invert sugar, an equimolar mixture of fructose and glucose caused by inversion (hydrolysis) of sucrose, significantly decreases the solubility of sucrose in water, as shown in Table 4–2.[32] This may have commercial importance in sugar-based foods that are heated since inversion occurs more readily at higher temperatures. Products where sugars are cooked for significant time include hard candy and fondant. Table 4–2 also shows that the maximum concentration that can be achieved in solution when two sugars are added together is higher than the solubility of either sugar. This has important ramifications for texture and stability (eg, microbial) of foods.

Another commercially important example of mixed sugar effects is the addition of corn syrup to products containing sucrose in order to control crystallization, as is quite common in many confectionery products (ie, hard candies, fondants and creams, caramels, etc). The addition of corn syrup causes the solubility of sucrose to decrease, the extent of which is dependent on the amount of corn syrup added. Tjuradi and Hartel[51] published solubility data showing the effect of corn syrup solids on solubility of sucrose (Figure 4–17). Ice creams and caramels are examples where mixed sugars occur in foods, and controlling crystallization is critical to product quality. In these products, there is a mixture of lactose and sucrose (and perhaps corn syrup as well). The presence of lactose depresses the solubility of sucrose, and vice versa.[54,61] Figure 4–18 shows the effect of sucrose on the solubility of lactose in solution. Further discussion of the effects of various components on solubility of sugars can be found in Hartel and Shastry.[54]

Ice

It is well known that the addition of small molecules (salts, sugars, etc) causes a depression in the freezing point of water. This colligative effect is based on the number

Table 4–2 Effect of Invert Sugar on Solubility of Sucrose at 30°C

Invert Sugar Solids (g/100 g water)	*Sucrose Solubility (g/100 g water)*
0	213.6
20	206.7
40	200.1
80	188.1
160	167.3
200	159.3
300	147.3
400	140.4

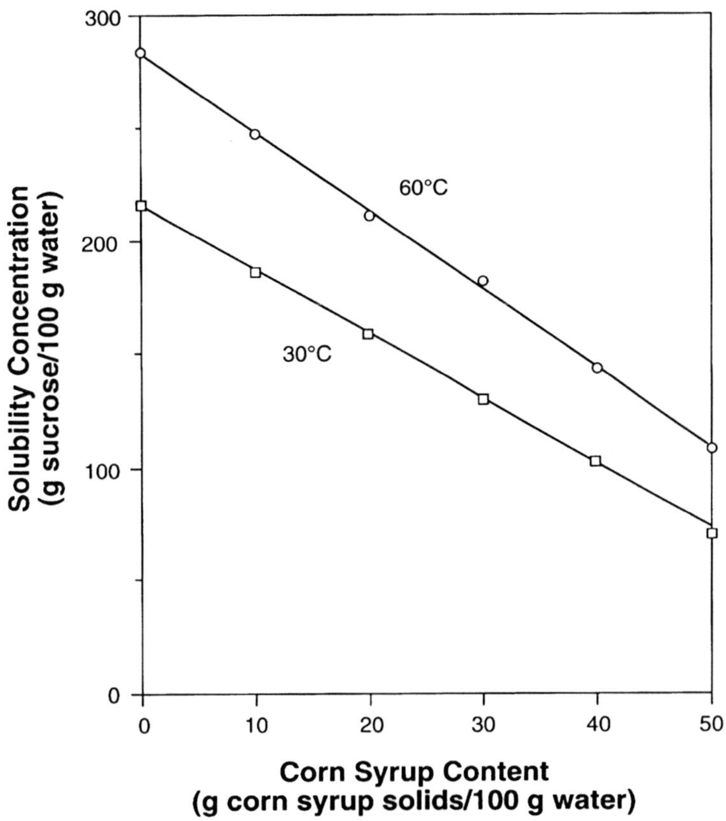

Figure 4–17 Solubility of Sucrose in the Presence of 35 DE (Dextrose Equivalent) Corn Syrup at 30°C and 60°C.

and size of molecules added to the water, and it can be calculated using Equations 4.3 and 4.4.

The decrease in freezing point as concentration increases has important ramifications on texture and stability of frozen foods. When a product is frozen slowly, the heat of fusion released on freezing is sufficient to increase the temperature in the vicinity of the ice crystal to the melting point. As the solution freezes, however, the concentration in the solution increases (freeze concentration) and the freezing point temperature decreases according to the solutes present. Thus, slow freezing leads to a freezing curve (temperature

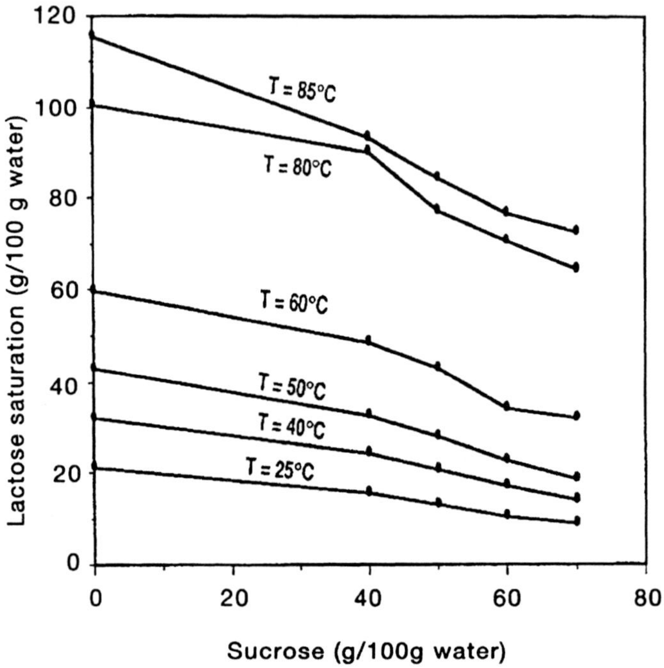

Figure 4–18 Solubility of Lactose in the Presence of Sucrose.

versus time) that essentially follows the freezing point depression curve once nucleation has been initiated (Figure 4–19). For example, ice cream may be frozen slowly so that the process essentially follows the path shown in Figure 4–19. As temperature is decreased, additional ice is formed so that the product temperature is approximately the same as the freezing point of the remaining solution phase. Actually, the temperature must be slightly below the freezing point so that a driving force for crystallization always remains during freezing. In the case of slow cooling, the amount of water frozen into ice at any temperature can be calculated from a mass balance. The ice cream industry uses a plot of percent water frozen at any temperature (Figure 4–20) to indicate the amount of ice in ice cream stored at any temperature.[62] Note that this calculation assumes that the equilibrium phase volume of ice is present at any temperature.

Sakai and Hosokawa[63] compared five different methods for calculating ice phase volume at equilibrium for salt and sugar solutions. The simple mass balance approach using the freezing point depression curve noted above works well for dilute solutions.

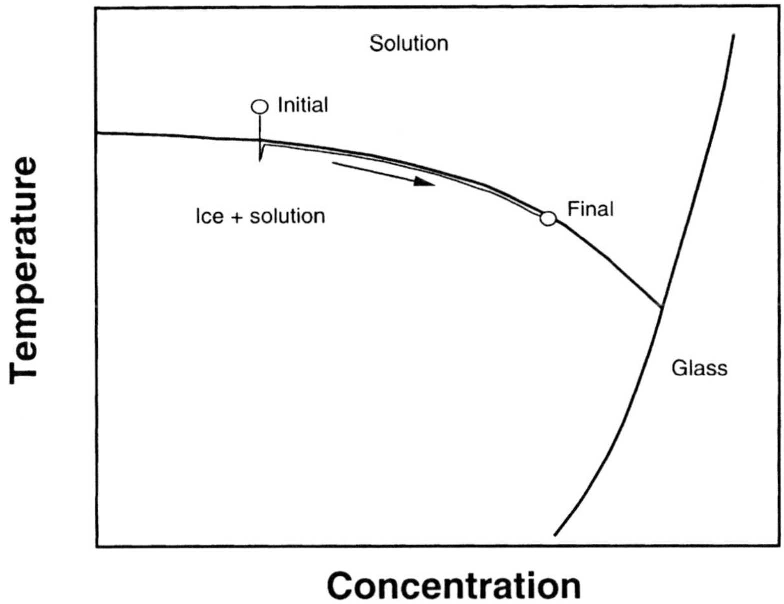

Figure 4–19 Path (Shown Schematically) on Phase/State Diagram for Slow Freezing.

However, the inaccuracies in drawing the freezing point curve at higher concentrations led to a loss of accuracy in the ice content. The best results were found when a method based on an enthalpy balance as the ice in the food melts was used. This method required measurement of the enthalpy curve (from DSC, for example) of the food in the range of interest.

Lipids

Natural lipids exhibit complex phase behavior, and simple binary phase diagrams are not often used due to the large number of triacylglycerol species that are normally present. For example, cocoa butter (one of the simplest natural fats) contains three main triglyceride species (they make up between 70 and 80% of cocoa butter) with several other triacylglycerols as important impurities. Milk fat, on the other extreme, contains over 400 triacylglycerol species, none of which is present at greater than 5 mol percent.[64] Thus, even the simplest natural lipids are beyond our capabilities of dealing with simple phase diagrams.

For mixtures of natural lipids, some measure of compatibility between different fats is needed, and simple phase diagrams are not appropriate due to the large number of

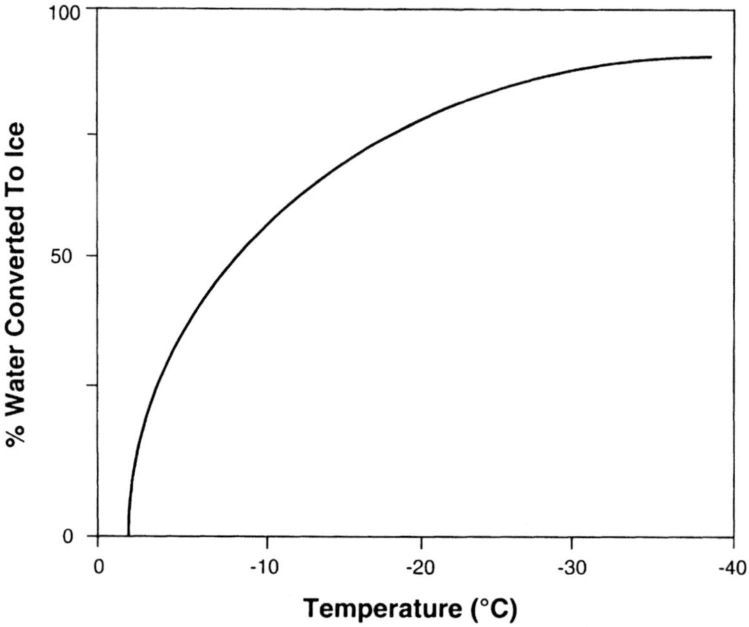

Figure 4–20 Percentage of Water Frozen at any Temperature for Slow Freezing of a Typical Ice Cream.

molecular constituents in each fat. For example, when milk fat is added to cocoa butter in chocolate or to palm kernel oil in compound coatings, the phase behavior plays an important role in determining the final product texture (hardness). However, there are more than 400 molecular species that are likely to interact when these fats crystallize together. One approach to this problem is to treat the two fats as single entities, not materials composed of many molecular species, and develop binary phase diagrams for the mixtures of the two fats. Timms[26] studied mixtures of milk fat and cocoa butter in this way. The two fats were crystallized together at different mass ratios and the samples stored to produce the most stable polymorphic form and the maximum amount of crystalline material. The fats were considered to be fully crystallized when no change in calorimeter (differential scanning calorimeter) curves or X-ray spectra were observed. Solidus and liquidus lines (representing the phase boundaries) were obtained by slow melting of the sample in a DSC. The solidus boundary was defined as the onset temperature of the first melting peak and the solidus temperature as the endset of melting. These data were corroborated with X-ray spectroscopy to delineate regions where different polymorphs existed. An example of such a phase diagram is shown in Figure 4–21. Although this approach has potential to help us understand phase behavior of fats, it has not been used widely due to the time required to generate the stable forms and the difficulty in interpreting the DSC scans.

A simpler technique that allows one to determine phase compatibilities utilizes NMR to measure solid fat content (SFC) of fats at different temperatures. This melting profile is a form of phase diagram that represents the amount of crystal content at any temperature. Figure 4–22 shows the changes in NMR melting curves as cocoa butter and anhydrous milk fat are mixed at different ratios. The primary advantage of this technique is the relative ease of experimental measurement, since SFC curves may be generated in only a few hours. In this technique, a fat is tempered (controlled cooling and heating) to produce a condition close to equilibrium crystallization. Tempering profiles for fats are recommended by both International Union of Pure and Applied Chemistry (IUPAC)[65] and American Oil Chemist's Society (AOCS)[66] and generally involve cooling to promote initial crystallization and then warming to promote equilibration prior to sample analysis by NMR. Differences between the two tempering methods are primarily temperatures used and the holding times at elevated temperature, as seen schematically in Figure 4–23. Some fats (milk fat, etc) solidify rapidly enough that shorter times are sufficient (AOCS method), whereas other fats (cocoa butter in particular) require extended times (IUPAC method) to ensure full crystallization and formation of the most stable polymorph.

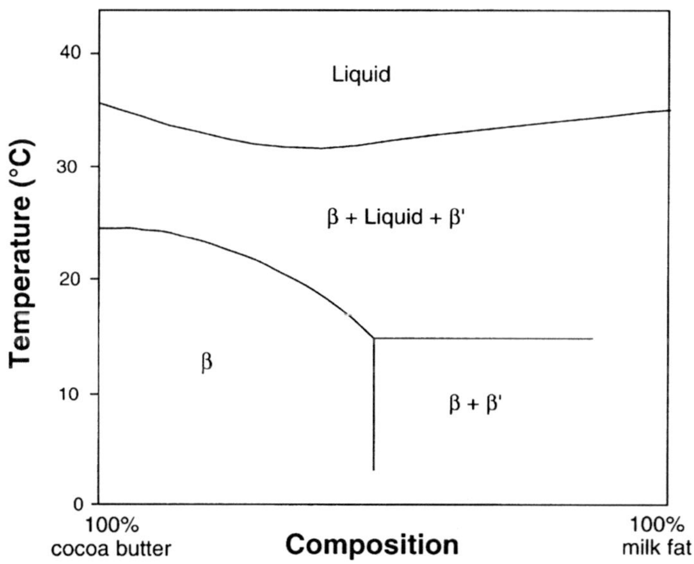

Figure 4–21 Phase Diagram for Mixtures of Milk Fat and Cocoa Butter.

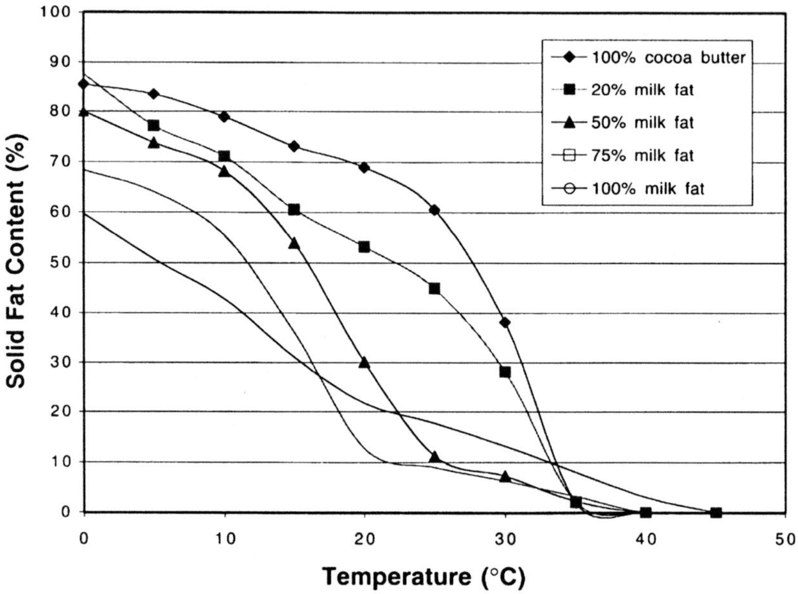

Figure 4–22 Solid Fat Content (Measured by Nuclear Magnetic Resonance Spectroscopy) Curve for Mixtures of Cocoa Butter and Anhydrous Milk Fat (AMF).

To generate a type of phase diagram for mixtures of two fats, SFC curves are generated for several mixtures of the two fats (from 100% of fat A to 100% of fat B), as seen in Figure 4–22 for cocoa butter and anhydrous milk fat. The SFC data are then recalculated for each mixture level to give the SFC at specific temperatures (or sometimes as temperature for a given SFC). Lines of constant SFC are plotted[26] for different temperatures against the level of mixture of the two fats, as seen in the isosolid diagram of Figure 4–24. Alternatively, lines of constant temperature can be plotted as functions of the SFC for given mixture levels.[67]

The isosolid diagram shows whether two fats exhibit phase compatibility.[67] If the isosolid lines are straight between the pure fats (Figure 4–25A), the two fats are completely compatible from a phase behavior standpoint. This means that the triacylglycerols of the one fat crystallize together with the triacylglycerols of the other fat. This behavior is seen, for example, between mixtures of cocoa butter and a cocoa butter equivalent, which is designed to contain essentially the same triacylglycerols as cocoa butter. When the isosolid diagram for the mixture of two fats exhibits straight lines but have dissimilar melting profiles, the two fats are still compatible but show significant softening (Figure 4–25C).

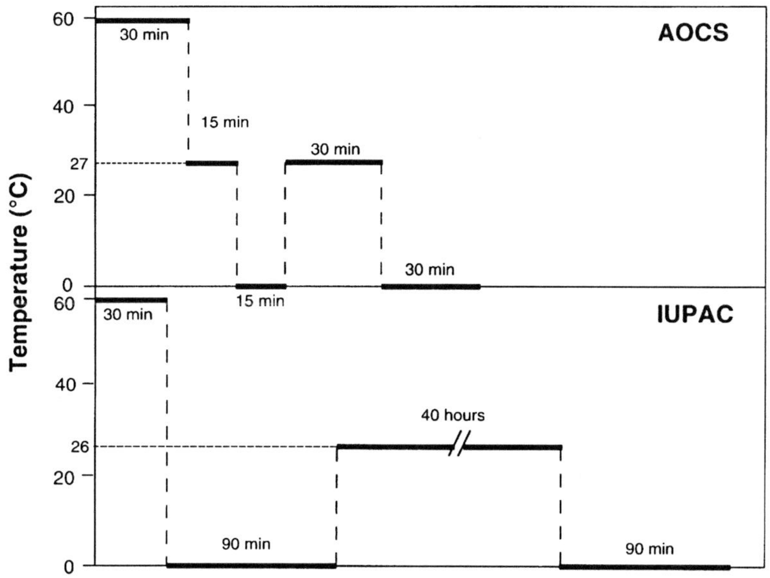

Figure 4–23 Comparison of Tempering Profiles Recommended by AOCS and IUPAC.

In this case, the triacylglycerols of the two fats mix well enough together that there is only a dilutional softening between them. This behavior occurs, for example, between mixtures of cocoa butter and peanut oil. When the triacylglycerols of the two fats impede the crystallization of each other, a eutectic softening occurs. In this case, the isosolid lines fall at temperatures well below the straight line connecting the melting profiles of the two pure fats (Figure 4–25B). Such a eutectic softening is seen between cocoa butter and many vegetable fats (like palm kernel oil). Even the addition of milk fat to cocoa butter causes a significant eutectic softening at levels of about 30% addition of milk fat (Figure 4–24). Below 30% addition, there is only a dilutional softening of the cocoa butter due to addition of milk fat triacylglycerols. An in-depth analysis of isosolid diagrams can show the levels of incompatibilities between different compositional elements of fats. For example, the intermediate-melting components of milk fat have a different effect on cocoa butter than the low-melting components.[68] Isosolid diagrams for mixtures of milk fat with confectionery fats have been studied by Williams et al[69] and Ali and Dimick.[70]

Marangoni and Lencki[46] studied the phase behavior of mixtures of three distinct component fractions of milk fat, based on the classification of Timms.[71] Binary isosolid diagrams showed that no eutectics were formed in any of the binary mixtures, although monotectic, partial solid solution behavior was found in mixtures with low-melting frac-

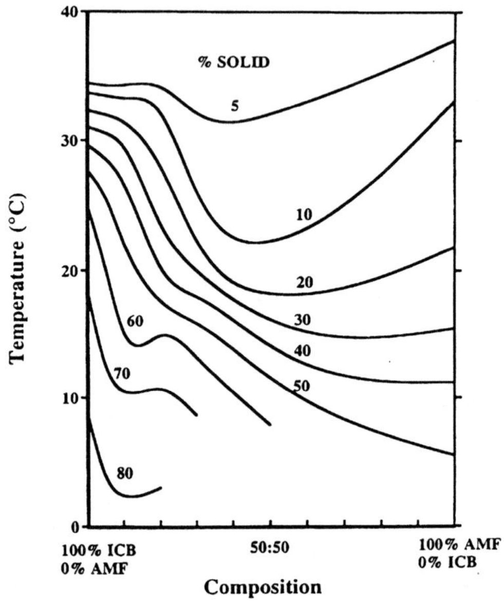

Figure 4–24 Isosolid Diagram for Mixtures of Anhydrous Milk Fat (AMF) and Cocoa Butter (ICB).

tion. Ternary phase diagrams for mixtures of the three fractions also were drawn as if each fraction was a pure component.

GLASS TRANSITIONS

Various materials, including some polymers and many food components, transform into a noncrystalline solid-like state under certain conditions. For example, at high concentration of sugar and low temperature, the reduced molecular mobility leads to formation of an amorphous material, rather than a crystal. In the amorphous state, the material is a very viscous fluid that does not flow over long periods of time. At some point with increased concentration or reduced temperature, the amorphous material goes through a transition to a glassy state. The structure of a glassy material is significantly different from the crystalline lattice, even though a glass acts like a solid material in many ways. The energy associated with the transition from a highly viscous fluid to a glass is significantly less than the energy associated with formation of the crystal lattice (latent heat of fusion) from solution. This difference can be used to distinguish glasses from crystals by using a DSC. Another technique used to distinguish between glasses and crystals is X-ray diffraction. An example of the different X-ray spectra produced from sugar

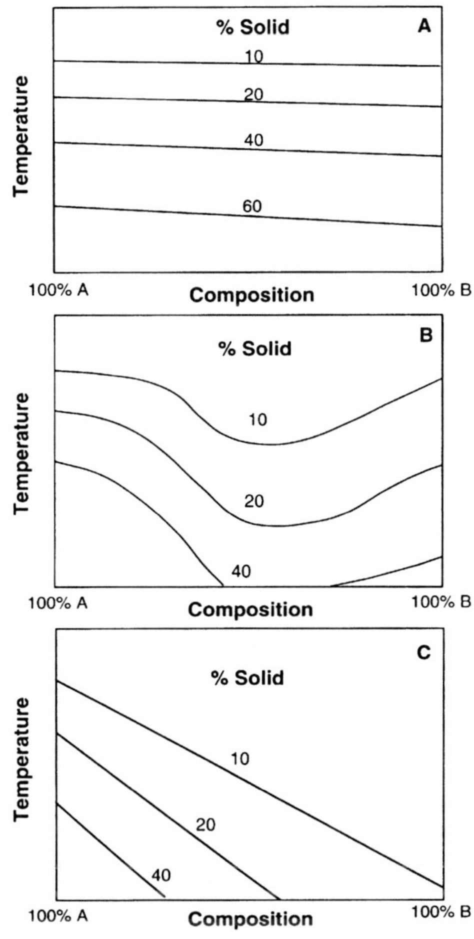

Figure 4–25 Isosolid Diagrams (Highly Schematic) for Mixtures of Two Fats: (A) Compatible Fats with Similar Melting Points; (B) Incompatible Fats Showing Strong Eutectic Softening; and (C) Compatible Fats with Different Melting Points Showing Dilutional Softening.

crystal and sugar glass was shown previously in Figure 3–17. In an amorphous glass, there is no long-range order and an X-ray spectrum gives only a broad, diffuse scattering pattern. In a glass, the molecules have a random orientation with solvent molecules distributed throughout a close-packed, solute molecule matrix. Thus, X-rays scatter randomly when passed through a glass. Crystals, with a characteristic lattice structure, scatter X-rays in well-defined ways based on the molecular spacings. Thus, the X-ray spectra of crystals contain many sharp intensity peaks that can be used to differentiate the type and geometry of the molecular spacings in the lattice structure.

Many materials exhibit a continuum of flow characteristics depending on their composition or external conditions. Many foods follow these general patterns, acting like a liquid under certain conditions yet exhibiting characteristics of a solid under other conditions. Typically, there is a gradual transition between the two, where the solid-like characteristics gradually increase as temperature decreases and/or concentration increases, as seen in Figure 4–26. Using a sweetened fruit juice as an example, let us consider conditions that lead to the glass transition. Dilute juice at almost any temperature has flow characteristics of a liquid, although the viscosity increases as temperature goes down. Likewise, a more concentrated sugar juice has higher viscosity at all temperatures. At high temperatures, this concentrated juice flows easily as a liquid. As temperature decreases, the viscosity increases dramatically. If the water content is low enough (10 to 15%), it will set into a semi-solid material, like a fruit leather product, at room temperature. This material is in the rubbery or leathery region according to Figure 4–26. If the juice is highly concentrated (less than 5% water), it still flows easily as a very viscous liquid at very high temperatures, but rapidly becomes more viscous at lower temperatures. If cooled rapidly to room temperature, it will pass through the glass transition zone and the resulting product has the hard, brittle characteristic of a glass, although it really is a very concentrated liquid in this form. As suggested in Figure 4–26, there is a range of physical properties that can be obtained depending on the state (temperature and concentration) of the material.

Many food ingredients (sugars, complex carbohydrates, proteins, sugar alcohols, organic acids, and even water) exhibit glass transitions, and this has become an important concept in food processing.[72] Understanding the glassy state is important in controlling crystallization in many food products. For example, hard candies are sugar glasses typically made from some mixture of sucrose and corn syrup. The amount and type of corn syrup present determine many of the characteristics of hard candy. These characteristics include sweetness and mouthfeel, hygroscopicity, and the likelihood of crystallization (graining defect) of the candy during storage. The glass transition temperature plays an important role in these properties. Another common food product where glass transition is critical to controlling crystallization is spray-dried milk powders. Milk contains sufficient lactose that when it is concentrated and dried, the lactose forms into the glassy state. If handled improperly (warm temperatures and high relative humidity), the lactose glass can convert to the crystalline form, which negatively affects the physical characteristics (eg, flowability) of the powder.

The glassy state can be attained in several ways. Rapidly cooling a highly concentrated solution causes solidification into the glassy state, as in production of hard candy from a concentrated sugar syrup. By cooling quickly, the sucrose molecules do not have sufficient time to rearrange and orient into a crystal lattice structure. The glass transition occurs when the viscosity of this highly concentrated solution is so great (typically thought to be $> 10^{12}$ to 10^{14} kPa) that no liquid characteristics are observed over reasonable time scales (eg,

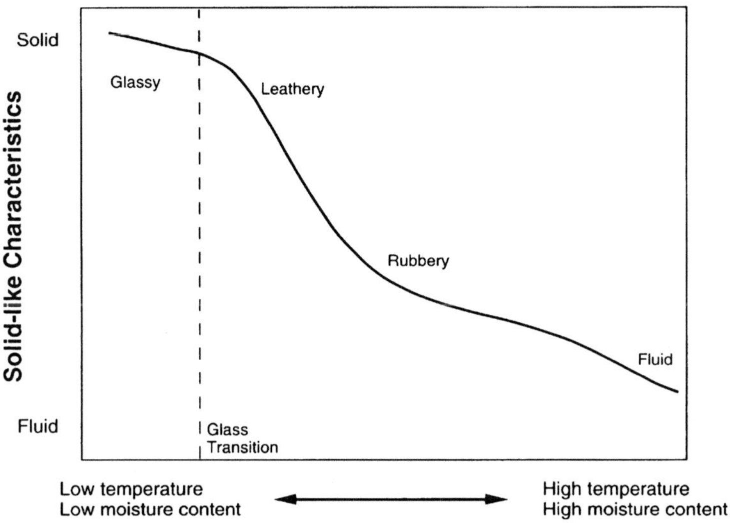

Figure 4–26 Solid-Liquid Characteristics of Glass-Forming Materials as Temperature and/or Water Content Changes.

the shelf life of the product). Glasses also can be produced by drying a solution rapidly, as in spray drying. Here, small droplets are formed in an atomizer, and these droplets are exposed directly to warm, dry air. Since each drop has a large surface area, the droplet dries quickly and rapidly approaches the glass transition point. A third possibility for producing a glass is from a melt of the material. When cotton candy is made, sucrose goes through several transitions but ultimately ends up as a glass. Crystalline sugar is melted in the center of the cotton candy spinner and the fluid melt forced out the small holes as the spinner rotates. When the fine strands of sugar melt contact room air, they cool rapidly and set into a glassy state.

As a material passes through the glass transition point, whether by cooling or increasing concentration, its viscosity increases dramatically. The WLF relationship, developed by Williams, Landel, and Ferry,[73] describes this increase in viscosity at temperatures near the glass transition temperature (T_g). The WLF equation can be written in the following form[23]:

$$\ln\left(\frac{\mu}{\mu_s}\right) = \frac{-C_1(T - T_s)}{C_2 + (T - T_s)} \tag{4.18}$$

where T is temperature; μ is viscosity; μ_s is the viscosity at some reference temperature, T_s; and C_1 and C_2 are constants that describe the relationship. Typically, T_s is given as the glass transition temperature, T_g, of the system. Other equations also have been proposed to fit the relationship between viscosity (or any relaxation phenomena) and temperature in the vicinity of the glass transition.[23]

Other thermal or physical parameters also change as a material undergoes a glass transition,[23] and these changes can be used to characterize the glass transition region. These other techniques include specific heat (C_p), mechanical moduli, dielectric constant, and specific volume. On a DSC, a change in C_p is observed as a shift in the baseline as the material is slowly heated through the glass transition region. This slight shift in the baseline in a DSC often is used to identify the T_g of a material. As might be expected, such a small shift often might be missed due to other transitions (melting, etc) within the product. For example, it often is difficult to measure the T_g in frozen foods because the enthalpic peak for melting of ice obscures the small glass transition shift. Similar problems may be encountered in foods containing lipids that transform from solid to liquid in the region of the glass transition. Often it is helpful to perform a derivative function on a DSC scan to highlight the glass transition, as seen in Figure 4–27.

Changes in mechanical properties (expansion coefficient, specific volume, etc) as a material goes through a glass transition also can be used to characterize T_g. Dynamic mechanical thermal analysis (DMTA) measures the changes in mechanical properties (storage modulus) as the material passes through the glass transition zone. A fairly sharp decrease in storage modulus (which represents the elastic nature of a material) occurs as the material changes from the solid-like glassy state to a more fluid state above T_g. However, the point of this sharp decrease in storage modulus is dependent on the frequency of the forces applied to the material in the DMTA. Thus, values of T_g obtained from DMTA are not always consistent with those obtained from DSC.

NMR and electron spin resonance (ESR) spectroscopy have become important techniques for characterizing glass transitions based on changes in molecular relaxations,[23] as described in Chapter 3. Dielectric thermal analysis utilizes the sharp change in dielectric properties in the vicinity of T_g to characterize the glass transition.[23] A rapid increase in dielectric constant occurs as the material goes from the glassy state to the more fluid-like state above T_g since dielectric constant is related to molecular mobility. The response with this technique is dependent on the frequency of the electric field applied to the food,[74] in a similar manner as the frequency dependence in DMTA. Comparison of T_g values found by the different techniques is often difficult since different measures of increasing molecular mobility are being measured.

An important characteristic that distinguishes glasses from liquids or crystals is the order of the relaxation time of the mechanical properties, which is related to the molecular mobility and viscosity.[23] In liquids, both solute and solvent molecules can move about relatively freely. Relaxation times of the mechanical properties are very short in liquids, whereas in glasses, relaxation times are much higher (may be on the order of 10^{12} years). Thus, the relative mobility of molecules in glasses is extremely low, and it is difficult for the molecules to rearrange into more ordered states. For this reason, food materials maintained

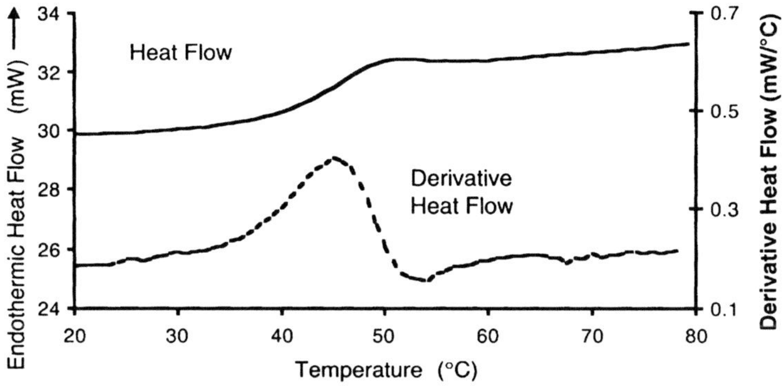

Figure 4–27 Differential Scanning Calorimetry Heating Curves To Identify the Glass Transition. Top line (solid) is normal scan; bottom line (dashed) is derivative curve.

in the glassy state are quite stable—processes that depend on molecular mobility (eg, diffusion) occur very slowly. However, there is a continuum of relaxation times so that the closer a glass is stored to its glass transition point, the more mobile the molecules become. If a glassy material is stored above its glass transition point, the molecules have sufficient mobility that they will eventually rearrange and form crystals, given enough time. In this "rubbery" or "leathery" amorphous phase, crystallization may occur over long time periods, although the time scale at which it occurs depends on the kinetics of the system. Thus, foods that are maintained in this amorphous or rubbery region exhibit varied degrees of stability.

State Diagram

The glass transition point of a food depends on many parameters, including water content (or concentration of solute) and type of solute. There is a well-defined relationship between temperature and composition, as described by the glass transition curve. The glass transition curve can be overlaid on a phase diagram to indicate regions of stability, metastability, and instability in foods. Such a modified phase diagram, or state diagram, is shown in Figure 4–28 for a simple binary system (sucrose in water). In the state diagram, the equilibrium solid-liquid relationships for both solvent and solute can be found in addition to the glass transition curve. Roos and Karel[72] discuss the application of the state diagram to food processing situations.

The state diagram indicates regions of varying stability. Below and to the right of the glass transition curve is a region of high stability where the material is in the glassy state. This is not an equilibrium condition, but rather a metastable state of high stability. In general, reactions or processes that are governed by a mass transfer or diffusion step either

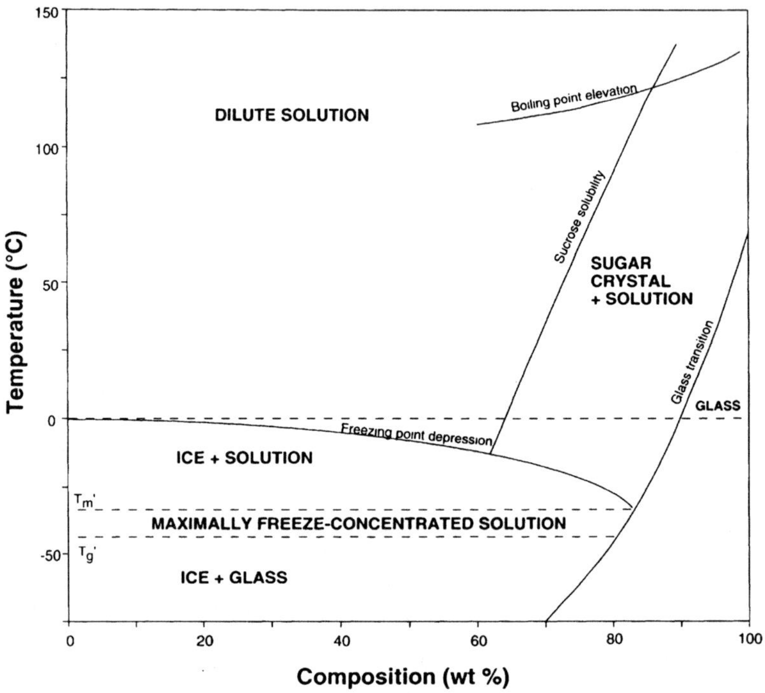

Figure 4–28 State Diagram for Sucrose-Water System Showing Regions of Crystallization.

do not occur or occur very slowly in the glassy state. Crystallization, for example, requires significant molecular movement, whether rotational or translational; thus, crystallization generally does not occur while a material is in the glassy state. This is particularly true for larger-molecular-weight materials (sugars, starches, proteins, etc). In this sense, foods that are maintained in the glassy state are stable and will not crystallize. For example, a hard candy stored with temperature and moisture content low enough that the product remains in the glassy state will not grain or crystallize even though held for long periods of time. The packaging layer must be sufficient to protect the candy from picking up moisture, and the temperature must remain below T_g of the product. Typically, the T_g of hard candy varies between about 25°C and 45°C depending on the final moisture content and the composition. Products with lower T_g generally have shorter shelf life.

Crystallization of solute occurs in the region bounded by the solute solubility and the glass transition curves. This is the region of solute supersaturation, as defined by the metastable and labile zones discussed previously. At conditions of temperature and concentration close to the solubility curve, crystallization rates increase with increasing supersaturation. However, as a solution moves closer to the glass transition curve, either by concentration or cooling, the viscosity increases dramatically and crystallization rates decrease due to diffusion limitations. As the glass transition curve is reached, crystallization is effectively prohibited over any reasonable time scale due to the limited mobility of molecules. Again, this is a dynamic (kinetically constrained) state and not an equilibrium state.

The region above the solvent equilibrium (freezing point depression) curve and to the left of the solute solubility curve is the region of undersaturated solution. Here, crystals dissolve and no crystallization can occur. The equilibrium line for the solvent, or freezing point curve, defines the conditions where solvent (water) can crystallize. Below the freezing point depression curve, ice forms in solution. If solute concentration increases above that defined by the eutectic point on the phase diagram, either a eutectic solid (solute and solvent) is formed or solvent continues to crystallize in a supersaturated solution of solute. The former situation occurs with salts, whereas the latter typically occurs with sugars.

Several transitions occur at very low temperatures (–30 to –50°C) in food products and biologic materials. Apparently, there is not full agreement among researchers on the exact cause of these transitions (usually seen by DSC) so different nomenclature exists in the current literature.[23] Rasmussen and Luyet[75] identified two transitions during melting of biologic materials and called them T_{am} (antemelting) and T_{im} (incipient melting). Roos and Karel[76] considered these transitions to indicate the point where the maximum amount of ice can be formed (maximally freeze concentrated solution) and the temperature for onset of melting, identified as T_g' and T_m', respectively.

The point that denotes the maximally freeze concentrated solution is called T_g'. This is the temperature at which the greatest amount of ice has formed and the remaining unfrozen phase is in the glassy state. T_g' values are typically found by analyzing DSC curves of frozen materials. To ensure that the maximal amount of ice has formed, it is common to anneal the samples at a temperature just above the suspected T_g'. The sample is then cooled below T_g' and scanned to obtain the true value. T_g and T_g' values for several sugars are listed in Table 4–3. Tables also are available for glass transition temperatures of various starches, sugars, and polymers of importance to the food and pharmaceutical industries.[13,23,77] The importance of T_g' is related to the stability of frozen foods. Levine and Slade[78] suggested that the rate of ice crystal growth and recrystallization approaches zero at T_g' due to the high viscosity. Thus, this is the temperature at which frozen foods are stabilized against ice crystallization for extended periods of time.

A state diagram also can be represented for polymeric molecules,[23] as seen for starch in Figure 4–29. In this case, there is no freezing point depression due to the large molecular weight of the solute. The state diagram is thus represented by ice formation at the freezing point of water (0°C) and the glass transition curve of the starch. It has proven difficult to measure T_g of dry starch, due to chemical changes upon drying, although it has been estimated to be between 227 and 243°C. T_g' values for starch fall between –5 and –7°C.[23]

Table 4–3 Glass Transition Temperatures (T_g), T_g for the Maximally Freeze Concentrated State (T_g'), and the Corresponding Concentration at That Point (C_g') for Several Sugars and Sugar Alcohols

Compound	*T_g (°C)*	*T_g' (°C)*	*C_g' (% wt/wt)*
Fructose	5	–57	82.5
Glucose	31	–57	80.0
Lactose	101	–41	81.3
Maltose	87	–42	81.6
Sucrose	62	–46	81.7
Raffinose	70	–36	84.1
Sorbitol	–9	–63	81.7
Maltitol	39	–47	82.9
Xylitol	–29	–72	80.2

Factors Affecting Glass Transition Temperature

Water content of a material plays a significant role in determining T_g. Water is an excellent plasticizer of many food components (starch, protein, sugar), and its presence generally reduces T_g significantly. For example, the T_g of pure sucrose is between 65 and 70°C; however, addition of only 2% water to a sucrose glass lowers the T_g to 50°C; at 5% water, the T_g is already below room temperature.[23] This strong dependence of T_g on water content can be seen by the sharp slope of the glass transition curve on the state diagram (Figures 4–28 and 4–29). Thus, very little moisture uptake is required to make a sugar glass unstable by lowering T_g below room temperature.

As a general rule, the T_g of a material is based on its molecular weight, with higher molecular weight resulting in higher T_g. However, this rule is not always strictly followed. For example, sucrose, maltose, and lactose have the same molecular weight, but significantly different T_g values[13,23] for dry material (65 to 70°C for sucrose, 87 to 92°C for maltose, and about 101°C for lactose). In addition, the hydrogenated form of the sugar molecule has significantly lower T_g. For example, glucose has a T_g of 30 to 35°C, whereas sorbitol has a T_g between –9 and –4°C.[23] Thus, molecular structure plays an important role in T_g of a material.

The T_g of mixtures of two components in a binary system can be found from the Gordon and Taylor equation below.[23]

$$T_g = \frac{X_1 T_{g1} + kX_2 T_{g2}}{X_1 + kX_2} \tag{4.19}$$

where X_1 and X_2 are mole fractions of the two components (1 and 2), T_{g1} and T_{g2} are the glass transition temperatures of the two components, and k is an interaction parameter

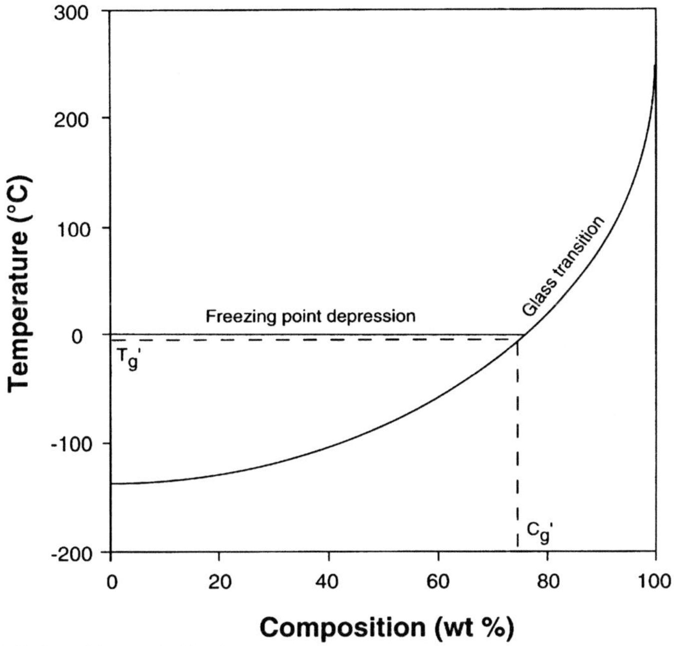

Figure 4–29 State Diagram for Starch.

---

related to the differences in heat capacity at T_g between the two components. Knowledge of the T_g and k values of the pure components allows one to calculate the T_g of any mixture; k values for simple sugars with water can be found in the literature.[23] For example, the glass transition curve for mixtures of sugars and water can be predicted reasonably well from the Gordon-Taylor equation with a T_g for amorphous water of $-135°C$.[23]

The T_g of high-molecular-weight molecules also increases as average molecular weight increases. Fox and Flory[79] give this relationship as

$$T_g = T_g^{\infty} - \frac{K}{Mn} \tag{4.20}$$

where T_g^{∞} is the limiting value of T_g at high molecular weight,[23] K is a constant, and Mn is the number-averaged molecular weight. This equation has been shown to hold for linear-chain polymers including maltodextrins,[80] but not for molecules with substantial branching.[81] Figure 4–30 shows the T_g of mixtures of dried sucrose and corn syrup components (glucose polymers) with different average molecular weights.[81] In general, a linear relationship was found between the T_g of the mixture and average molecular weight of the corn syrup component. The discontinuities in this linear relationship were due to the presence of branched chain oligosaccharides (from amylopectin) in these particular components. Thus, the molecular conformation (straight chain or branched) influences T_g for material with the same *Mn*.

In complex foods, it is often difficult to define completely the state diagram due to the complex mixture of ingredients. However, in some cases, a single component may dominate the state diagram of a food. For example, Jouppila and Roos[82] studied milk powders and found that lactose was the main component that determined the T_g of the powder. Fat content of the milk had no effect and neither did the proteins and salts normally found in milk. In this case, the physico-chemical behavior and physical properties of milk powders were predicted based on the lactose content.

Often, the choice of ingredients in a product is based on the desired texture and stability of the final product; glass transition principles play a role in this choice. In hard candies, for example, significant amounts of corn syrup are added to help stabilize the product against crystallization (graining) and change the physical properties (eg,

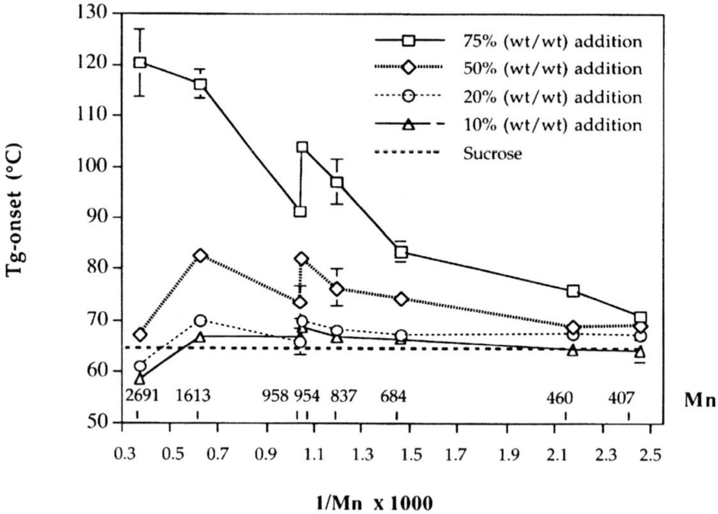

Figure 4–30 Effect of Molecular Weight of Starch Hydrolysates (Corn Syrup Components) on Glass Transition Temperature (T_g -Onset) of Sucrose Blends.

machinability) of the sugar mass. Since corn syrups contain a wide range of saccharide polymers with a wide range of molecular weights, it is important to select the correct corn syrup. The goal for hard candy manufacture is to ensure that the T_g is well above room temperature or even some maximum storage temperature (up to 35°C to 40°C in some locations), through proper formulation. A low-conversion (low-dextrose equivalent [DE]) corn syrup contains more long-chain polysaccharides and increases T_g of a mixture with sucrose at any level. On the other hand, a high-conversion (high-DE) corn syrup contains significant amounts of glucose (T_g around 30°C) and this may cause the T_g of a mixture with sucrose to decrease. Hard candies made with high-DE corn syrup will be less stable than those made with low-DE corn syrup due to the lower T_g. Table 4–4 shows how the T_g of sucrose-corn syrup mixtures changes with the amount of corn syrup added to two types of corn syrup.[83] Although it is important to note the effect of corn syrup addition on T_g, there are other factors that also influence product stability. Hygroscopicity is affected by choice of corn syrup, with lower-DE corn syrup resulting in glasses that are more hygroscopic (as well as lower T_g). Another factor to consider is the effect of corn syrup on inhibiting sucrose crystallization. At high levels of corn syrup, sucrose crystallization is effectively prevented and, even at lower levels, the rate of sucrose crystallization is decreased. So even if a sugar glass falls below its T_g, graining may still not occur if the corn syrup prevents sucrose crystallization. In this case, it is more likely that the product will become sticky since the material will tend to flow as it moves farther from the glass transition point.

SUMMARY

Control of crystallization requires an understanding of the phase behavior of the system. Specifically, the equilibrium conditions given by the phase diagram provide knowledge of the extent of supersaturation during processing and allow one to calculate the yield of crystalline material (equilibrium phase volume). However, the phase diagram is insufficient to specify the nature of the crystalline structure obtained under real conditions, where kinetic constraints may be more important than equilibrium conditions. Thus, further

Table 4–4 Glass Transition Temperature (Onset; °C) for Mixtures of Sucrose with Corn Syrup

*Addition Level** *%*	*20 DE*	*42 DE*
0	63.9 ± 1.4	63.9 ± 1.4
10	66.8 ± 0.4	64.1 ± 2.2
20	65.8 ± 2.6	66.9 ± 0.7
50	73.6 ± 3.0	69.1 ± 2.0
75	91.6 ± 1.3	70.9 ± 1.6
100	139.3 ± 0.8	79.0 ± 2.9

DE, dextrose equivalent.
* Addition on a dry solids basis (g corn syrup solids/100 g total solids).

information about crystallization kinetics is necessary to define crystallization in foods. One kinetic limitation of particular importance in many foods occurs at high concentrations and low temperatures and leads to the glass transition in many food components. In the glassy state, crystallization does not occur due to limited molecular mobility. In a state diagram, the glass transition conditions are overlaid on the phase diagram for the system to provide additional information about the state of the system at various conditions.

REFERENCES

1. Richardson SJ, Baianu IC, Steinberg MP. Mobility of water in sucrose solutions determined by deuterium and oxygen-17 nuclear magnetic resonance measurements. *J Food Sci.* 1987;52(3):806.
2. Mathlouthi M. Amorphous sugar. In: Mathlouthi M, ed. *Sucrose: Properties and Applications.* Glasgow: Blackie; 1995:75–100.
3. Allen AT, McDonald MP, Nicol WM, Wood RM. An investigation of athermal concentration gradients in supersaturated solutions of sucrose. In. *Particle Growth in Suspensions.* New York: Academic Press; 1973:239–246.
4. Mullin JW, Leci CL. Evidence of molecular cluster formation in supersaturated solutions of citric acid. *Philosophical Magazine.* 1969;19:1075–1077.
5. McMahon PM, Berglund KA, Larson, MA. Raman spectroscopic studies of the structure of supersaturated KNO_3 solutions. In: Jancic SJ, deJong EJ, eds. *Industrial Crystallization '84.* London: Elsevier; 1984:229–239.
6. Cerreta MK, Berglund KA The structure of aqueous solutions of some dihydrogen orthophosphates by laser Raman spectroscopy. *J Crystal Growth.* 1987;84:577.
7. Chakraborty R, Berglund KA. Steady state fluorescence spectroscopy of pyranine as a trace extrinsic probe to study structure in aqueous sugar solutions. *J Crystal Growth.* 1992;125:81–96.
8. Yedur SK, Berglund KA. Use of fluorescence spectroscopy in concentration and supersaturation measurements in citric acid solutions. *Applied Spectroscopy.* 1996;50:866–870.
9. Mathlouthi M, Cholli A, Koenig JL. Spectroscopic study of the structure of sucrose in the amorphous state and in aqueous solution. *Carbohydr Res* 1986;147.1.
10. Larsson K. Molecular arrangement in glycerides. *Fette Seifen Anstrichmittel.* 1972;74:136.
11. Hernqvist L. Crystal structures of fats and fatty acids. In: Garti N, Sato K, eds. *Crystallization and Polymorphism of Fats and Fatty Acids.* New York: Marcel Dekker; 1988.
12. Hernqvist L. On the structure of triglycerides in the liquid state and fat crystallization. *Fette Seifen Anstrichmittel.* 1984;86:297–300.
13. Fennema OR. Water and ice. In: Fennema OR, ed. *Food Chemistry.* 3rd ed. New York: Marcel Dekker; 1996.
14. Fennema OR, Powrie, WD, Marth EH. *Low-Temperature Preservation of Foods and Living Matter.* New York: Marcel Dekker; 1973.
15. Kennedy CJ. Formation of ice in frozen foods and its control by physical stimuli. In: Reid DS, ed. *The Properties of Water in Foods: ISOPOW 6.* London: Blackie; 1998:327–365.
16. Bohenek M, Myerson AS. Diffusion coefficient and viscosity in crystal growth in microgravity. In: Myerson AS, Green DA, Meenan P, eds. *Crystal Growth of Organic Materials. American Chemical Society Conference Proceedings Series* Washington, DC: American Chemical Society; 1996:273–278.
17. Shallenberger RS, Birch GG. *Sugar Chemistry* Westport, CT: AVI; 1975.
18. Pigman W, Isbell HS. Mutarotation of sugars in solution I. *Adv Carbohydr Chem Biochem.* 1968;23:11.
19. Nickerson TA. Lactose. In: Webb BH, Johnson AH, Alford JH, eds. *Fundamentals of Dairy Chemistry*, 2nd ed. Westport, CT: AVI; 1974:273–324.
20. Haase G, Nickerson TA. Kinetic reactions of alpha and beta lactose. I. Mutarotations. *J Dairy Sci.* 1966;49.127.
21. Troy HC, Sharp PF. α and β lactose in some milk products. *J Dairy Sci.* 1930;13 140.
22. Patel KN, Nickerson TA. Influence of sucrose on the mutarotation velocity of lactose. *J Dairy Sci.* 1970;53:1654.
23. Roos YH. *Phase Transitions in Foods.* San Diego: Academic Press, 1995.
24. Fletcher NH. *The Chemical Physics of Ice.* London. Cambridge University Press; 1970.

25. Mullin JW. *Crystallization.* 3rd ed. Oxford: Butterworth-Heinemann; 1993.
26. Timms RE. Phase behaviour of fats and their mixtures. *Prog Lipid Res.* 1984;23:1–38.
27. Rossell JB. Phase diagrams of triglyceride systems. *Adv Lipid Res.* 1967;5:353–408.
28. Rossell JB. Interaction of triglycerides and of fats containing them *Chemistry and Industry.* 1973;17:832–839.
29. Sato K. Newest understandings of molecular structures and interactions of unsaturated fats and fatty acids. *Progress Colloid Polymyer Science.* 1998;108:58–66.
30. Lutton ES. Technical lipid structures. *J Am Oil Chem Soc.* 1972;49:1–9.
31. Young FE, Jones FT. Sucrose hydrates: the sucrose-water phase diagram. *J Phys Chem.* 1949;53:1334.
32. Pancoast HM, Junk WR. *Handbook of Sugars* Westport, CT: AVI; 1980.
33. Timms RE. Crystallization of fats. In: Hamilton RJ, ed. *Development in Oils and Fats.* London: Blackie; 1995:204–223.
34. Kloek W, Walstra P, Van Vliet T. Crystallization kinetics of fully hydrogenated palm oil in sunflower oil mixtures. *J Am Oil Chem Soc.* 2000;77(6):643–652.
35. Koyano T, Hachiya I, Sato K. Phase behavior of mixed systems of SOS and OSO. *J Phys Chem.* 1992;96(25):10514–10520.
36. Wesdorp LH. *Liquid-Multiple Solid Phase Equilibria in Fats.* Delft: Technical University of Delft, The Netherlands; 1990. PhD dissertation.
37. Roussett P, Rappaz M, Minner E. Polymorphism and solidification kinetics of the binary system POS-SOS. *J Am Oil Chem Soc.*1998;75(7):857–864.
38. Heldman DR, Singh RP. *Food Process Engineering.* 2nd ed. Westport, CT: AVI; 1981.
39. Heldman DR. Food freezing. In: DB Lund, Heldman DR, eds. *Handbook of Food Engineering.* New York: Marcel Dekker; 1992:277–315.
40. Bradley RL Jr, Smith K. Finding the freezing point of frozen desserts. *Dairy Rec* 1983;84(6):114–115.
41. Schwartzberg HG. Effective heat capacities for the freezing and thawing of food. *J Food Sci.* 1976;41:152–156.
42. Smith KE, Bradley RL Jr. Effects on freezing point of carbohydrates commonly used in frozen desserts. *J Dairy Sci.* 1983;66:2464–2467.
43. Blond G. Freezing in polymer-water systems and properties of water. In: Simatos D, Blond G, eds. *Properties of Water in Foods in Relation to Quality and Stability.* Dordrecht, The Netherlands: Martinus Nijhof; 1985:531–542.
44. Tavare NS. *Industrial Crystallization: Process Simulation Analysis and Design.* New York: Plenum Press; 1995.
45. Bubnek Z, Kadlec P. Sucrose solubility. In: Mathlouthi M, ed. *Sucrose: Properties and Applications.* Glasgow: Blackie;1995:101–125.
46. Marangoni AG, Lencki RW. Ternary phase behavior of milk fat fractions. *J Agric Food Chem.* 1998;46(10):3879–3884.
47. Boistelle R. Fundamentals of nucleation and growth. In: Garti N, Sato K, eds. *Crystallization and Polymorphism of Fats and Fatty Acids.* New York: Marcel Dekker; 1988:189–226.
48. Mazur P. Physical and chemical basis of injury in single-celled microorganisms subjected to freezing and thawing. In: Meryman HT, ed. *Cryobiology.* London: Academic Press; 1966.
49. Schwartzberg HG. Food freeze concentration. In: Schwartzberg HG, Rao MA, eds. *Biotechnology and Food Processes Engineering.* New York: Marcel Dekker; 1990:127–202.
50. Hartel, RW. Mechanisms and kinetics of recrystallization in ice cream. In: Reid DS, ed. *The Properties of Water in Foods: ISOPOW 6.* London: Blackie; 1998:287–319.
51. Tjuradi P, Hartel RW. Effects of corn syrup oligosaccharides on sucrose crystallization. *J Food Sci.* 1995;60(6): 1353–1356.
52. Bertran JF, Ballester L. The heat of crystallization and activity coefficients of sucrose in saturated water solutions. *Int Sugar J.* 1971; 71:40–43.
53. Van Hook A. The thermodynamic activity of concentrated sugar solutions. *Zuckerindustrie.* 1987;112(7):597–600.
54. Hartel RW, Shastry AV. Sugar crystallization in food products. *Crit Rev Food Sci Nutr.* 1991; 30:49–112.
55. Mullin JW, Sohnel O. Expressions of supersaturation in crystallization studies. *Chem Eng Sci.* 1977;32:683–686.
56. Van Hook A, Bruno AJ. Nucleation and growth in sucrose solutions. *Faraday Discuss.* 1949;5:112.
57. Van Hook A, Nucleation in supersaturated solutions. In: Honig P, ed. *Principles of Sugar Technology.* New York: Elsevier; 1959:2–113.

58. Vasatko J, Smelik A. The crystallization of sucrose, D-glucose and D-fructose from supersaturated solutions of metastable state. *Zucker*. 1968;21(6):144.
59. Singh G. Analysis of crystallization systems with applications to continuous fractional crystallization of fatty acid glycerides. *Am Inst Chem Eng*. Symposium Series no. 153. 1976;72:100–109.
60. Herrington, BL. Some physico-chemical properties of lactose. VI. The solubility of lactose in salt solutions; the isolation of a compound of lactose and calcium chloride. *J Dairy Sci*. 1934;17.805.
61. Nickerson TA, Moore EE. Solubility interrelations of lactose and sucrose. *J Food Sci*. 1972;37:60.
62. Bradley RL Jr. Protecting ice cream from heat shock. *Dairy Rec*. October 1984;120–121.
63. Sakai N, Hosokawa A. Comparison of several methods for calculating the ice content of foods. *J Food Eng*. 1984; 3:13–26.
64. Gresti J, Bugaut M, Maniongui C, Bezard J. Composition of molecular species of triacylglycerols in bovine milk fat. *J Dairy Sci*. 1993;76:1850–1869.
65. International Union of Pure and Applied Chemistry, Method 2.150. In: Paquot C, Hautfenne A, eds. *Standard Methods for the Analysis of Oils, Fats and Derivatives*. 7th ed. Boston: Commission on Oils, Fats and Derivatives; 1987.
66. AOCS Method Cd 16-81. In: *Official Methods of the American Oil Chemist's Society*. 4th ed. Champaign, IL: Am Oil Chem Soc; 1989.
67. Bigalli G. Practical aspects of the eutectic effect on confectionery fats and their mixtures. *Manufacturing Confectioner*. 1988;68(5):65.
68. Metin, S. *Crystallization and Kinetics of Blends of Cocoa Butter and Milk Fat Fractions*. Madison, WI: University of Wisconsin-Madison; 1997. PhD thesis.
69. Williams SD, Ransom-Painter KL, Hartel RW. Mixtures of palm kernel oil with cocoa butter and milk fat in compound coatings. *J Am Oil Chem Soc*. 1997;74(4):357–366.
70. Ali AR, Dimick PS. Melting and solidification characteristics of confectionery fats: anhydrous milk fat, cocoa butter and palm kernel stearin blends. *J Am Oil Chem Soc*. 1994,71(8):803–806.
71. Timms RE. The phase behaviour and polymorphism of milk fat, milk fat fractions and fully hardened milk fat. *Aust J Dairy Technol*. 1980;35(2):47–53.
72. Roos YH, Karel M. Applying state diagrams to food processing and development. *Food Technol*. 1991;45(12):66, 68–71, 107.
73. Williams ML, Landel RF, Ferry JD. The temperature dependence of relaxation mechanisms in amorphous polymers and other glass-forming liquids. *J Am Chem Soc* 1955;77:3701–3707.
74. Noel TR, Ring SG, Whittam MA. Dielectric relaxations of small carbohydrate molecules in the liquid and glassy states. *J Phys Chem*. 1992;96:5662–5667.
75. Rasmussen D, Luyet B. Complementary study of some nonequilibrium phase transitions in frozen solutions of glycerol, ethylene glycol, glucose, and sucrose. *Biodynamica*. 1969;10:319–332.
76. Roos YH, Karel M. Amorphous state and delayed ice formation in sucrose solutions. *Int J Food Sci Technol*. 1991; 26:553–566.
77. Slade L, Levine H. Beyond water activity: recent advances based on an alternative approach to the assessment of food quality and safety. *Crit Rev Food Sci Nutr*. 1991;30:115–360.
78. Levine H, Slade L. A food polymer science approach to the practice of cryostabilization technology. *Comments Agric Food Chem*. 1989;1:315–396.
79. Fox TG, Flory PJ. Second-order transition temperatures and related properties of polystyrene. I. Influence of molecular weight. *J Applied Physics*. 1950;21:581–591.
80. Roos YH, Karel M. Water and molecular weight effects on glass transitions in amorphous carbohydrates and carbohydrate solutions. *J Food Sci* 1991;56:1676–1681.
81. Gabarra P, Hartel RW. Corn syrup solids and their saccharide fractions affect crystallization of amorphous sucrose. *J Food Sci*. 1998;63(3):523–528
82. Jouppila K, Roos YH. Glass transition and crystallization in milk powders. *J Dairy Sci* 1994;77:2907–2915.
83. Gabarra P. *The Effects of Corn Syrup Solids and Their Oligosaccharide Fractions on the Glass Transition and Crystallization of Amorphous Sucrose*. Madison, WI: University of Wisconsin-Madison; 1996. MS thesis.

CHAPTER 5

Nucleation

Nucleation, or the formation of the crystalline phase, is probably the most important step for controlling crystallization in foods. The correct extent of nucleation at the appropriate time in the process is required to produce the desired target crystalline structure (number, size, distribution, shape, and polymorph). Therefore, it is important to understand the mechanisms of nucleation, the kinetics or rate at which it occurs, and how operating parameters and product formulation influence these rates.

Nucleation is typically classified into either primary or secondary mechanisms, with primary nucleation involving either homogeneous or heterogeneous mechanisms. Figure 5–1 provides a breakdown of these nucleation mechanisms. Homogeneous nucleation involves formation of a crystal lattice structure based on accretion of molecules into a stable size, whereas heterogeneous nucleation involves a foreign impurity that promotes formation of a nucleus. Secondary nucleation can occur in several ways, with the most common being contact nucleation. In this case, contacts between a crystal and something else, like another crystal, a stirrer, or a solid wall, cause formation of nuclei under conditions where nucleation would not normally have occurred (in the metastable zone).

NUCLEATION THEORIES

Homogeneous Nucleation

Primary homogeneous nucleation is the classical mechanism of formation of nuclei. Although homogeneous nucleation rarely occurs in practical situations, it is informative to understand this mechanism of crystal formation. In classical nucleation theory[1,2] molecules in solution come together, either singly or in units, with a certain probability of staying together as clusters of molecules. These clusters are continually being formed and dissociated in a supersaturated solution, as discussed in Chapter 4. This molecular accretion process occurs based on localized energy fluctuations[3] and eventually a stable nucleus forms, depending on the supersaturation and temperature. Classical homogeneous nucleation theory was initially developed for formation of droplets from a subcooled vapor. However, nucleation from solution is generally considered to follow along the same lines.

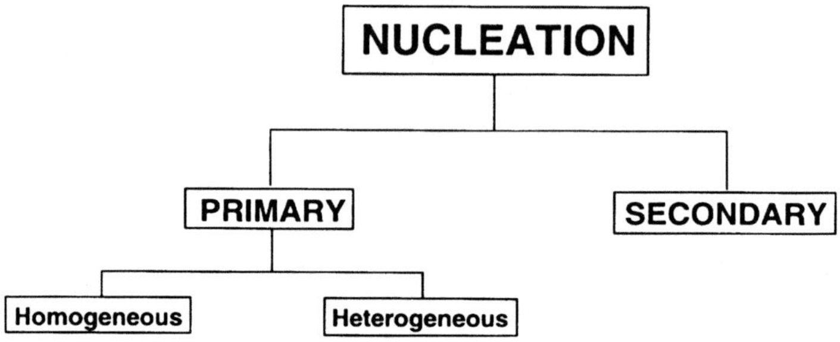

Figure 5–1 Mechanisms of Nucleation.

The exact nature of the molecular clusters, or pre-nuclei embryos, is unknown. Clustering is most likely a random event where molecules come together, interact, and then dissociate again on some time scale. If conditions are correct (sufficient supersaturation or supercooling), the rate of association is greater than the rate of dissociation and a cluster of critical size is formed. At some point, a cluster must convert from a random grouping of molecules into a crystal lattice. Some evidence[4] suggests that the ordering or clustering of molecules in a supersaturated solution may not take the same structure as the ordering found in the crystal lattice. That is, the molecular organization in these clusters must be broken and rearranged prior to formation of the crystal lattice. However, our understanding of solution structure is far from complete, and it is still an active area of research.

Homogeneous nucleation theory has been developed in great detail elsewhere.[1] For our purposes, a simple approach showing the main events and parameters is sufficient. For a spherical nucleus to form, the work of formation can be broken down into a surface term and a bulk term. The free energy change for formation of the crystal surface, or development of an interfacial tension, is positive and proportional to the surface area (size squared) and the interfacial tension of a macroscopic crystal. The free energy change for formation of the bulk of the crystal is negative, due to the latent heat of fusion released upon phase change, and proportional to the volume (size cubed). The net free energy change during nucleation is the sum of these terms, as shown in Figure 5–2. The free energy of formation goes through a maximum at some critical size, r_c, corresponding to the critical nucleus size. Once a cluster attains this size, it is more likely to continue to grow and appear as a stable crystal. Thus, there is an energy barrier that must be overcome in order for nuclei to form. This energy must come from the supersaturated solution and from microscale fluctuations in the solution energy levels.

By writing the energy terms for surface and volume influences, the critical nucleus size, r_c, can be calculated according to classical homogeneous nucleation theory.[2] This is given for crystallization from a solution as

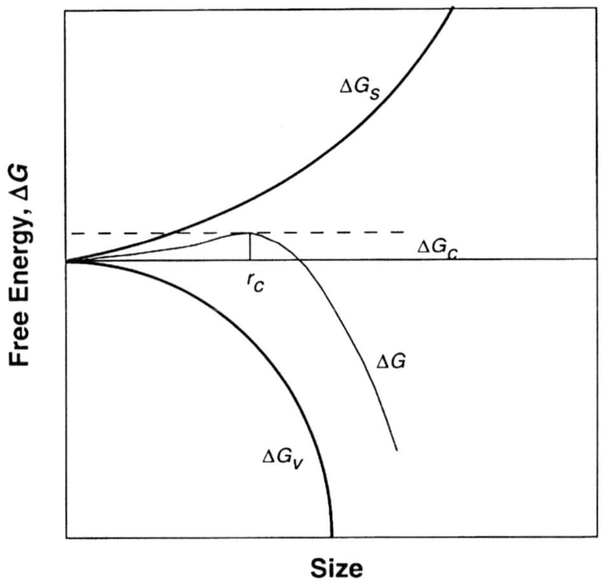

Figure 5–2 Free Energy (ΔG) Changes Associated with Formation of a Stable Nucleus. ΔG_s, free energy of surface formation; ΔG_v, free energy of volume formation; ΔG_c, critical free energy for nuclei formation; r_c, critical size.

$$r_c = \frac{2\sigma_s v}{kT \ln(S)} \tag{5.1}$$

where σ_s is the interfacial tension, v is the molecular volume, k is Boltzman's constant, T is absolute temperature, and S is the supersaturation ratio. S should more accurately be the ratio of activities of the supersaturated and saturated solutions (as discussed in Chapter 4), although the concentration ratio is often used. For crystallization from a melt, the critical size is

$$r_c = \frac{2\sigma_s v T_f}{\Delta H_f (T_f - T)} \tag{5.2}$$

where T_f is the freezing point and ΔH_f is the latent heat of fusion.

The critical free energy, ΔG_c, required for nuclei formation is then given, for solution crystallization, as

$$\Delta G_c = \frac{16\pi\sigma_s^3 v^2}{3k^2T^2 \ln^2(S)} \tag{5.3}$$

The classical nucleation rate equation can then be written as

$$J = A\exp\left\{-\frac{\Delta G_c}{kT}\right\} = A\exp\left\{-\frac{16\pi\sigma_s^3 v^2}{3k^3T^3 \ln^2(S)}\right\} \tag{5.4}$$

From Walton,[3] the equation for homogeneous nucleation in melt systems can be written as

$$J = A\exp\left\{-\frac{16\pi\sigma_s^3 v^2 T_f^2}{3kT\Delta H_f^2 (T_f - T)^2}\right\} \tag{5.5}$$

From Equation 5.4, the temperature and supersaturation are the major operational variables influencing nucleation in a solution, according to the classical homogeneous nucleation theory. The nature of the system also is important, as determined by the characteristic interfacial tension between liquid and solid, the molecular volume, and the equilibrium conditions (freezing temperature or solubility). As the supersaturation increases, the rate of nucleation increases dramatically due to the functional nature of S in Equation 5.4. Generally, the rate of nucleation is quite low until a critical supersaturation has been reached, at which point nucleation rate increases dramatically.[5] The time required for nucleation to occur decreases precipitously as supersaturation increases slightly above the critical point; this kinetically determined lag time leads to the concept of metastability that has been observed in many systems.

The critical radius for nuclei formation is dependent primarily on temperature and supersaturation, as seen in Equations 5.1 and 5.2. During freezing of ice, the critical radius decreases in size as temperature is lowered below the freezing point. The size of the critical cluster for nuclei formation by homogeneous nucleation is estimated to be between 20 Å and 120 Å at temperatures of –30 and –3°C, respectively.[6] Van Hook[7] calculates that the size of a critical nucleus for sucrose crystallizing from water is about 20 Å in radius, suggesting that between 80 and 100 molecules are needed to form a stable cluster.

The interfacial tension between solution and crystal is a parameter that is difficult to find directly, although it can have a significant effect on nucleation. For example, the reason that lower-stability polymorphs nucleate before the more stable polymorphs is often attributed to differences in interfacial tension. For lipids, the α-polymorph often nucleates before the β′-1 or β-polymorphs[8] due to the lower surface tension of the α-polymorph. In

Equation 5.5, the α-polymorph would have the lowest supersaturation at any temperature, but since it has lower interfacial tension than the other forms, the α-polymorph nucleates first. However, measurement of the interfacial tension is very difficult. In fact, interfacial tension values are often calculated from Equation 5.5 based on nucleation data.[9] Van Hook and Frulla[10] estimate the interfacial tension between water and a sucrose crystal to be 6 erg/cm^2, whereas Hobbs[11] estimates the interfacial tension between a sucrose solution and an ice crystal to be between 15 and 35 mJ/m^2.

For viscous fluids, volume diffusion effects also may influence the rate of nucleation. These effects are usually accounted for with a free energy associated with molecular diffusion, $\Delta G'_v$, during development of the energy changes in nucleation.[3,12,13] This is given as

$$J = A\exp\left\{-\frac{\Delta G_c}{kT} + \frac{\Delta G'_v}{kT}\right\} \tag{5.6}$$

At high solute concentrations, or high viscosities, the viscous barrier to nucleation becomes more important than the supersaturation driving force and nucleation is inhibited. At sufficiently high viscosities or supersaturations, nucleation no longer occurs and this leads to the glass transition region on the state diagram.

Although the classical accretion model of nucleation has been the theory most widely used, others have suggested alternative models for nucleation. Rasmussen and Liang[14] proposed a model of nucleation based on spontaneous decomposition within fluid regions undergoing energy fluctuations that are large in extent but small in degree. That is, spontaneous decomposition (sometimes called spinodal decomposition) in a region of fluid undergoing a fluctuation results in formation of a new phase cluster (called an iceberg for water-ice) in equilibrium with residual undercooled liquid. Nucleation rate is dependent on a fluctuation that is large enough in both degree and extent so that a crystalline particle is formed. Upon cessation of the energy fluctuation in the nucleating region, the crystalline particle must still be stable in the supercooled parent phase. Computer simulations of the nucleation process[15] may ultimately allow a better understanding of the exact mechanism of nuclei formation under the wide range of commercially important conditions.

Heterogeneous Nucleation

As mentioned previously, classical homogeneous nucleation is thought to occur quite rarely in real-life situations. Generally, the presence of foreign nucleating sites induces nuclei formation at much lower supersaturations or subcoolings than those required for homogeneous nucleation. These nucleating sites may be dirt or dust particles, or rough and pitted surfaces of the vessel containing the supersaturated solution. These sites are thought to orient solute molecules in such a way that formation of the crystal lattice is promoted by reducing the energy required to form a stable surface and decreasing the number of molecular interactions required to form a stable nucleus. Lattice matching parameters have been used to describe the similarity between two structures (the nucleating site and the crystal that forms on it). A substrate that induces too much strain within a nucleus, due to poor lattice matching, will not be an effective nucleator. The general rule is that better matching between lattices promotes heterogeneous nucleation,[3,16] although this trend is not always strictly followed.[2,7] Walton[3] suggests that it is the matching capability of the nucleating site that is critical and that sites at dislocations or impurity structures may act differently from the bulk material. For example, the ability of silver iodide to promote nucleation of water (eg, cloud seeding) has been related to the matching of the crystal lattice parameters between the two systems.[7,16] Furthermore, Walton states that active sites on substrates "are subject to ill-defined aging effects which render reproducible data difficult to obtain."[3(p 289)] The subject of the variability of nucleation events will be taken up further in a later section. Our understanding of the exact processes involved in heterogeneous nucleation is quite limited.

In general, heterogeneous nucleation is treated by modification of homogeneous nucleation theory.[2,3] The rate equation for heterogeneous nucleation is given as

$$J = A\exp\left\{-\frac{\Delta G_c'}{kT}\right\} \tag{5.7}$$

where $\Delta G'_c$ is the critical free energy required for heterogeneous nucleation. It is often given as

$$\Delta G_c' = \phi \Delta G_c \tag{5.8}$$

The factor, ϕ, is determined by the interrelation between the nucleating solid, the crystal lattice, and solution. The factor, ϕ, is often correlated to the interfacial tensions existing between the foreign solid, the crystallizing species, and the solution.[2] This relationship has been given as

$$\phi = \frac{1}{4}(2 + \cos\theta)(1 - \cos^2\theta) \tag{5.9}$$

and

$$\cos\theta = \frac{\sigma_{sl} - \sigma_{cs}}{\sigma_{cl}} \tag{5.10}$$

where θ is the angle (of wetting) between crystal and nucleating solid, and σ_{sl}, σ_{cs}, and σ_{cl} are interfacial tensions between solid-liquid, crystal-solid, and crystal-liquid, respectively. When $\theta = 0°$, $\Delta G_c' = 0$ and nucleation occurs readily, whereas when $\theta = 180°$, nucleation should not occur at all due to the lattice mismatch.[17] It often is difficult to quantify the extent of matching between nucleating sites and crystal lattices, so heterogeneous nucleation is generally considered from an empirical perspective.

Figure 5–3 shows several scenarios of interaction between solid particles and solution. This figure attempts to show schematically how the contact angle between solution and solid can influence heterogeneous nucleation. In Figure 5–3A, there is no solid surface and the entire molecular cluster must be constituted of crystallizing species before nucleation begins. In Figures 5–3B through 5–3D, varying extents of wetting of the surface by the crystallizing molecule occurs. In each case, the number of molecules of the crystallizing species to cover the surface decreases based on the contact angle. Thus, at lower contact angles, fewer molecules are needed to provide a critical size for a stable nucleus and nucleation occurs at lower driving forces. In this case, the foreign solid particle provides a portion of the energy needed to overcome the energy barrier to nucleation.

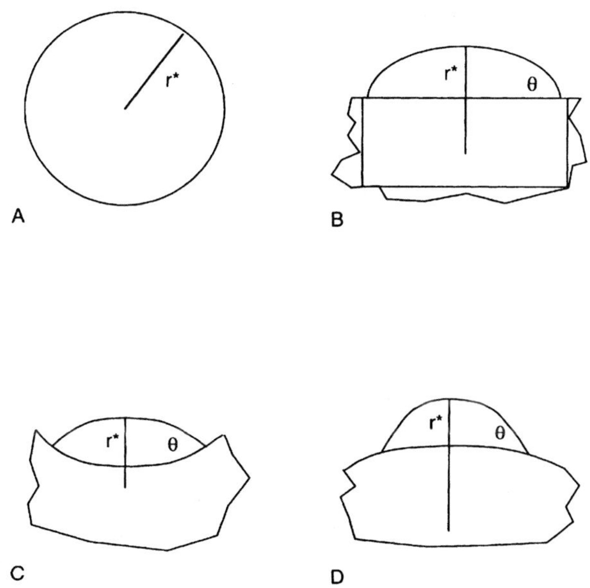

Figure 5–3 Potential Interactions between a Solid Particle and Solution Showing Different Angles of Wetting: (A) Homogeneous Nucleation—No External Surface; (B) Heterogeneous Nucleation on Flat Surface; (C) Concave Surface; and (D) Convex Surface. *Note:* r* represents the radius of curvature of a stable nucleus.

Fennema et al[6] reviewed the general requirements of nucleants to promote formation of ice. These requirements were given as: (1) low solubility in water, (2) reasonably close matching of the lattice parameters between ice and the nucleant, (3) capability of sufficient hydrogen bonding to interact with water molecules, and (4) sufficiently small size. Numerous materials fit these qualifications, including silver iodide, silicate minerals (kaolite, covellite, magnetite, etc), steroids (eg, cholesterol), and proteins. Some bacteria (eg, *Pseudomonas syringae*) have ice-nucleating characteristics based on the water structuring capability of the protein in the cell wall.[18,19] Charoenrein et al[20] compare the effects of silver iodide and a *P syringae* preparation on heterogeneous nucleation. Preparations of this bacteria have been used to promote ice nucleation in frozen foods.

NUCLEATION KINETICS

There are two aspects of nucleation kinetics to consider. The first is the rate of nuclei formation (in number of nuclei formed per unit volume per unit time) as evidenced by spontaneous production of stable crystals and described above by the classical nucleation theories. The second aspect of nucleation kinetics relates to the time required for nucleation to occur, which might be considered as the rate of formation of embryos or clusters. Nucleation data are generally either given as rates of nuclei production at specified conditions or as the lag, or induction, time required for onset of nuclei production. Although these parameters are related, they are probably not measures of exactly the same thing. Despite this fact, many consider induction time to be the inverse of the nucleation rate. Both parameters are important in controlling crystallization processes.

Rate of Nucleation

According to Equation 5.4, the rate of nucleation (number of nuclei formed per unit volume and time [#/mL-s]) in solution systems depends on the supersaturation and temperature. Note that supersaturation also depends on temperature since solubility is generally a function of temperature (Chapter 4). Thus, it is often difficult to separate the effects of temperature from supersaturation in solution systems. The functional dependency for supersaturation or subcooling in Equation 5.4 predicts that, at low supersaturations, the nucleation rate is very small and at high supersaturations, the nucleation rate is very large. Nucleation rate goes through a substantial increase as supersaturation increases through some critical point.[5] Below this critical supersaturation, nucleation rate is very low, whereas above this critical point, nucleation rate is very high. This functionality has led to the idea of metastability in supersaturated solutions, where solutions with low supersaturation are effectively stable against nucleation for long periods of time.

For many food systems (ie, sugars, carbohydrates, organic acids, etc) the nucleation rate for a solution of constant concentration goes through a maximum as temperature decreases. At higher temperatures within the supersaturated region, just below the saturation point, spontaneous nucleation is low since the driving force is still small. As temperature decreases, the supersaturation increases from a combination of mechanisms. Lowering the temperature itself increases the driving force for nucleation, as seen in Equation 5.4. However, lowering temperature also, in general, increases supersaturation through the depression in solubility concentration. Thus, the driving force for crystallization increases rapidly as temperature is decreased. This increased driving force results in a dramatic increase in nucleation rate. However, at some point, the diffusion term in Equation 5.6 becomes dominant, and the nucleation rate decreases. A general form of a nucleation rate curve is shown in Figure 5–4. The exact point where nucleation begins to decrease again is not well defined, and very little quantitative data for food systems exist to document this decrease in nucleation rate. Nevertheless, the existence of this maximum has been documented, at least qualitatively, in several food processing operations. For example, fondant manufacturers know that if they cool the sugar mass below some optimal temperature prior to crystallization, their product does not have as many crystals. When crystallization occurs at the optimal temperature, the peak temperature of maximal nucleation rate depends primarily on the nature of the system (composition of ingredients) being studied. When temperature goes below the glass transition temperature, T_g, for the system, nucleation is essentially reduced to zero and the material becomes a glassy solid.

A comparison of the relative rates of nucleation for homogeneous versus heterogeneous nucleation also is shown in Figure 5–4. Homogeneous nucleation does not begin until some reasonably high driving force, or a critical supersaturation, is attained. Beyond this point, however, nucleation increases dramatically with small increases in driving force. Heterogeneous nucleation occurs initially at lower supersaturation driving forces and increases at a slower rate than for homogeneous nucleation. An emulsification technique of Rasmussen and McKenzie[21] often is used to distinguish between homogeneous and

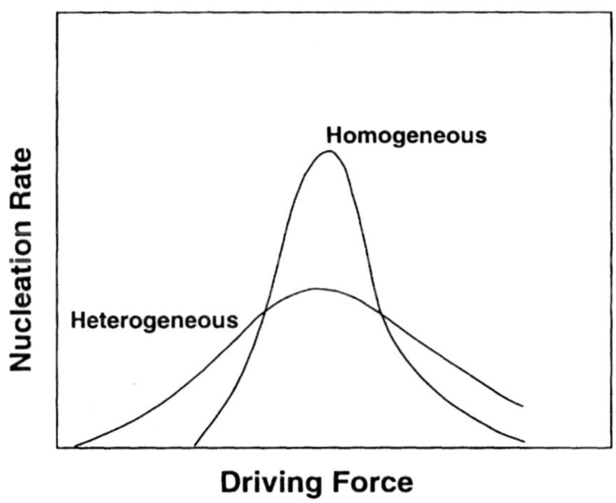

Figure 5–4 General Form of Nucleation Rate Curve with Increasing Driving Force, Showing a Comparison of Homogeneous and Heterogeneous Nucleation.

heterogeneous nucleation. In water, for example, homogeneous nucleation occurs approximately 41°C below the freezing point.[6] However, heterogeneous nucleation in pure water occurs approximately 12°C below the freezing point, depending on the nature of the nucleating sites.[22] At extremely low temperatures, viscosity effects limit the rate of nucleation. Below T_g nucleation is no longer thought to occur. In water, the temperature where nucleation rate begins to decrease again has not been clearly determined,[6] although it is thought to be below −100°C. Note that T_g for pure water is often quoted as −135°C.[23]

Little data showing the optimal temperature for maximum nucleation in food systems are available. Figure 5–5 shows this trend for crystallization of citric acid, a highly viscous solution crystallization system.[24] In these experiments, supersaturation was increased by decreasing the temperature of a solution with a specified concentration. Figure 5–5 shows an optimum temperature at which maximal nucleation occurred, and this temperature was correlated by Mullin and Leci[24] to a rapid increase in solution viscosity. Once temperature was lowered below this point, the viscosity was sufficiently high that nucleation was inhibited and fewer crystals were formed. The optimal temperature for maximal nucleation is dependent on the degree of supersaturation, the extent of metastability possible in the

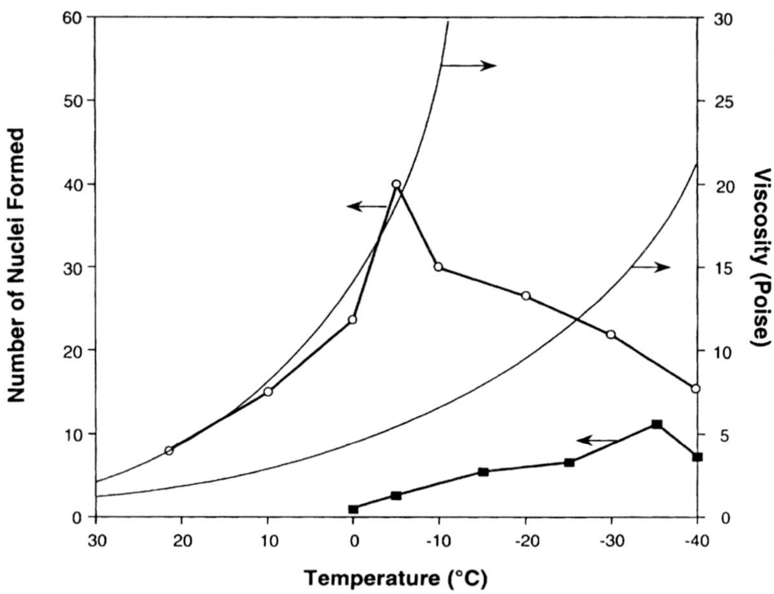

Figure 5–5 Nucleation Rate Curve and Viscosity for Citric Acid Monohydrate Cooled Rapidly to Different Temperatures. Open circles: 7.0 g citric acid monohydrate/g "free" water (saturation temperature = 85°C); closed squares: 4.6 g citric acid monohydrate/g "free" water (saturation temperature = 62°C).

system, the glass transition temperature, and any processing factors that affect nucleation (eg, agitation rate).

In many circumstances in food processing, it would be quite useful to know this optimal nucleation temperature. For example, fondants are a highly crystallized sucrose syrup where a large number of extremely small crystals (< 10 µm) are desired to maintain a smooth product. In fondant processing, the temperature at which the syrup is nucleated is a critical parameter, since sucrose nucleation should follow a curve similar to that seen in Figure 5–5. If nucleation is induced at temperatures other than the optimal temperature, fewer crystals will be formed than the maximum number, and the texture of the fondant will be unsatisfactory. This is explained in greater detail for fondant processing in Chapter 7.

Despite the importance of nucleation rate data to controlling quality and texture of many food products, there are very few quantitative data available for complex foods. The problem arises in that there are many factors that influence these rate data, and the experimental measurements are extremely difficult. The factors that influence nucleation include formulation (other ingredients and impurities) and processing conditions (heat and mass transfer rates). For example, it is relatively easy to concentrate a sucrose syrup to 85 to 90% without crystallization by using a rapid evaporation technique (thin-film evaporator). In this case, the kinetics of heat and mass transfer (during evaporation) are significantly faster than the kinetics of nucleation. However, a slower concentration technique (rotary evaporator in the laboratory) will almost always result in crystallization by the time concentrations of 90% are attained. In this case, the longer time required in the rotary evaporator results in greater likelihood of crystallization at a given condition. That is, the kinetics of nucleation are on the same order of magnitude as the kinetics of evaporation. Thus, each food product experiencing a set of processing conditions exhibits a certain nucleation rate profile (similar to that of Figure 5–5).

Induction Time

Induction times measure the time required for the first detectable evidence of nuclei formation. Effectively, induction times indicate a combination of the true time required for nuclei to form plus a measurement time depending on the sensitivity of the experiment. Induction time (τ) measurements can then be broken into two terms:

$$\tau = \tau_n + \tau_g \tag{5.11}$$

where τ_n is the true induction time for nuclei formation and τ_g is the time required for a nucleus to grow to sufficient size to be detected. Detection techniques (see Chapter 3) include visual observation, refractive index, turbidimetric measurement, light scattering, and calorimetry. The validity of an induction time measurement depends on the technique. If visual observation of crystals is required to denote lag time, the growth of nuclei to detectable size must be taken into account. If nuclei form with a size of a few nanometers and the first detectable crystals are 50 μm in size, obviously the growth time plays an important role in induction time measurements. Much of the early data on induction time measurements suffered from this experimental disadvantage. Visual induction time measurements can be enhanced by use of microscope techniques to identify the time when nuclei are first formed. However, the sampling volume observed in the microscope must be fairly large to be sure that the first nuclei are observed. Change in concentration of a supersaturated solution can be used to indicate onset of nucleation. This can be accomplished with a refractive index detector situated directly in the solution. Calorimetry has been used to detect heat changes that occur during nucleation. If there is a large latent heat release and if a large amount of nucleation occurs instantaneously, this is an excellent method of obtaining induction time data. Another common technique for measuring induction time is turbidity measurement. Here, the sample may be held at controlled temperature and agitation conditions and absorbance at high wavelength used to denote when sufficient particles are formed to scatter light. In all of these measurements, it is generally the onset of massive nucleation that is observed. However, localized supersaturation gradients within the solution may lead to formation of a few nuclei prior to the catastrophic event. Formation of a small number of nuclei is very difficult to detect.

Induction time and nucleation rate are functions of the conditions of temperature and pressure, and they may vary with processing conditions. Figure 5–6 shows the trend for induction time of sucrose nucleation. The induction time is initially high at low supersaturations, reaching infinity at saturation (since no nucleation can occur). Within the metastable zone, the induction time is finite and decreases with increasing supersaturation. Eventually a supersaturation is reached where nucleation occurs instantaneously. This is often defined as the critical supersaturation or supersolubility point. From this curve, it is clear that metastability of solutions depends on the time scale over which those solutions must remain uncrystallized. If crystallization is to be delayed for periods of weeks to months, supersaturations must be lower than if crystallization must only be delayed for days. This is an important consideration in the shelf life of many food products, where crystallization may be a defect (eg, ungrained caramel or lactose in ice cream).

Typical data for the effective metastability zone for sucrose solutions were shown in Figure 4–16. In sugar refining, crystallization occurs in a batch evaporator where fresh feed is input at a rate that counterbalances the rate of evaporation of water such that supersaturation is maintained constant. The refiner tries to grow sucrose seed crystals to product size in the minimum amount of time (at the highest supersaturation). However, if the supersaturation is too high, spontaneous nucleation occurs and the result is a wide distribution of sizes that does not centrifuge very well. Thus, the refiner must keep the supersaturation below the metastable zone limit to avoid this "false grain." The time scale for the refining batch is on the order of 5 to 6 hours so that solutions that are stable to nucleation for this period of time are called metastable. However, the time scale for nucleation of a supersaturated solution in a caramel during storage is significantly longer, perhaps equivalent to the desired shelf life of the product, and metastable zone width is effectively smaller. Thus, the metastable zone width is dependent on the time scale

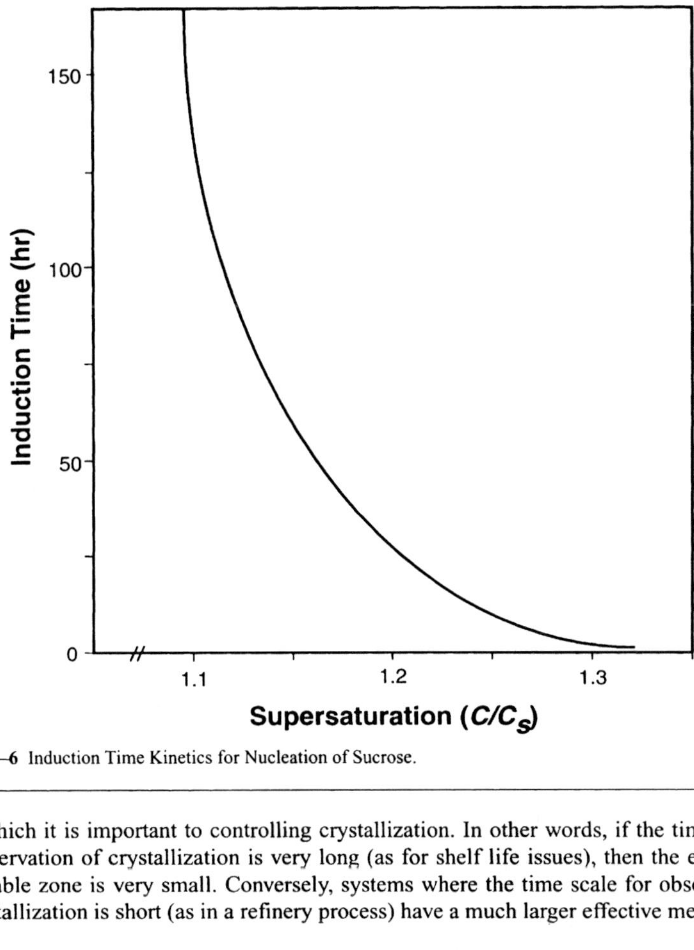

Figure 5–6 Induction Time Kinetics for Nucleation of Sucrose.

over which it is important to controlling crystallization. In other words, if the time scale for observation of crystallization is very long (as for shelf life issues), then the effective metastable zone is very small. Conversely, systems where the time scale for observation of crystallization is short (as in a refinery process) have a much larger effective metastable zone.

In viscous systems, higher supersaturations result in increased lag times as diffusion becomes more limiting.[25] Lag times increase up to the glass transition temperature (and corresponding supersaturation), where no further nucleation occurs. Past the glass transition point, the time required for nucleation to occur becomes effectively infinite. Thus, it is possible to make a sugar glass (or hard candy) even though a very high supersaturation is generated to attain the low water content. However, these high concentration syrups (upward of 98% total solids) do not crystallize in the process because the induction times for nucleation are sufficiently long that the material is stabilized in the amorphous or glassy state before nucleation occurs. In Figure 5–7, data for induction time of lactose crystallization at different temperatures and concentrations are shown. At the lower concentrations, induction time generally decreases as temperature decreases. However, at higher concentrations, there is a slight increase in induction time at the lower temperatures, especially at the highest concentration. The data in Figure 5–7 suggest that somewhere between 25°C and 30°C would be the optimal temperature to obtain the most rapid nucleation for lactose at the highest concentration.

Crystallization of lactose in ice cream also follows a similar behavior based on the two competing mechanisms. As temperature decreases, the supersaturation increases, which initially causes induction time to decrease. When temperature decreases far enough to cause viscous limitations, the induction time increases despite the greater supersaturation. The induction time for the onset of lactose nucleation was measured for storage of ice cream at different temperatures[26] and is shown in Figure 5–8. Between –5°C and about –12°C, the induction time decreased due to the increase in supersaturation. However, between –12 and –20°C, the induction time for onset of lactose nucleation increased, due to the decrease in molecular mobility as temperature approaches T_g.

The natural variability in the nucleation event can be observed as a relatively large scatter of the measurements for induction time. At equivalent supersaturation ratios, and

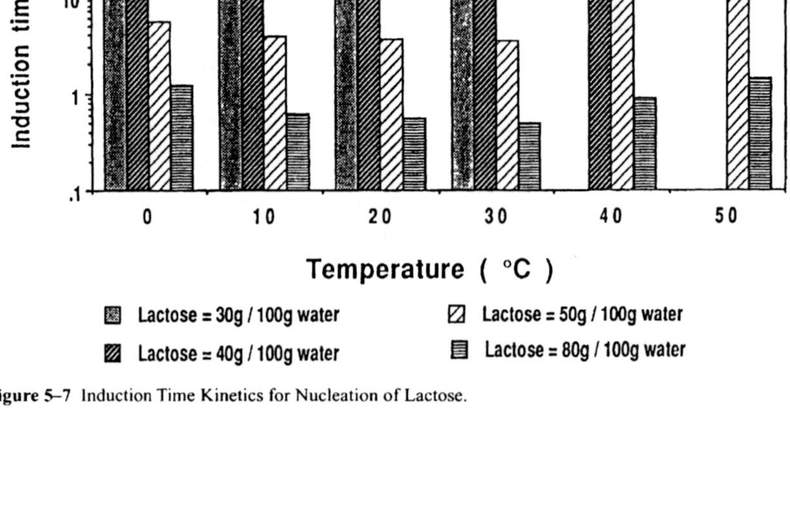

Figure 5–7 Induction Time Kinetics for Nucleation of Lactose.

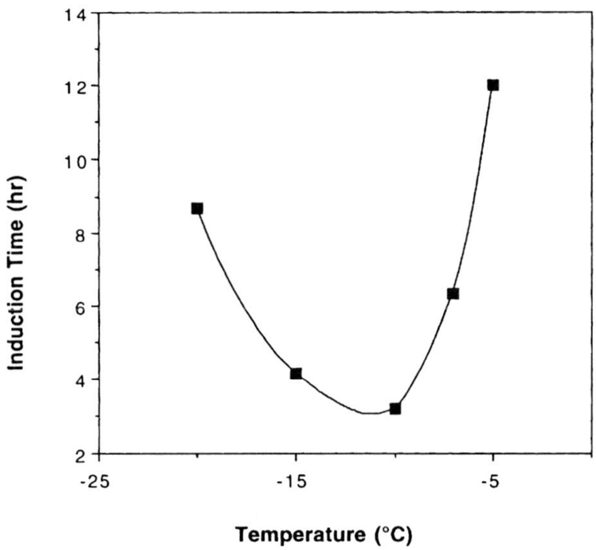

Figure 5–8 Induction Time for Onset of Lactose Nucleation in Ice Cream Stored at Different Temperatures.

with all other macroscopic conditions held the same, a wide scatter of induction times is typically observed even under such carefully controlled conditions. For example, the data generated by van Hook[7] for sucrose solutions shows a range of induction times from 5 to 30 hours for solutions at a supersaturation ratio of 1.21. Part of this variability may be due to the measurement technique (visual observation of induction time), as discussed previously. However, the natural variability in the nucleation event leads to problems with trying to control nucleation in many systems. Despite generating nearly identical processing conditions, there still may be a large variability in onset time for nucleation. The atomic and molecular interactions that lead to formation of a stable nucleus vary slightly each time it occurs, even though the conditions may be nearly identical. There also may be some variability in the nature of the activation sites (heterogeneous nuclei) present in the system. Vonnegut and Baldwin[27] suggested that the activity of nucleation sites may vary with conditions and even change within a single sample over time. The induction time for nucleation of ice, catalyzed by silver iodide, in a small water sample was repeatedly measured to determine the scatter in nucleation times. The sample was heated to ensure melting and cooled rapidly to nucleation temperature, and the induction time was measured as indicated by a change in conductivity of the sample. Induction times for the same sample were measured over 800 times, with induction times varying from a few seconds to over 300 seconds. Interestingly, Vonnegut and Baldwin[27] showed that there was a period of sampling (trial numbers 210 to 260) when the sample froze instantaneously nearly (but not completely) all of the time. Since the experiments were conducted in exactly the same way each time by a mechanical system, their explanation for this variability is that nucleation is a stochastic event, related to the activity of the silver iodide surfaces. Similar results on variability of induction time for freezing of ice in distilled water (although a small amount of salt was added to provide a conductivity signal) were found by Wang and Vonnegut.[28]

An interesting example of the variability or random nature of the nucleation event can be seen in the results of an experiment on metals, where nucleation temperature was repeatedly measured in a calorimeter by cycling the temperature of a small aluminum droplet above and below the melting point.[29] As temperature of the molten system was lowered, the onset temperature for nucleation was denoted by an enthalpic peak on the calorimeter. After crystallization, the sample was remelted by increasing temperature above the melting point and then cooled once again. This was repeated over 700 times to generate a statistical distribution of crystallization temperatures. A normal (Gaussian) distribution in the crystallization temperature over a range of 10°C was observed, which clearly documents the variability in nucleation even within the same sample. That is, the same droplet exhibited freezing temperatures over a range of 10°C (with an equilibrium melting temperature of 660°C).

There are many examples of process systems where crystallization is repeatable the majority of the time, but where every once in a while the system remains uncrystallized. When this occurs during processing, the product must be scrapped because the desired texture and quality were not obtained. Thus, many processors choose to avoid nucleation in the first place by adding the appropriate quantity of seed crystals to generate the crystalline structure. For example, fudge can be made either by crystallizing the batch in the process or by simply adding powdered or fondant sugar to the batch to obtain the appropriate crystalline structure in the product. However, seeding is not always possible, as in the case of fondant or ice cream, since the desired crystals are smaller than any readily available seed crystals.

NUCLEATION FROM THE AMORPHOUS STATE

In many food products, the amorphous or glassy state is desired for product quality, and subsequent crystallization within these products leads to the end of shelf life. Some examples include hard candies and lactose, or milk powders. In hard candy, the glassy state is desired for flavor distribution and dissolubility in the mouth. When the sugar in a hard candy crystallizes, the texture changes, the flavor escapes, and the dye molecules rearrange around the crystals to leave a mottled appearance. When lactose in a milk powder crystallizes, the flow properties of the powder change and the dissolution rate of the powder in water decreases substantially. Thus, understanding the mechanisms and kinetics of nucleation from the amorphous state is important in many food processes.

In hard candy and milk powder, the sugars are stable to crystallization as long as the product remains in the glassy state, with storage at a temperature below the glass transition temperature. However, there are two scenarios where such a product becomes unstable. If the temperature of the product increases to a point above T_g, the product is unstable and crystallization occurs based on the kinetics relevant at that temperature and supersaturation. The second scenario occurs when temperature is constant, but the product picks up moisture from the environment. Since sugar glasses are quite hygroscopic, this happens quite readily unless the product is well protected from ambient humidity (packaged in a water barrier). Upon moisture sorption, the T_g of the sugar glass decreases until eventually T_g is below ambient temperature. Again, crystallization occurs based on the kinetic rates relevant at those conditions.

In condensed phases, like amorphous materials, the mechanism of nucleation may be different from those pertinent in solution environments. However, only limited data are available for nucleation from the amorphous state in foods despite interest in other fields (ceramics, metals, pharmaceuticals, etc).

Techniques to study nucleation rates, in general, are extremely difficult, and the problem is not simplified in condensed states.[30] One approach is to use the differential scanning calorimeter (DSC) to monitor phase changes in a glass as it is heated under controlled conditions (see Chapter 3 for more details). The temperature at onset of crystallization is characteristic of the point where molecular mobility has increased to the point where nuclei formation can occur. However, the dynamic nature of this experiment allows limited fundamental information about the nucleation event. Isothermal studies on amorphous materials allow determination of an induction time for onset of nucleation, where spectroscopic techniques, such as X-ray diffraction, can be used to characterize the appearance of crystals. Other research[31] used a two-step technique to quantify nucleation from the condensed phase. The amorphous material is allowed to nucleate at the temperature of interest for a specified time period; then, temperature is quickly increased to a point where no further nucleation occurs, but growth of any nuclei present continues. After a suitable growth period, the number of crystals present in a specified volume are counted, and the nucleation rate is calculated as the number of crystals found in that volume over the time spent at nucleation temperature. Based on proper choice of temperatures and times, a reasonably accurate measure of nucleation rate can be obtained.[31] This technique has been widely used to study nucleation in amorphous ceramic materials, but not in food systems.

One of the earliest studies on crystallization of food glasses was undertaken by Makower and Dye.[32] In this work, the impact of moisture uptake on stability of a sugar glass was studied. Spray-dried sugar glasses were stored in air with different relative humidities (RHs) to allow the glass to pick up moisture. The change in water content was monitored gravimetrically as the samples picked up moisture from the air, dependent on the level of RH. At each RH, the amorphous sugar picked up water and reached a water content in equilibrium with the environment. At sufficiently high RH, the amorphous sugar picked up sufficient water so that the sucrose molecules had sufficient mobility to crystallize. The time scale before onset of crystallization was dependent on the RH of the storage environment, with higher RH leading to more rapid onset of crystallization. Interestingly, once crystallization occurs in these amorphous sugars, the water dynamics change and they begin to lose water. Crystallization leads to a concentration of water in the noncrystallized portion; thus, water migrates from the product back into the air. The peaks in water content profiles therefore can be used to indicate onset of nucleation. Makower and Dye[32] verified that this decrease in water content coincided with an increase in crystallinity of the material. A schematic diagram of the moisture sorption and desorption profile for this situation is shown in Figure 5–9.

The ability of an amorphous material to nucleate may be dependent on the process used to produce the amorphous product. Mathlouthi[33] shows that freeze-dried sucrose has microcrystalline regions, as verified by electron diffraction spectroscopy and scanning electron microscopy (SEM) visualization. Sugar glasses quenched from the melt (or spray dried) do not have this microstructure, and the sugars remain in the glassy state. These differences may impact the ability of these products to crystallize when stored above T_g. If there are small microcrystalline regions in a freeze-dried product, nucleation is not necessary since, in effect, the material has been seeded with small crystallites. Thus, a freeze-dried product generally crystallizes more readily than a spray-dried product, all else being the same.

Several researchers have studied the onset of crystallization of lactose systems in either pure systems or in milk powder. Roos and Karel[34] characterized the onset of crystallization of a pure lactose glass held at different temperatures. When the glass was held below T_g, nucleation did not occur. For a given water content in the glass, nucleation occurred more rapidly as temperature was increased above T_g. The WLF (Williams, Landel, and Ferry; see Chapter 4) kinetic equation defined for relaxation time of a kinetic process

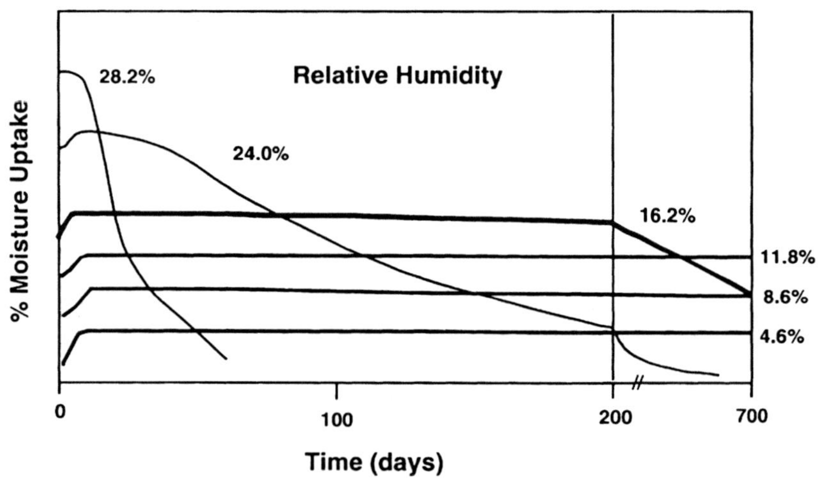

Figure 5–9 Moisture Sorption and Desorption Profile for an Amorphous Sugar Product Stored in Air with High Relative Humidity.

was able to predict the change in onset time with temperature. Jouppila and Roos[35,36] characterized the change in lactose crystallization of milk powders held at different temperatures and relative humidities. As the amorphous powder picked up water from the air, T_g decreased and the onset of crystallization was observed by using X-ray diffraction. The onset time for crystallization decreased as temperature increased above T_g. Interestingly, the presence of fat in a milk powder had essentially no effect on lactose crystallization. Based on these measurements, Jouppila and Roos[35,36] compare the effect of water on T_g with the moisture sorption isotherm to yield a critical moisture content for onset of crystallization. Their comparison between the sorption isotherm and the effect of water content on T_g is shown schematically in Figure 5–10. The point where the two curves intersect is taken as the point of instability of an amorphous product.

NUCLEATION OF LIPID POLYMORPHS

Although nucleation of lipids follows the same general principles, the complexity of lipids in their natural state causes some interesting differences. In particular, the formation

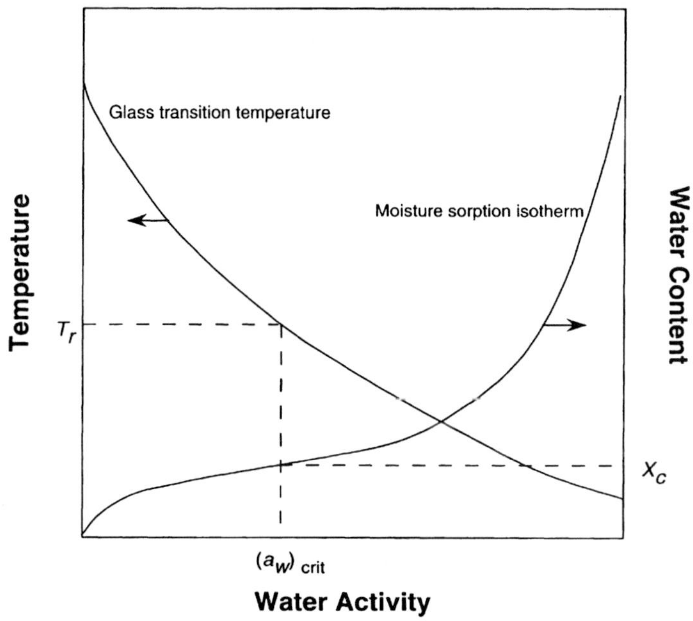

Figure 5–10 Schematic Diagram Depicting the Stability of an Amorphous Sugar Glass Exposed to Humid Air. The water activity [$(a_w)_{crit}$] at which the glass transition temperature reaches room temperature (T_r) represents the critical moisture content (X_c) for stability.

of different lipid polymorphs causes study of lipid crystallization to be more difficult. In addition to the kinetics of nucleation, one also must know which polymorphic form is most likely to nucleate. When different triacylglycerols are cooled from the molten state, different polymorphic forms nucleate first. Hernqvist[37] determined the formation of the first polymorph as different trisaturated triacylglycerols were cooled slowly (Figure 5–11). In the case of tristearin (C18:0), the α-polymorph was the first to form despite the existence of a substantial undercooling of the two more stable polymorphs (β and β′). However, the shorter-chain (C16:0, C14:0, and C12:0) trisaturated triacylglycerols nucleated into the β′ polymorph under these cooling conditions.

Nucleation of the lower energy polymorph rather than the more stable form has been explained based on free energy effects. The final energy state of the lower-stability polymorphs is higher than that of the more stable polymorphs, as shown in Figure 5–12.[8,38] However, the energy barrier to form the more stable polymorphs is higher than for the least stable polymorph. Thus, nucleation of lipid polymorphs is kinetically controlled and typically, the lower-energy polymorph forms first. Boistelle[17] relates this effect to differences in interfacial tension between polymorphs. Since the less stable polymorphs have a looser structural arrangement, it is logical to think that they have lower interfacial tension than the more stable polymorphs. In this case, according to Equation 5.5, the relative effects of interfacial tension and subcooling of each polymorph can be seen. For different polymorphs, the effect of interfacial tension must be substantially greater than the

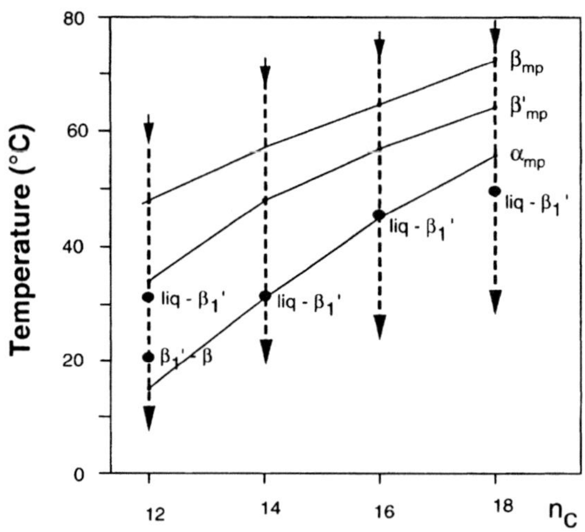

Figure 5–11 Onset Temperature of Nucleation and Polymorphic Form of Monoacid Triacylglycerols with Different Chain Lengths at Slow Cooling Rate of 0.4°C/min. α_{mp}, β'_{mp}, and β_{mp} represent melting points of the different polymorphs.

differences in subcooling between polymorphs; thus, the polymorph with lower interfacial tension (σ_s) has a higher nucleation rate.

Induction time data for tripalmitin in the different polymorphic forms was measured by Sato and Kuroda,[39] as shown in Figure 5–13. In this case, the molten lipid was cooled rapidly from 80°C to crystallization temperature and induction time measured by detection of crystals with a polarizing microscope. The induction time required for nucleation of the α-polymorph was significantly less than that of the more stable polymorphs. In addition, the induction time for the β′-polymorph was lower than that for the β-polymorph. Again, these differences can be explained by differences in interfacial tension. These nucleation kinetics suggest the general behavior for nucleation rate of lipids as shown in Figure 5–14. At temperatures above the melting point of the β′-polymorph, only the β-polymorph nucleates, but the nucleation kinetics are extremely slow. Below the melting point of the β′-polymorph, the melt is subcooled with respect to both the β′- and β-polymorphs. However, the β′-polymorph is more likely to nucleate, despite the lower subcooling, due to the more rapid kinetic rate.[38,40] Van Putte and Bakker[41] studied crystallization of lipids containing saturated fatty acids from palm oil using a population balance approach to obtain nucleation and growth rates separately. The rate of nucleation of the β′-polymorph was significantly higher than that of the β-polymorph, although the general trend shown in Figure 5–14 was demonstrated.

Nucleation rate of a pure lipid should follow the expression given in Equation 5.5, where nucleation rate is an exponential function dependent on the inverse square of

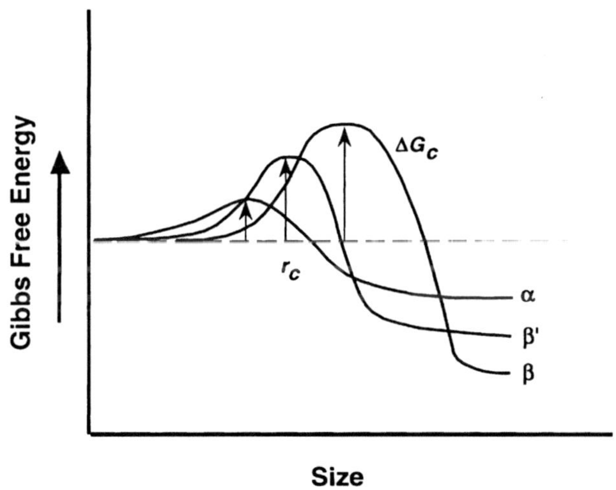

Figure 5–12 Schematic Diagram of Free Energy Change during Nucleation for Different Polymorphs.

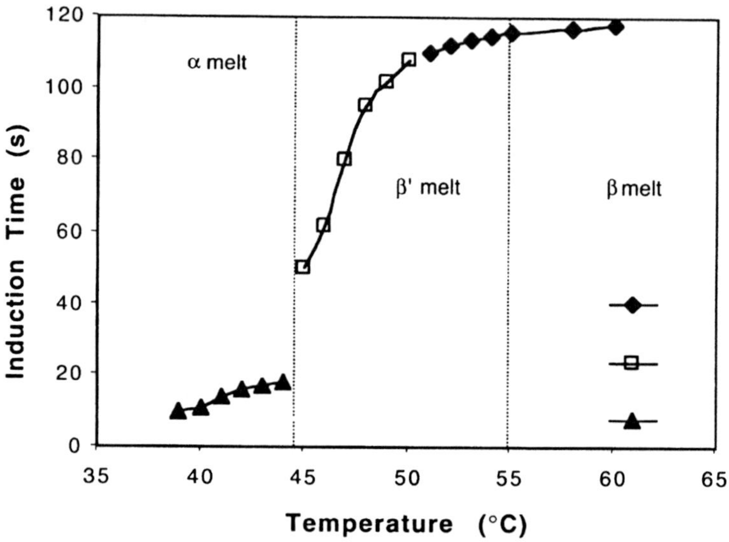

Figure 5–13 Induction Time Kinetics for Nucleation of Different Polymorphic Forms of Tripalmitin. Melting temperatures of each polymorph noted by straight line.

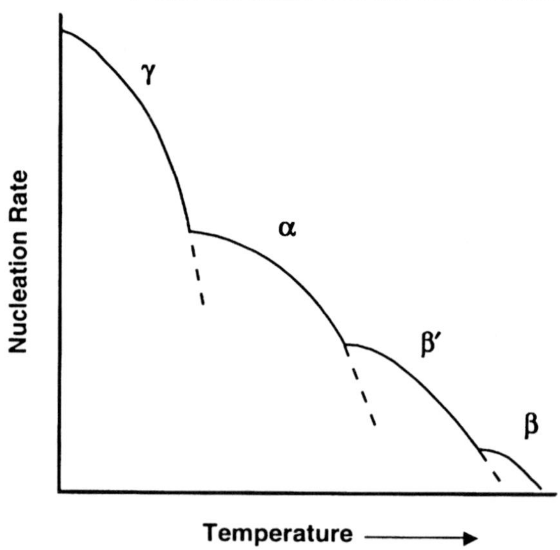

Figure 5–14 Schematic Diagram Showing General Form of Nucleation Rate of Lipids.

subcooling ($T_f - T$). Kellens et al[42] measured the nucleation rate of tripalmitin by counting the number of nuclei formed in the microscope field over a certain time. For nucleation of the β′-polymorph (melting point of 58.5°C), an increase in temperature from 45 to 52°C caused a significant decrease in nucleation rate. When these data were plotted according to Equation 5.7, with log of nucleation rate plotted against the inverse of the square of the subcooling, a straight line was obtained between 45 and 50°C. Interestingly, above 50°C, a straight line with a different slope was obtained, indicating that the β-polymorph was forming under these conditions. This conclusion was corroborated by DSC measurements and the habits of the crystal images observed by microscopy. Ng[43] found that the induction time for onset of nucleation of tripalmitin (PPP) crystallized from a solvent of triolein (OOO) was related to temperature by a semi-logarithmic function. The logarithm of induction time was linear with temperature, at least over a small temperature range, in contrast to the results of Kellens et al.[42] From the analysis of Ng,[43] an interfacial tension for the β-polymorph of tripalmitin was found to be 9.45 erg/cm^2.

Nucleation of natural fats can be quite complex due to the varied nature of molecular species present. However, it is possible to consider nucleation to be a solution-based process in some cases. That is, there may be a small percentage of a high-melting component that nucleates first upon cooling, which is followed by massive compound crystal formation as lower-melting triacylglycerols join into the crystal lattice. Thus, the rate of nucleation in natural fats is very compositional dependent. For example, Metin[44] studied nucleation rates of cocoa butter and milk fat by using a DSC in isothermal mode to monitor the exotherm produced upon crystallization. The results for nucleation at three different temperatures are shown in Table 5–1. One might expect that cocoa butter would nucleate more readily since it has a much simpler molecular composition. However, milk fat, with the much wider extent of molecular species, nucleates significantly faster than cocoa butter over this temperature range. This can be explained as nucleation of a very high-melting component of milk fat (primarily long-chain saturated triacylglycerols) from the remaining liquid oil. Interestingly, the nucleation rate of these components in milk fat is not a strong function of temperature in this study.

Other researchers have taken a similar approach in studying lipid nucleation. For example, Kloek et al[45] studied crystallization of a fully hardened palm oil from sunflower oil by considering this system to be a binary mixture of a solute (hardened palm oil) in a solvent (sunflower oil). Note that both components are composed of multiple molecular species of triacylglycerols. However, due to the differences in melting point, which is attributable to differences in the nature of the triacylglycerols (fully saturated fatty acids for hardened palm oil and mostly unsaturated fatty acids for sunflower oil), the hardened palm oil can be considered to crystallize separately. Their analysis generally followed Equation 5.7 for heterogeneous nucleation, although they also considered the entropy change upon nucleation. Induction times for onset of nucleation of the α-polymorph were determined by using an isothermal nuclear magnetic resonance (NMR) technique. By making the assumption that the inverse of the induction time is proportional to the nucleation rate, Kloek et al[45] were able to show that their data followed the general trend predicted by Equation 5.7. However, they were not able to quantify nucleation completely due to the lack of understanding of heterogeneous nucleation mechanisms.

Table 5–1 Induction Time (Measured by Isothermal Differential Scanning Calorimetry) for Onset of Nucleation of Cocoa Butter, Anhydrous Milk Fat (AMF), Milk Fat Fractions, and Their Mixtures with Cocoa Butter

	*Induction Time (min)**		
Fat Type	*15°C†*	*20°C*	*25°C*
Cocoa butter	9.4 ± 1.2	58.0 ± 8.0	N‡
AMF§	2.6 ± 0.4	2.1 ± 0.5	4.5 ± 0.4
5% AMF in cocoa butter	13.2 ± 0.8	72.6 ± 4.5	N
10% AMF in cocoa butter	15.3 ± 0.9	72.9 ± 4.4	N
HMF‖	1.4 ± 0.2	0.9 ± 0.1	0.4 ± 0.5
5% HMF in cocoa butter	12.7 ± 3.0	60.5 ± 0.2	N
10% HMF in cocoa butter	20.8 ± 3.5	64.4 ± 7.2	41.7 ± 0.1
MMF**	N	N	N
5% MMF in cocoa butter	17.9 ± 1.2	63.5 ± 6.5	N
10% MMF in cocoa butter	29.0 ± 2.9	N	N
LMF††	N	N	N
5% LMF in cocoa butter	11.9 ± 1.5	53.2 ± 9.0	N
10% LMF in cocoa butter	20.6 ± 1.4	N	N

* Induction time given as mean of three replicates ± standard deviation.
† Isothermal crystallization temperature.
‡ N indicates sample did not crystallize within 3 hours.
§ AMF, anhydrous milk fat (clear point = 37°C).
‖ HMF, high-melting milk fat fraction (clear point = 49°C).
** MMF, middle-melting milk fat fraction (clear point = 23°C).
†† LMF, low-melting milk fat fraction (clear point = 12°C)

Nucleation from Emulsions

Nucleation of emulsions has been used to study the difference in homogeneous and heterogeneous nucleation.[21] If a solution containing a certain number of active nucleation sites can be dispersed into sufficiently small droplets so that some of the droplets do not contain heterogeneous nucleation impurities, homogeneous nucleation can be studied. Thus, a population of droplets formed in this way will show two freezing points. The first freezing point occurs at the heterogeneous nucleation condition, dependent on the nature of the active sites present. The second freezing point occurs at the homogeneous nucleation temperature. Reid et al[22] used this technique to study the effects of various additives (silver iodide and nucleating bacteria) on freezing of water.

However, when materials crystallize in emulsions, the size of the emulsion droplet is an important factor that determines onset of nucleation. Walstra[46] notes that dispersion of a fat into an emulsion form may have several consequences on nucleation. First, the supercooling needed for onset of nucleation is generally higher (lower temperatures needed) than for nucleation of the same fat in bulk form. Kinetics of nucleation in an emulsified fat have been worked out by Walstra and van Beresteyn.[47] The probability of nucleation within an individual droplet is decreased as compared with nucleation of the bulk fat. The finer the dispersion of droplets, the less likely that a heterogeneous nucleation site becomes incorporated within the droplet. Those droplets with no active site must undergo nucleation either through the homogeneous nucleation mechanism or by some surface-induced action (induced by the presence of surfactant molecules at the droplet interface). Thus, in general, emulsions with smaller droplets require greater supercooling for onset of nucleation. Walstra[46] also suggests that the composition of lipid crystals formed in an emulsion of a natural fat may be different from those formed in bulk phase. This is attributed to differences in heat transfer that lead to formation of crystals at different temperatures than would be observed in the bulk fat. Crystal size and arrangement also may be different due to the physical limitations for growth provided by the droplet surface. Under some conditions, crystals that are oriented at the surface may be seen, where platelet-like crystals follow the surface curvature. It has been claimed that nucleation of the high-melting components of the fats initiates at the surface with concentration of the low-melting lipid components toward the center of the droplet.[48] This gives a stratified crystallization structure, although the nature of the crystalline microstructure within an emulsion droplet is dependent on the thermal treatments used to crystallize the fat. Walstra and van Beresteyn[47] showed that the initial polymorphic form that nucleated in an emulsion was different from the one that nucleated in bulk fat. In emulsions, more stable forms tended to nucleate as compared to bulk fat. In the milk fat system studied by Walstra and van Beresteyn[47] it was primarily the α form that nucleated in the bulk fat, whereas both α- and β′-polymorphs formed in the emulsion.

Kaschiev et al[49] studied the kinetics of crystallization of a palm oil emulsion in water. The effects of polydispersed emulsion (varied size distributions) was modeled theoretically assuming that a single nucleus formed in each droplet and experimentally verified based on results obtained by using an ultrasonic velocity measurement of crystallization.[50] The results showed that the critical supercooling was not a function of the polydispersity of the emulsion. However, the half time for crystallization of the emulsion was considerably shorter than for the corresponding monodisperse emulsion (all droplets the same size). Kaneko et al[51] studied the effects of various impurities on crystallization of hexadecane in aqueous emulsions. The rate of crystallization was retarded in the emulsion system, as compared with bulk crystallization. However, addition of a surface-active agent to the emulsion promoted nucleation, since the onset temperature for crystallization of the emulsions droplets was higher when the surfactant had been added. No effect on crystallization temperature was observed when the surfactant was added to the bulk. It was speculated that the surfactant promoted heterogeneous nucleation of hexadecane at the surface of the droplet.

Since nucleation in emulsified systems is important commercially, these differences with crystallization from the bulk phase may be quite important. For example, careful control of crystallization of cream prior to whipping or churning is very important to obtain the desired physical properties of whipped cream or butter, respectively.[48] Proper crystallization of emulsified milk fat in ice cream also is critical to obtain the desired product quality.[52]

SECONDARY NUCLEATION

In many systems, including sucrose, ice, citric acid, lipids, and many salts, there is a sizeable metastable zone of supersaturation where spontaneous, heterogeneous nucleation does not occur for reasonably long times (Figure 4–16). If seeds are added to such a supersaturated solution, they will grow based on the kinetics of the system, but no new nuclei will form spontaneously. However, under certain conditions within this region of concentration and temperature, new nuclei may be formed due to the presence of the existing crystal surface. The process of formation of nuclei due to the presence of existing crystals is called secondary nucleation, which may occur by several mechanisms.[53] Note that secondary nucleation occurs under conditions that would not normally result in spontaneous, heterogeneous nucleation. The metastable zone is often subdivided to show this region of secondary nucleation, sometimes called the forced crystallization zone.[54] This is shown in Figure 5–15 for sucrose.

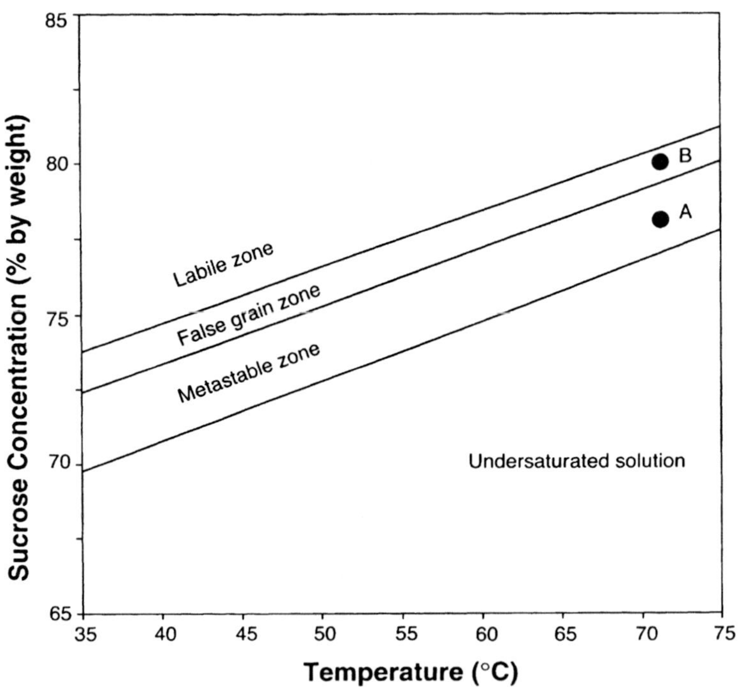

Figure 5–15 Metastable Zone Showing Region of Secondary Nucleation (False Grain Zone) for Sucrose-Water System.

Secondary nucleation is particularly important in suspension crystallization, as in refining of sugars and fractionation of fats, where agitation of the seed crystals leads to secondary nucleation. In sucrose refining, for example, the goal is to grow a distribution of seed crystals to product size in as short a time as possible. This is accomplished by maintaining supersaturation at the highest possible point (for a given operating temperature) in the metastable zone (see point A in Figure 5–15). At this point, growth rate is most rapid and occurs without the undesired formation of new nuclei. If a process upset occurs such that supersaturation moves into the forced grain zone (point B in Figure 5–15), secondary nucleation, or false grain, occurs and a population of small crystals is introduced into the batch. This distribution of crystal sizes (large and small) is difficult to separate from the liquid in the centrifuge and causes processing problems.

There are several potential mechanisms for formation of secondary nuclei.[53,55,56] Probably the most important mechanism is contact secondary nucleation, where nuclei are generated by contacts between crystals and either other crystals, the crystallizer walls, or the stirrer in an agitated tank (Figure 5–16). Alternatively, sufficient fluid shear may cause small surface irregularities on the crystal to be sheared off and dispersed into the solution where they grow into stable nuclei. In some cases, irregular growth of existing crystals,

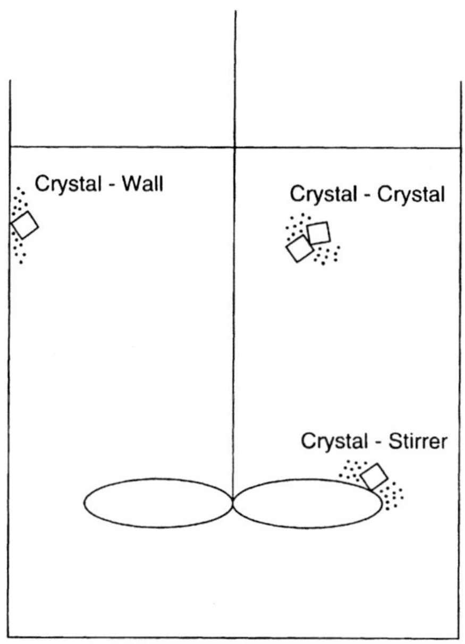

Figure 5–16 Potential Sources of Contact Nucleation in a Suspension Crystallizer.

combined with changing environmental conditions of heat and mass transfer, may lead to melting or dissolution of portions of a crystal with the remaining portion able to survive as a new nucleus. Ice and lipid crystals, for example, often grow in irregular, needle-like or dendritic form. The base of these needle-like appendages may provide a site where the radius of curvature makes it more unstable than the bulk of the crystals, and this thin neck melts or dissolves away, leaving new crystals (nuclei) adjacent to the original source crystal.[40] This is shown schematically in Figure 5–17.

The exact mechanisms of contact nucleation remain somewhat unclear, despite substantial research in this area.[56] It is likely that contact secondary nucleation occurs due to one of two possible causes. The first potential mechanism is microattrition, where microscopically small bits and pieces of crystal are dislodged from the lattice under conditions of intense energetic contacts. Survival of these bits depends on their size and the supersaturation of the surrounding solution. However, microscopic observation of crystal surfaces following controlled contact nucleation often indicates no observable damage to the crystal surface structure, putting the importance of this mechanism in question. The second possible mechanism involves disruption of a thin layer around the growing crystal (see Chapter 6 for further details of this layer). This adsorption layer is thought to represent a transition zone between molecules in solution and molecules in the lattice. Here, it is likely that groups of molecules orient into some form of precrystalline structure prior to being incorporated into the lattice. If these precrystalline embryos are dislodged and dispersed into the solution due to a contact, they will grow into stable nuclei depending on the critical size and supersaturation at that temperature.

There is some experimental evidence for this latter mechanism. In many crystallizers, the energy of contact between crystals and crystallizer surfaces does not appear to be sufficient to generate enough nuclei to explain the observed results. However, low-energy contacts, equivalent to a gentle sliding contact between a crystal and a solid surface, are sufficient to generate hundreds of nuclei under appropriate conditions.[57] In fact, simply sliding a growing crystal across a glass surface is sufficient to produce hundreds of secondary nuclei in many systems (sucrose or lactose, for example). When an inert surface (a metal rod) is scratched along the surface under the same conditions, no new nuclei are generated, verifying that this is indeed a secondary nucleation mechanism. These results suggest that one important mechanism for contact nucleation is disruption of the adsorbed layer at the surface of a growing crystal.

Experiments utilizing enantiomeric differences in molecules have been used to distinguish between surface and solution-based mechanisms of secondary nucleation. Denk and Botsaris[58] studied secondary nucleation from sodium chlorate crystals, which can crystallize in different enantiomeric forms (D and L). They used a seed crystal of one form to study secondary nucleation under conditions that would yield the other enantiomeric form when crystallized by primary nucleation. For example, the D form was placed in a supersaturated solution that would normally crystallize in the L form if there was no secondary nucleation. Their purpose was to determine whether secondary nuclei come from the seed crystal itself or are induced by some solution mechanism. At high supersaturations, where primary nucleation was occurring, they found an equal mixture

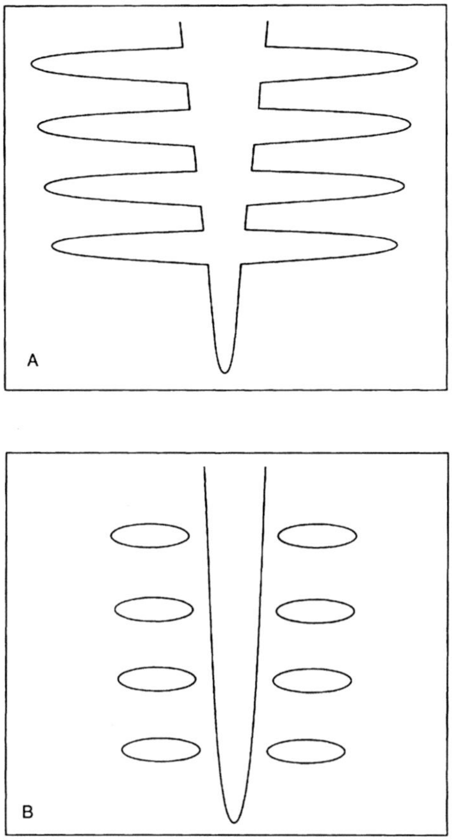

Figure 5–17 Secondary Nucleation Due to Preferential Dissolution of Irregular or Dendritic-Shaped Crystal.

of the two enantiomers was formed. At intermediate supersaturations, nearly all of the nuclei were the same enantiomer as the seed crystal. However, at low supersaturations, the percentage of nuclei with the same enantiomeric form as the seed crystal fell to 60%. Since two different species were crystallized during the same contact, it is not clear that the mechanism of secondary nucleation can be attributed completely to microattrition. Secondary nuclei of the same enantiomeric form as the seed crystal could indeed arise from the surface of the seed during contact, but the presence of the other enantiomeric form suggests a solution mechanism. Similar results were found for secondary nucleation of glucose under sliding conditions.[59] They found that two crystal forms (α-glucose monohydrate and α-glucose anhydrous) were formed during secondary nucleation under certain conditions. Thus, the exact mechanisms of secondary nucleation under all conditions could not be determined from this study.

Wissing et al[55] studied secondary nucleation generated by contacting a potassium dihydrogen phosphate crystal with a probe at a contact energy controlled by height of fall. The surface of the crystal at the point of contact was closely observed by optical microscopy. The falling probe disturbed the volume diffusion field around the growing crystal and caused an increase in the velocity of the advancing macrosteps on the crystal surface. Turbulence in the solution surrounding the crystal caused an increase in supersaturation at that point. Within several minutes after the contact, the step pattern on the surface of the growing crystal restabilized. If the contact energy was less than some critical value, E_c, only a few nuclei (10 to 100) were generated for each contact. No damage to the crystal was observed when contact energy was below E_c, and the nuclei generated were visible immediately after contact (larger than a few microns in size). When the contact energy was greater than E_c, severe impact damage to the growing crystal was observed accompanied by generation of vast numbers (over 1,000) of secondary nuclei. Under these conditions, holes in the growing crystal surface were observed. Secondary nuclei generated were visible immediately with size as large as 20 μm. The damage to the crystal surface healed as the crystal continued to grow. These results suggest that two mechanisms of secondary nucleation exist. At low contact energies, there is a disruption of the surface field around a growing crystal that leads to nuclei formation. However, at higher contact energies, microattrition is the cause of significant numbers of secondary nuclei as the surface of a growing crystal is chipped away.

Secondary nucleation is affected by factors such as agitation rate, the number and size of existing crystals, the surface hardness of the crystallizer vessel and agitator, as well as supersaturation, temperature, and the presence of additives/impurities. Typically, an expression similar to the form of Equation 5.12 is used to describe the rate of secondary nuclei formation, B (#/mL-min).

$$B = k_{SN}(S)^i(m)^j(N)^l \tag{5.12}$$

where k_{SN}, i, j, and l are constants; m is either the second or third moment of the existing crystal size distribution; S is supersaturation ratio; and N is stirrer rate (agitation energy input). Typically, secondary nucleation is a strong function of supersaturation (given as either S or ΔT), with greater numbers of nuclei being generated at higher supersaturations (assuming that supersaturation remains below the labile zone limit where nucleation occurs spontaneously). In ice, secondary nucleation rate was found to be a second order function ($i = 2$) of ΔT at low supercoolings (0.01 to 0.20°C) and a first order function ($i = 1$) of ΔT at higher supercoolings (0.25 to 1.0°C).[60,61] When the third moment of the distribution ($m = m_3$) is used to correlate the surface area available for secondary nucleation, the exponent, j, is usually equal to one.[56] The number of nuclei formed generally increases with the energy of contact and the hardness of the contacting surface. Another important factor is the presence of impurities that influence crystal growth. In general, materials that inhibit crystal growth also depress rates of secondary nucleation. Perhaps this is due to a smoothing effect on the surface of the growing crystal and fewer embryos are dispersed into the solution upon contact.

Secondary nucleation has been shown to occur in a wide variety of crystallizing systems, from salts to fats. The importance of contact secondary nucleation in the food industry is probably greatest in those systems where crystallization is controlled for separation of a certain component material, for example, refining of sugars, freeze concentration of fluid foods (juices, etc), and fractionation of fats. The importance of secondary nucleation in sugar refining was mentioned previously. Operation of a sucrose crystallizer is carefully controlled to limit any secondary nucleation as this would result in a crystalline product with a size distribution that leads to inefficient filtration. Secondary nucleation also is important in refining of other sugars (lactose, glucose, etc.). Fats also are believed to exhibit secondary nucleation, where the formation of many new nuclei form in the vicinity of a crystal once it has formed.[46] Hachiya et al[62] claim that secondary nucleation is the cause of crystal formation when low levels of high-melting seed crystals are added to promote crystallization of cocoa butter.

Generation of secondary nuclei also has been found during agitation of ice crystals. Garabedian and Strickland-Constable[63] found that an ice crystal slid across a glass plate did not produce secondary nuclei, as found for sugars.[57] However, a single crystal circulating in a stirred tank produced numerous secondary nuclei, presumably due to collisions with the stirrer. Hartel and Chung[64] found similar results and characterized a minimum ΔT needed for onset of secondary nucleation. Numerous studies[60,61] have quantified secondary nucleation in water, sugar solutions, or fruit juices. It has been speculated that secondary nucleation may be an important mechanism for production of ice crystals in a scraped-surface freezer, as in production of ice cream.[65,66]

PARAMETERS INFLUENCING NUCLEATION

In order to control and predict nucleation in foods and other products, it is necessary to quantify rates of nucleation under different conditions. However, it is rare that we have this kind of data for systems of specific interest in complex foods since the nucleation rate curves are extremely difficult to obtain. In addition, factors other than supersaturation and temperature may influence the induction time and nucleation rate curves. Processing factors such as agitation rate and energy input and the presence of impurities or additives (soluble or insoluble) may have significant impact on nucleation rates and induction times.

Supersaturation

The effects of supersaturation or subcooling on nucleation rates have been discussed already. Briefly, as driving force increases initially above the critical point necessary for onset of nucleation, nucleation rates increase and induction times decrease. In this region of low driving force, small increases in supersaturation or supercooling result in very large increases in nucleation rate. However, in systems that exhibit mobility limitations, a point is reached where nucleation becomes inhibited by decreasing molecular mobility (increasing viscosity) as driving force is further increased. At even higher driving forces (lower temperatures and/or high concentrations), molecular mobility is decreased sufficiently that nucleation is inhibited and the rates decrease rapidly. When the glass transition point is surpassed, nucleation no longer takes place.

Temperature

Temperature is intimately related to driving force in both solution and melt systems. In melt systems, lowering temperature below the metastability zone induces rapid onset of nucleation. Nucleation rate increases and induction time decreases rapidly with slight decreases in temperature in the region near the melting point (melts) or solubility condition (solutions). However, at some temperature, a maximum in the rate of nucleation is observed as supersaturation, and mobility effects counterbalance one another. When temperature decreases to the point where mobility effects become rate limiting, nucleation rate decreases. At these lower temperatures, the rate of nucleation is low, and more time is required before nucleation occurs.

The nucleation rate of sucrose at low supersaturations (from −10 to 40°C) was found to have an activation energy of 4.4 kcal/mol.[10] Since this value is close to the activation energy for the change in viscosity over this temperature range, it is likely that mass transfer plays an important role in primary nucleation of sucrose.

Materials in the glassy state do not nucleate over any finite period of time and can be considered stable to crystallization. However, if their temperature is increased or if their concentration is decreased through moisture sorption, they become prone to crystallization with a stability that is time dependent. The induction time for nucleation from an amorphous material depends primarily on temperature and water content, although the method of preparing the glassy state also may affect onset of nucleation. Higher temperatures and increased water content, leading to greater molecular mobility, make these materials more prone to nucleation.

Viscosity

For the most part, the effect of viscosity on nucleation is controlled by the concentration and temperature of the system. As viscosity increases due to lower temperature and higher concentration, nucleation processes become more limited. This was shown in Figure 5–4, where nucleation rate decreased at higher driving forces. In this case, viscosity is governed directly by changes in concentration or temperature. Thus, it can be said that as viscosity increases due to a decrease in temperature and an increase in concentration, nucleation is inhibited. However, nucleation is not controlled per se by viscosity of the liquid phase, contrary to popular belief.

There are cases where viscosity of the bulk solution may be modified by addition of small amounts of ingredients (gums, proteins, starches, etc) with little effect on nucleation. For example, addition of small amounts of gums to an ice cream mix increases the bulk viscosity of the mix by over an order of magnitude. However, the addition of small levels of hydrocolloids, which have a major effect on viscosity, has been shown to have no effect on nucleation of ice.[67] Two mechanisms may lead to an effect of viscosity on nucleation. The first is the true effect of solution viscosity on mobility of the molecules of the crystallizing species. If diffusivity of the crystallizing species is reduced by addition of another component (eg, hydrocolloid), then nucleation will be inhibited. The second effect is that increased viscosity can alter the rate of heat transfer, which can influence nucleation processes. In the case of hydrocolloids added to influence freezing,[67] apparently neither mechanism was important under the conditions studied, and the hydrocolloids had no net effect on ice nucleation. That is, mobility of water molecules was sufficient to promote nucleation despite the presence of hydrocolloids, and apparently heat transfer rates during the freezing process were not influenced by increased viscosity.

Cooling Rate

In situations where the driving force for crystallization is generated by cooling of the material, the cooling rate may impact when nucleation occurs. This is certainly true in melt systems[40] where the material is cooled below its melting point to initiate crystallization, but it also is true for solution systems that are supersaturated by cooling below the solubility temperature (for example, the citric acid system shown in Figure 5–5). A rapid rate of cooling generally leads to nucleation at a lower temperature than for slow cooling. A difference of 2 to 3°C in temperature of onset of nucleation was reported by Mullin[2] for salt solutions cooled slowly (5°C/hour) or rapidly (30°C/hour). Similar results often are found when using the DSC to study induction time for nucleation (in lipids, for example), where slower cooling rates generally lead to nucleation at higher temperatures. Note that agitation rate also plays a role here since heat transfer is promoted at higher agitation rates, but higher agitation rate also can promote nucleation. Thus, Singh[68] found that higher cooling rates led to crystallization of partially hydrogenated soyabean oil at higher temperatures in a stirred tank.

Milk fat cooled slowly on a microscope slide has significantly different crystalline microstructure when cooled rapidly. Confocal microscope images of these samples were shown in Figure 3–7. The slowly cooled sample (0.2°C/min) forms spherulitic crystals, whereas the rapidly cooled sample (15°C/min) forms random crystalline strands that arrange in a network. Herrera and Hartel[69] studied the effect of cooling rate on crystallization kinetics and lipid crystalline structure in a model milk fat system. Mixtures of high- and low-melting components of milk fat were cooled from 60°C and allowed to crystallize under agitation at 25°C for several hours. Polarized light microscope images of the crystalline microstructure formed at different cooling rates are shown in Figure 5–18. The rate of cooling from 60°C to 25°C also had a significant effect on crystal size distribution, as seen in Figure 5–19.

A

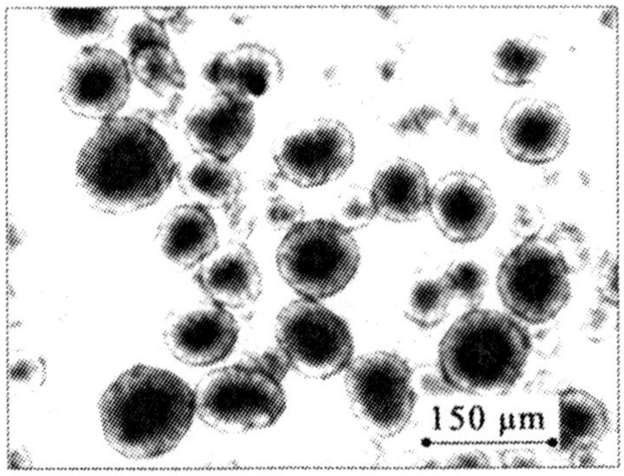

B

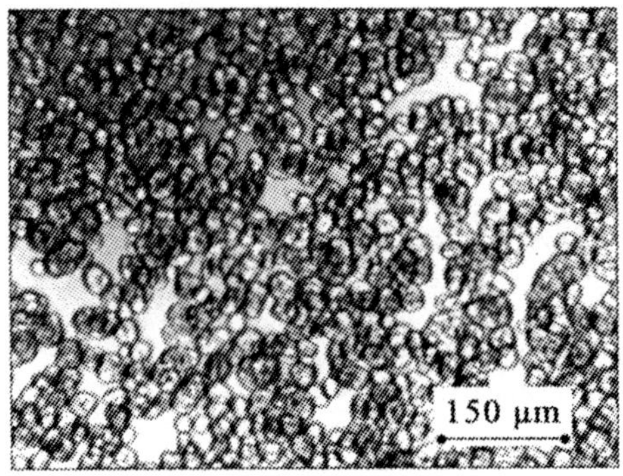

Figure 5–18 Effect of Cooling Rate on Crystalline Microstructure (Polarized Light Microscopy) of a Milk Fat Model System (50% High-Melting and 50% Low-Melting Component of Milk Fat) Crystallized at 25°C and Agitation Rate of 50 rpm. Cooling Rate: (A) 0.2°C/min and (B) 5.5°C/min.

The rate of cooling to a supersaturated state can be an extremely important process parameter in some systems. In some products, it is desirable that nucleation of all crystals occurs at approximately the same time so that all nuclei grow to be crystals of about the same size. If some nuclei form early in the process, they will grow larger than the majority of crystals formed later in the process. These large crystals can have a negative impact on product quality. For example, in fondant processing, rapid cooling of the sugar mass is desired so that all crystals form at the same time. If some crystals form during cooling prior to the main nucleation event, these crystals will be significantly larger than the average; a fondant with coarse texture will result. Thus, the cooling rate in fondant processing should be rapid enough that no crystals form until the main event triggered by agitation in the beating vessel (see Chapter 7). If cooling of the sugar mass is slow, nucleation may potentially occur at a higher temperature than desired and continue as temperature is lowered. A wider distribution of crystal sizes is then obtained. A similar result was found

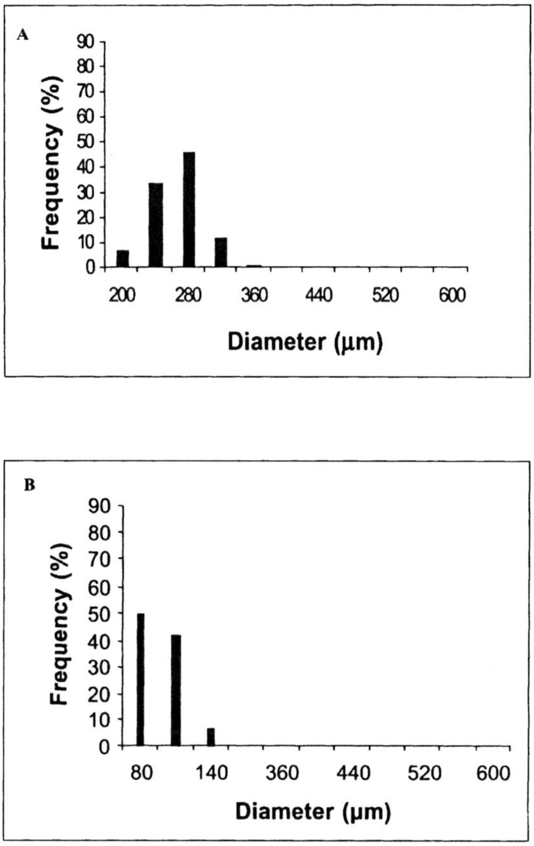

Figure 5–19 Effect of Cooling Rate on Crystal Size Distribution of a Milk Fat Model System (50% High-Melting and 50% Low-Melting Component of Milk Fat) Crystallized at 25°C and Agitation Rate of 50 rpm. Cooling rate: (A) 0.2°C/min and (B) 5.5°C min.

for crystallization of a milk fat model system.[69] Slower cooling resulted in a distribution with larger mean size and wider size distribution.

Agitation Rate

Agitation is widely considered to promote the rate of nucleation. The energy input due to agitation is thought to help overcome the energy required for formation of nuclei.

In general, greater agitation or energy input into a supersaturated solution hastens the onset of nucleation and is thought to increase the rate of nuclei production. Also, increased agitation rate is thought to decrease the extent of the metastable zone.[70] These effects are most likely due to the effect of agitation on localized supersaturation gradients that leads to regions where local supersaturation exceeds the critical value for nuclei formation. In the manufacture of fondant, a highly supersaturated sugar syrup is exposed to high agitation levels to promote maximum formation of nuclei. If the highly concentrated syrup is not agitated, nucleation occurs very slowly and the product does not have the many small crystals desired for smooth texture. Even if the agitation (beating) occurs with lower energy input; this can influence the extent of nucleation and the smoothness of the resulting fondant.[71]

Although it is widely accepted that agitation enhances nucleation, not all research studies have documented this result in all cases. For example, van Hook and Bruno[12] observed no increase in nucleation rate of a supersaturated sucrose solution as agitation rate was increased when the supersaturation ratio (S) was below 1.4. Above $S = 1.4$, however, nucleation rate occurred more rapidly as agitation level was increased. Mullin and Raven[70] explained this effect based on two competing mechanisms—diffusion and attrition.

Certainly, agitation rate has a major effect on secondary nucleation in stirred tanks. According to Equation 5.12, an increase in agitation rate (either energy input or rpm) leads directly to an increase in secondary nucleation rate, with an exponent between 0 and 1. For secondary nucleation in fruit juices and sugar solutions, Omran and King[60] found that the exponent of agitation rate in Equation 5.12 was 1 at lower agitation rates, but then nucleation rate became independent of agitation ($j = 0$) for higher agitation rates. Other researchers[72] have found similar behavior with values of j somewhere between 0 and 1 for secondary nucleation of ice in stirred tanks.

Herrera and Hartel[69] found that higher agitation rates led to smaller mean size and narrower distribution for crystallization of a milk fat model system. In this case, the influence of agitation rate was not separated from that of cooling rate, since higher agitation rate leads to faster cooling. Thus, the combination of higher agitation and faster cooling resulted in a narrow size distribution with small mean size.

External Fields or Physical Phenomena

The input of external energy sources, particularly ultrasonic energy, often has been claimed to enhance nucleation in various systems.[73] Numerous examples of freezing of supercooled water have been given where systems that would normally be metastable for long periods of time were found to nucleate when an external force (either sound wave or shock wave) was applied.[74] Turner and van Hook[75] showed that water could only be supercooled by about 2°C when an ultrasound generator was turned on, whereas supercoolings of 5 to 10°C were obtained when the ultrasound was turned off. In addition, van Hook and Frulla[10] observed an increase in nucleation rate (higher seed densities) when an ultrasonic field was applied to a supersaturated sucrose solution.

However, there remains some question about the exact mechanism of this effect, and exactly when and where such an effect occurs. Mechanistically, it is argued that the physical energy reduces any diffusional limitation to nucleation and thereby enhances the process. Alternatively, the collapse of air bubbles due to passage of a shock wave is suggested as the main cause of nucleation under the influence of external energy. Kennedy[74] provides a theoretical basis for the effect of ultrasonic energy and collapse of cavitation bubbles on nucleation of ice. Pressure and temperature changes in the vicinity of the collapsing bubble were described according the theory of Hickling.[76] In this theory, an adiabatic increase in pressure occurs due to the collapse of the bubble. Although temperature increases in this case, the phase boundary at higher pressure occurs at higher temperatures as well. Thus, the combination of increased pressure and temperature causes the water in the vicinity of the collapsed bubble to be at a position on the phase diagram (in pressure and temperature) below the homogeneous nucleation temperature for that pressure. Nucleation must occur quite rapidly; then, upon cooling back to the original state, the nucleus must remain stable and grow into a viable crystal.

The application of ultrasonic energy to promote nucleation generally requires sufficient energy input that cavitation occurs. Kennedy[74] estimated the minimum energy input needed for cavitation to be on the order of 1 W/cm^2. However, it also has been suggested that ultrasonic energy of sufficient power to cause cavitation also is likely to generate substantial numbers of heterogeneous nuclei caused by the cleaning effect at the walls (recall that ultrasonic baths are used for cleaning surfaces). Release of entrapped particles from microcrevices in the wall may add sufficient nucleation sites to promote more rapid nucleation.

Additives and Impurities

In food formulations, ingredients other than the crystallizing species can have profound influence on mechanisms and kinetics of nucleation. Ingredients that influence crystallization may be specifically added to the formulation or present as impurities in other ingredients. Sometimes other ingredients are added to foods to help control crystallization. The addition of corn syrup to sucrose solutions in making hard candies is an example of an ingredient added to slow down nucleation of sucrose. Addition of seed materials is an example where the additive promotes crystallization. Certain materials can be added to a supersaturated solution to promote nucleation, although there also are many other materials that inhibit nucleation processes. Perhaps one way to classify these materials is based on their characteristics in solution. Insoluble materials added to a supersaturated solution are likely to promote nucleation through addition of catalytic sites that promote heterogeneous nucleation. On the other hand, both soluble and insoluble materials may impede the ability of molecules of the crystallizing species to come together and form a stable nucleus. In general, insoluble components have the potential to promote nucleation, whereas soluble components are more likely to inhibit nucleation. In emulsion systems, nucleation may be enhanced by addition of a surface-active component that helps order the molecules of the crystallizing species within the emulsion droplet.

Intentional addition of insoluble materials (or seeding) often is used in practice to promote nucleation and crystallization. Crystals of the desired species may be added to a supersaturated solution to seed the material and help control crystallization. Alternatively, solid particles of different material than the crystallizing species may be added, either knowingly or not, that induce nucleation of the desired species. Seeding with another solid material promotes nucleation only when that material acts as an efficient heterogeneous nucleation site. For example, cloud seeding with silver iodide crystals promotes formation of ice crystals that eventually fall as rain. Addition of dusty material in certain foods may promote nucleation during storage, as with lactose crystallization in ice cream during storage due to the addition of certain ingredients (nuts, etc). These dust particles act as nucleating sites where eventually lactose crystals will form.

Many soluble impurities and additives inhibit nucleation. Typically, if a material inhibits nucleation, the induction time is increased and the rate of nuclei formation is decreased. Potential mechanisms by which additives may influence nucleation are through diffusion limitations, steric influences, or charge effects. Numerous studies have documented the effects of other components on nucleation in foods. Only a few examples of these studies are presented here for nucleation of sugars, ice, and lipids. However, in general, there is still a significant lack of quantitative information about the effects of formulation factors (additives and impurities) on nucleation in complex food products.

Sugars

The addition of other sugars to sucrose has been widely used in the past to control sucrose crystallization. Nucleation is controlled through increases in both the metastability limit of supersaturation and the induction time for onset of nucleation. The presence of invert sugar, at only a 0.5% level, in a sucrose solution extended the critical supersaturation for onset of nucleation to greater than $S = 1.8$.[7] Recall that in a pure sucrose solution, the critical supersaturation for spontaneous nucleation was about $S = 1.3$. Corn syrups (which contain glucose polymers) are well known to delay the onset of sucrose crystallization when incorporated at sufficiently high levels,[77,78] as shown schematically in Figure 5–20. This phenomenon is used in the candy industry to control sugar crystallization, where addition of between 20 and 50% corn syrup in a formulation can greatly inhibit sucrose crystallization. Note that the effect of an impurity on nucleation should be compared at equivalent supersaturation rather than at equivalent solution concentration (or constant total solids) since impurities also can affect the phase diagram and change the equilibrium conditions.

To compare the effects of an additive on nucleation rates, one must first separate any solubility effects. In the case of invert sugar or corn syrup, the solubility of sucrose is decreased as more of the other sweetener is added (see Chapter 4). In order to determine the true effect of corn syrup, for example, on the nucleation process, nucleation rates must be compared at conditions of equivalent supersaturation. Consider two scenarios. In the first case, a solution of 80% total solids is nucleated with different ratios of sucrose to corn syrup. Because the total sugar content (sucrose plus corn syrup) did not change, increased levels of corn syrup caused a decrease in the level of sucrose. This decrease in sucrose

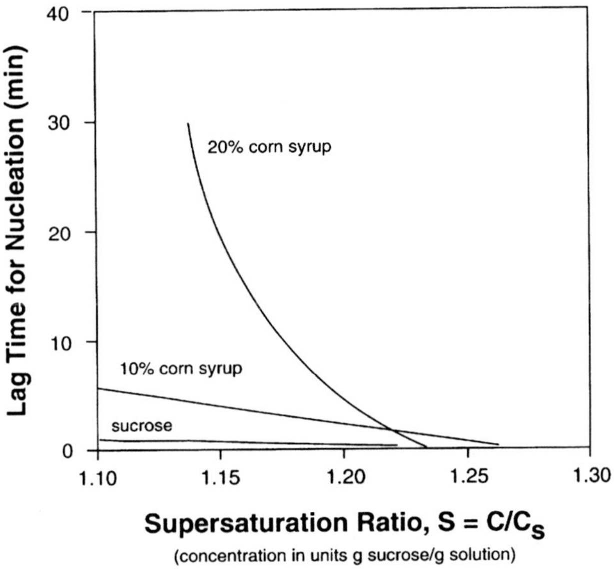

Figure 5–20 Effects of Corn Syrup on Induction Time for Sucrose Crystallization during Agitation in a Batch Crystallizer, Shown Schematically.

content, in addition to the depression of solubility of sucrose due to addition of corn syrup, results in a system at lower supersaturation (but the same initial total solids content). Based on Equation 5.4, the system with lower supersaturation will nucleate more slowly than the one with higher supersaturation. Thus, it is not possible to ascertain whether the corn syrup had an inhibition effect on nucleation separate from the effect on supersaturation. The second scenario documents the specific effect on nucleation kinetics. In this case, the supersaturation ratio of sucrose is maintained constant as more corn syrup is added to the solution. However, maintaining constant supersaturation requires addition of more sucrose as well, resulting in a solution with overall higher concentration. Thus, it is often difficult to sort out the competing effects of solubility and supersaturation, particularly as concentrations increase to very high levels.

Crystallization from the amorphous state also is influenced by the presence of impurities or other formulation components. This principle is widely used in both the food and pharmaceutical industries to stabilize sugar glasses that are prone to crystallize. In hard candy, the addition of corn syrup serves several purposes, not the least of which is to reduce the likelihood of graining. If crystallization occurs in the hard candy (a process called graining), dispersed colors and dyes are concentrated in noncrystalline areas. Thus, crystallization in hard candy results in uneven color and flavor distribution as well as more rapid diffusion of the flavor molecules out of the structure. Herrington and Branfield[79] clearly showed the effects of various corn syrups on inhibition of the rate of progress of a nucleation front as an amorphous sugar matrix crystallized. Other additives, such as gums, gelatin, and even fats, also may stabilize the sugar glass against crystallization. Van Scoik and Carstensen[80] documented the effects of various impurities, at the 5% addition level, on induction time for nucleation of sugar glasses as they picked up moisture from the environment. Corn syrup extended induction time by a factor of 7, whereas gelatin and acacia gum extended induction time by a factor of 15. Shamblin et al[81] showed that, at certain levels, additives (particularly polymeric materials) may significantly influence nucleation without having any effect on T_g. Thus, the effect of additives on sugar crystallization from the amorphous state is typically not due to the effect on glass transition. Jouppila and Roos[36] showed that the various components in milk (protein, salts, etc) had no effect on T_g of the amorphous phase, yet caused significant inhibition of nucleation of lactose (compared with a pure lactose system).

Ice

In the same manner as discussed above, the effects of additives on freezing generally can be categorized as to whether the additives are soluble or insoluble. Typically, insoluble components catalyze ice formation—or at least have the potential to provide active sites—and soluble components either have no effect or inhibit nucleation. For example, deionized water can be supercooled to a substantially greater degree than regular tap water[67,82] due to the presence of nucleating sites in tap water.

Several insoluble materials have been identified that dramatically influence ice nucleation. Silver iodide, a principal material for cloud seeding, provides a surface capable of inducing ice nucleation. In distilled water (with a small amount of dissolved salt), Wang and Vonnegut[28] found nucleation of ice to occur in a reasonable time scale (minutes) at temperatures of −13.5 to −16.5°C , whereas Vonnegut and Baldwin[27] found similar rates of nucleation at −3 to −9°C with addition of silver iodide particles. The promotion of nucleation by silver iodide is thought to be due to the similarity in the crystal structures of the two materials.[3] There also are certain species of bacteria, such as *Pseudomonas syringae*, that promote ice nucleation. The nucleating capability of these bacteria comes from the membrane cell wall that organizes the water molecules into a structure like that in the crystal lattice and thereby decreases the energy needed to form the lattice structure.[19] Note that there also are proteins that have "anti-freeze" effect. These anti-freeze proteins often are extracted from arctic fish and can be used to control crystallization of ice. However, the effect of these anti-freeze proteins is thought to be primarily on crystal growth rather than an on nucleation.[83]

Addition of soluble components in solution often causes inhibition of nucleation of ice. Charoenrein et al[20] summarized previous work that demonstrates the effects of solutes on heterogeneous nucleation of ice induced either by addition of silver iodide or *Pseudomonas syringae*. Addition of 15% sucrose in solution had little effect on nucleation rates, but addition of 50% sucrose in solution significantly influenced nucleation kinetics.

Sucrose lowered the temperature at which heterogeneous nucleation occurred as well as changing the slope of the nucleation rate-temperature dependence for heterogeneous nucleation. Typically, the change in nucleation temperature due to addition of solute is considered to be proportional to the change in freezing point caused by the solute.[20] This relationship is given as

$$\Delta T_n = k\Delta T_f \tag{5.13}$$

where ΔT_n is the depression of nucleation temperature and ΔT_f is the freezing point depression induced by the solute. The parameter, k, is a constant with a value typically between 1 and 2.[21]

In many frozen foods, hydrocolloids often are added to control ice crystallization, although the exact effects are not always clear. Muhr et al[67] and Blanshard et al[82] showed that many hydrocolloids (agar, gelatin, pectin, and several proteins) had no influence on rates of heterogeneous nucleation of ice. The only hydrocolloids found to have an effect were the sodium alginates studied. However, their conclusion was that the sodium alginate preparation used in this study contained small particles (of unknown organic origin) that acted as heterogeneous nucleation sites. Apparently, the effects of various hydrocolloids in controlling ice crystallization does not arise from an effect on nucleation.

Numerous studies have shown the influence of impurities on secondary nucleation of ice in stirred crystallizers. The presence of salts decreases the rate of secondary nucleation,[72] as do various polymers like dextrans and proteins,[84] pectin and gelatin,[60] and lactose.[84] However, the presence of other sugars (sucrose, fructose, and glucose) and juices (apple and orange) have been shown to enhance secondary nucleation.[60,61] These effects are most likely related to the structuring of the ice crystal surface during growth.

Lipids

In natural fats that are composed of a wide range of triacylglycerol species (eg, milk fat), considerable cocrystallization may take place,[46] where each triacylglycerol can be considered an impurity to the crystallization of other species. Since triacylglycerols are generally similar in structure, there is substantial affinity of one molecular species for the crystalline lattice of the others. Thus, it would be logical to expect that natural fats with fewer triacylglycerol species would nucleate more readily than those with wider variability in the triacylglycerol structure. However, this is not always the case, as other factors also must be considered. For example, cocoa butter contains primarily three triacylglycerols, whereas milk fat contains hundreds. Yet the induction times for milk fat are substantially faster than those for cocoa butter under the same conditions, as studied by isothermal calorimetry.[44] These results were shown in Table 5–1. Note that the driving forces for nucleation of the cocoa butter and milk fat may not be the same even though crystallization occurs at the same temperature and the two fats have similar clear points (the temperature where the last crystals have melted). Thus, comparison of two fats with similar melting point, even when crystallized at the same temperature (Table 5–1) may not necessarily compare nucleation rates at the same level of thermodynamic driving force.

When one fat is added to another, there often is a decrease in nucleation rate. For example, addition of milk fat to cocoa butter results in a slightly longer induction time for nucleation of cocoa butter,[44] as seen in Table 5–1. This inhibition of crystallization of cocoa butter through addition of milk fat has important commercial ramifications as milk chocolates must be processed (tempered) at lower temperatures to obtain the same crystalline structure as dark chocolate (no milk fat). Similar results have been found for the effect of milk fat on other confectionery fats.[85] The inhibition effect may be caused either by a change in the driving force for crystallization or by inhibition of nucleation itself. Since we do not currently have a means to express thermodynamic driving force for lipids, it is difficult to determine the effect of impurity addition on supersaturation driving force. At the levels (> 10%) of milk fat needed for significant inhibition of nucleation of cocoa butter crystallization, there also is a decrease in the solid fat content of the mixture below that of pure cocoa butter. This would suggest that there is a shift in phase equilibrium, although solid fat content does not necessarily define the phase equilibrium. It also is likely that some physical (steric) mechanism of inhibition applies to mixed lipids. Physically, one can imagine that the triacylglycerols in milk fat interfere with formation of the triacylglycerols in cocoa butter. Milk fat contains a wide range of fatty acid chain lengths (C4:0 to C20:0) and saturations (C16:1, C18:1, C18:2, etc), as compared with cocoa butter, which primarily contains two saturated fatty acids (C16:0 and C18:0) and one unsaturated fatty acid (C18:1). Thus, the general similarity between these molecules causes them to crystallize together, but their specific differences cause interference of formation of nuclei. This is an active area of research that should lead to greater understanding of complex lipid crystallization.

The more polar lipids (diacylglycerols, monoacylglycerols, free fatty acids, sterols, phospholipids, etc) in natural fats also may affect nucleation kinetics, although the mechanism of this effect is still under investigation. Walstra and van Beresteyn[47] speculated that micelles of monoacylglycerols may be the cause of nucleation in milk fat emulsions. There also is substantial evidence now that phospholipids play an important role in nucleation of cocoa butter. Simply removing the phospholipids by degumming results in substantial decrease in nucleation kinetics,[86] as measured by changes in turbidity curves for nucleation. The shape and appearance of cocoa butter crystals also are different when cocoa butter has been degummed to remove phospholipids. Savage and Dimick[87] postulated that the phospholipids form the initial seeds, based on a liquid crystal structure, that then cause nucleation of the cocoa butter. These seeds have significantly higher melting points than the cocoa butter and contain elevated levels of these phospholipids. Thus, at this point it is not clear if these components cause an effect as an insoluble substance (a liquid crystal) or as a dissolved material.

In milk fat, the polar lipids also play an important role in nucleation. Removal of these polar components results in a slight decrease in induction times for nucleation of milk fat, with the extent of inhibition related to the amount and perhaps composition of

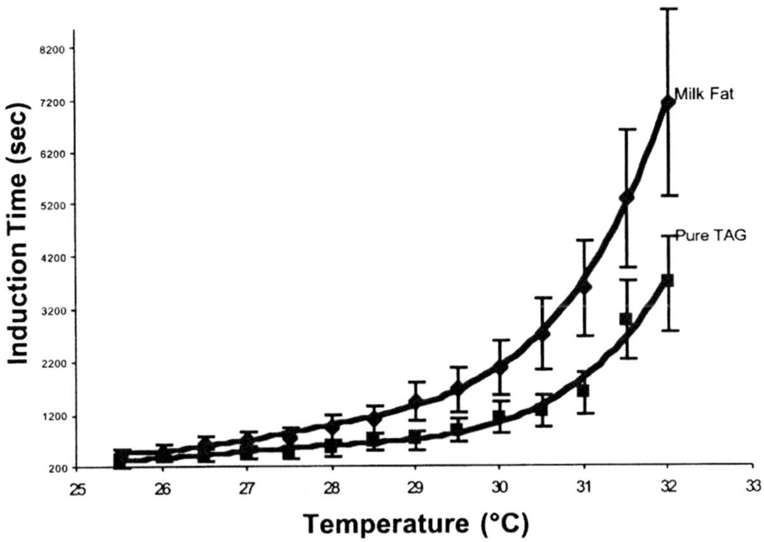

Figure 5–21 Effect of Polar Lipids on Induction Time of Milk Fat as Measured by Change in Turbidity upon Cooling to Crystallization Temperature.

these polar lipids,[88] as seen in Figure 5–21. When higher levels of these minor lipids were added, induction times increased substantially. Preliminary analysis suggests that the level of 1,3 diacylglycerols correlates with the observed inhibition of nucleation. However, the exact nature of this inhibition has not been fully identified as yet. This also is an active area of research that has great potential for modification of lipid crystallization rates and crystalline microstructures.

Another example of a solid material that catalyzes formation of lipid nuclei is the high-melting triglyceride BOB, which contains behenic (C22:0) and oleic (C18:1) fatty acids in a molecular form similar to that found in cocoa butter. Koyano et al[89] studied the effect of adding various (high-melting) seed materials on crystallization of cocoa butter in chocolate. It was found that low levels (less than 1%) of either SOS or BOB, both in the β-polymorph, were sufficient to enhance the crystallization rates of cocoa butter. These seed materials promote nucleation of cocoa butter directly into the stable (β) polymorph and thereby eliminate the need for the tempering process in chocolate manufacture. As long as the temperature of the chocolate remains below the melting point of the BOB crystals (53°C), this effect is realized. Once the BOB melts, however, the beneficial effect on nucleation is lost.

SUMMARY

Control of nucleation is probably the most important parameter in controlling crystallization and the nature of the crystalline phase in foods. However, our understanding of the mechanisms that control nucleation is still surprisingly limited. The variability experienced during normal operations limits the ability to control nucleation, although there are options for minimizing the effects of this variability. For example, processing conditions can be chosen that ensure a high level of nucleation all of the time. This condition might not produce the maximum amount of nuclei each time, but should provide sufficient nuclei to overwhelm any variability in crystallization due to raw materials.

Another option for circumventing the problem of variability in nucleation is to seed the system with the desired crystals rather than allowing nucleation to occur. Seeds with the desired characteristics (size, shape, polymorph, etc) must be available for this option to be successful. Seeding is often used in the food industry, although there are some cases where suitable seeds are not available.

There are a number of parameters that influence nucleation kinetics. Process conditions (heat and mass transfer rates) and formulation (composition) should be chosen that produce the desired crystalline structure. If many small crystals are desired to produce a smooth texture, nucleation should be controlled so that many nuclei form at about the same time. Nucleation that is spread out in a process (occurs over a wide time span) results in less than the maximal number of crystals and produces a distribution with a wider range of sizes. Those nuclei that formed first are able to grow significantly larger than those that formed later. On the other hand, if nucleation is to be avoided, processing conditions and formulations that minimize the likelihood of nucleation must be chosen. This generally involves rapid heating and cooling and selection of the appropriate ingredients to suppress nucleation.

REFERENCES

1. Zettlemoyer AC, ed. *Nucleation*. New York. Marcel Dekker; 1969.
2. Mullin JW. *Crystallization*. 3rd ed. Oxford: Butterworth-Heinemann; 1993.
3. Walton AG. Nucleation in liquids and solutions. In: Zettlemoyer AC, ed. *Nucleation*. New York: Marcel Dekker; 1969:225–308.
4. Cerreta MK, Berglund KA. The structure of aqueous solutions of some dihydrogen orthophosphates by laser Raman spectroscopy. *J Crystal Growth*. 1987;84:577.
5. Dunning WJ. General and theoretical introduction. In: Zettlemoyer AC, ed. *Nucleation*. New York: Marcel Dekker; 1969:1–68.
6. Fennema OR, Powrie WD, Marth EH. *Low-Temperature Preservation of Foods and Living Matter*. New York: Marcel Dekker; 1973.
7. Van Hook A. *Crystallization*. New York: Reinhold; 1961.
8. Sato K. Crystallization of fats and fatty acids. In: Garti N, Sato K, eds. *Crystallization and Polymorphism of Fats and Fatty Acids*. New York: Marcel Dekker; 1988:227–266.
9. Tavare NS. *Industrial Crystallization*. New York: Plenum Press; 1995.
10. van Hook A, Frulla F. Nucleation in sucrose solutions. *Industrial Engineering Chemistry*. 1952;44(6):1305.
11. Hobbs PV. *Ice Physics*. Oxford: Clarendon Press; 1974.

12. van Hook A, Bruno AJ. Nucleation and growth in sucrose solutions. *Faraday Discuss.* 1949;5:112.
13. Turnbull D, Fisher JC. Rate of Nucleation in Condensed Systems. *J Chemical Physics.* 1949;17(1):71–73.
14. Rasmussen DH, Liang M-T. A theory of nucleation of ice. In: *Nucleation and Crystallization in Liquids and Glasses.* Westerville, OH: The American Ceramic Society; 1993:85–91. Ceramic Transactions, Vol. 30.
15. Mountain RD. Computer simulations of crystal nucleation processes. In: *Nucleation and Crystallization in Liquids and Glasses.* Westerville, OH: The American Ceramic Society; 1993:55–64. Ceramic Transactions, Vol. 30.
16. Turnbull D, Vonnegut B. Nucleation catalysis. *Industrial Engineering Chemistry.* 1952;44(6):1292–1297.
17. Boistelle R. Fundamentals of nucleation and growth. In: Garti N, Sato K, eds. *Crystallization and Polymorphism of Fats and Fatty Acids.* New York: Marcel Dekker; 1988:189–226.
18. Lindow SE. Biological ice nucleation. In: Reid DS, ed. *The Properties of Water in Foods, ISOPOW 6.* London: Blackie; 1998:320–328.
19. Warren GJ, Wolber PK. Molecular aspects of microbial ice nucleation. *Mol Microbiol.* 1991;5:239–243.
20. Charoenrein S, Goddard M, Reid DS. Effect of solute on the nucleation and propagation of ice. In: Levine H, Slade L, eds. *Water Relationships in Foods.* New York: Plenum Press; 1991:191–198.
21. Rasmussen DH, Mckenzie AP. The effect of solute on the ice-solution interfacial free energy: calculation from measured homogeneous nucleation temperatures. In: Jellinek HHG, ed. *Water Structure at the Water-Polymer Interface.* New York: Plenum Press; 1972.
22. Reid DS, Foin AT, Lem CA. The effect of solutes on the temperature of heterogeneous nucleation of ice from aqueous solution. *Cryo-Letters.* 1985;6:189–198.
23. Roos YH. *Phase Transitions in Foods.* San Diego: Academic Press; 1995.
24. Mullin JW, Leci CL. Some nucleation characteristics of aqueous citric acid solutions. *J Crystal Growth.* 1969;5:75–76.
25. Herrington BL. Some physico-chemical properties of lactose. I. The spontaneous crystallization of supersaturated solutions of lactose. *J Dairy Sci.* 1934;17:501.
26. Livney YD, Donhowe DP, Hartel RW. Influence of temperature on crystallization of lactose in ice cream. *Int J Food Sci Technol.* 1995;30(3):311–320.
27. Vonnegut B, Baldwin M. Repeated nucleation of a supercooled water sample that contains silver iodide particles. *J Climate Applied Meterology.* 1984;23:486–490.
28. Wang MK, Vonnegut B. Repeated nucleation of supercooled water sample *J Recherches Atmospheriques.* 1984; 18:23–29.
29. Uttormark MJ, Zanter JW, Perepetzko JH. Repeated nucleation of an undercooled aluminum droplet. *J Crystal Growth.* 1997;177:258.
30. James PF. Nucleation in glass-forming systems—a review. In: Simmons JH, Uhlmann DR, Beall GH, eds. *Nucleation and Crystallization in Glasses.* Columbus, OH: American Ceramic Society; 1982:1–48.
31. James PF. Experimental studies of crystal nucleation in glasses. In: *Nucleation and Crystallization in Liquids and Glasses.* Westerville, OH: American Ceramic Society; 1993:3–12. Ceramic Transactions, Vol. 30.
32. Makower B, Dye WB. Equilibrium moisture content and crystallization of amorphous sucrose and glucose. *J Agric Food Chem.* 1956;4:72–77.
33. Mathlouthi M. Amorphous sugar. In: Mathlouthi M, Reiser P, eds. *Sucrose. Properties and Applications.* Glasgow: Blackie; 1995:75–100.
34. Roos YH, Karel M. Crystallization of amorphous lactose. *J Food Sci.* 1992; 57:775–777.
35. Jouppila K, Roos YH. Water sorption and time-dependent phenomena of milk powders. *J Dairy Sci.* 1994;77: 1798–1808.
36. Jouppila K, Roos YH. Glass transitions and crystallization in milk powders. *J Dairy Sci.* 1994;77:2907–2915.
37. Hernqvist L. On the structure of triglycerides in the liquid state and fat crystallization. *Fette Seifen Anstrichmittel.* 1984;86:296.
38. Hartel RW. Phase transitions in chocolates and coatings. In: Rao MA, Hartel RW, eds. *Phase/State Transitions in Foods.* New York: Marcel Dekker; 1998:217–252.
39. Sato K, Kuroda T. Kinetics of melt crystallization and transformation of tripalmitin polymorphs. *J Am Oil Chem Soc.* 1987;64:124.

40. Timms RE. Crystallization of fats. In: Hamilton, RJ, ed. *Developments in Oils and Fats*. London: Blackie; 1995: 204–223.
41. van Putte KPAM, Bakker BH. Crystallization kinetics of palm oil. *J Am Oil Chem Soc*. 1987;64(8):1138–1143.
42. Kellens M, Meeussen W, Reynaers H. Study of the polymorphism and the crystallization kinetics of tripalmitin: a microscopic approach. *J Am Oil Chem Soc*. 1992;69(9).906–911.
43. Ng WL. Nucleation behaviour of tripalmitin from a triolein solution. *J Am Oil Chem Soc*. 1989;66(8):1103–1106.
44. Metin S. *Crystallization Behavior and Kinetics of Blends of Cocoa Butter and Milk Fat or Milk Fat Fractions*. Madison, WI: University of Wisconsin-Madison; 1997. PhD dissertation.
45. Kloek W, Walstra P, von Vliet T. Crystallization kinetics of fully hydrogenated palm oil in sunflower oil mixtures. *J Am Oil Chem Soc*. 2000;77(4):389–398.
46. Walstra P. Fat crystallization. In: Blanshard JMV, ed. *Food Structure and Behavior*. New York: Academic Press; 1987:67-85.
47. Walstra P, van Beresteyn ECH. Crystallization of milk fat in the emulsified state. *Neth Milk Dairy J*. 1975;29:35.
48. Precht D. Fat crystal structure in cream and butter. In: Garti N, Sato K, eds. *Crystallization and Polymorphism of Fats and Fatty Acids*. New York: Marcel Dekker; 1988:305–361.
49. Kaschiev D, Kaneko N, Sato K. Kinetics of crystallization in polydisperse emulsions. *J Colloid Interface Sci*. 1998;208:167–177.
50. Hodate Y, Ueno S, Yano J. et al. Ultrasonic velocity measurement of crystallization rates of palm oil in oil-water emulsions. *Colloids Surfaces A*. 1997;128:217–224.
51. Kaneko N, Horie T, Ueno S, et al. Impurity effects on crystallization rates of n-hexadecane in oil-in-water emulsions. *J Crystal Growth*. 1999;197:263–270.
52. Adleman R, Hartel RW. Lipid crystallization and its effect on the physical structure of ice cream. In: Garti N, Sato K, eds. *Crystallization Phenomena in Fats and Lipids Systems*. New York: Marcel Dekker. In press.
53. Larson MA. Secondary nucleation: an analysis. *Chemical Engineering Communications*. 1981;12:161.
54. Hartel RW, Shastry AV. Sugar crystallization in food products. *Crit Rev Food Sci Nutr*. 1991;30:49–112.
55. Wissing R, Elwenspoek M, Degens B. In situ observation of secondary nucleation. *J Crystal Growth*. 1986;79: 614–619.
56. Randolph AD, Larson MA. *Theory of Particulate Processes*. 2nd ed. New York. Academic Press; 1988.
57. Shi Y, Hartel RW, Liang B. Formation and growth phenomena of lactose nuclei under contact nucleation conditions. *J Dairy Sci*. 1989;72(11):2903.
58. Denk EG, Botsaris GD. Fundamental studies in secondary nucleation from solution. *J Crystal Growth*. 1972;13/14: 493–499.
59. Elankovan P, Berglund KA. Contact nucleation from aqueous dextrose solutions. *Am Inst Chem Eng J*. 1987;33(11): 1844.
60. Omran AM, King CJ. Kinetics of ice crystallization in sugar solutions and fruit juices. *Am Inst Chem Eng J*. 1974;20(4): 795–803.
61. Stocking JH, King CJ. Secondary nucleation of ice in sugar solutions and fruit juices. *Am Inst Chem Eng J*. 1976;22(1): 131–140.
62. Hachiya I, Koyano T, Sato K. Seeding effects on crystallization behavior of cocoa butter. *Agric Biol Chem* 1989;53(2): 327–332.
63. Garabedian H, Strickland-Constable RF. Collision breeding of ice crystals. *J Crystal Growth*. 1972;22:188–192.
64. Hartel RW, Chung MS. Contact nucleation of ice crystals in fluid dairy products. *J Food Eng*. 1993;18(3):281–296.
65. Hartel RW. Ice crystallization during manufacture of ice cream. *Trends Food Sci Technol*. 1996;7(10):315–320.
66. Russell AB, Cheney PE, Wantling SD. Influence of freezing conditions on ice crystallization in ice cream. *J Food Eng*. 1999;39:179–191.
67. Muhr AH, Blanshard JMV, Sheard SJ. Effects of polysaccharide stabilizers on the nucleation of ice. *J Food Technol*. 1986;21:587–603.
68. Singh G. Analysis of crystallization systems with applications to continuous fractional crystallization of fatty acid triglycerides. *Am Inst Chem Eng Symposium*. 1976;72:100–109 Series No. 153.

69. Herrera ML, Hartel RW. Effect of processing conditions on crystallization kinetics of a model milk fat system. *J Am Oil Chem Soc*. 2000; 77(11):1177–1187.
70. Mullin JW, Raven KD. Influence of mechanical agitation on the nucleation of some aqueous salt solutions. *Nature*. 1962;195:35–38.
71. Lees R. *Factors Affecting Crystallization in Boiled Sweets, Fondants and Other Confectionery*. Leatherhead, UK: The British Food Manufacturing Industries Research Association; 1965. Scientific and Technical Survey, No. 42.
72. Evans TW, Margolis G, Sarofim AF. Mechanisms of secondary nucleation in agitated crystallizers. *Am Inst Chem Eng J*. 1974;20(5):950-958.
73. Kasputin AP. *The Effects of Ultrasound on the Kinetics of Crystallization* Moscow: Institute of Crystallography; 1962.
74. Kennedy CJ. Formation of ice in frozen foods and its control by physical stimuli. In: Reid DS, ed. *The Properties of Water in Foods, ISOPOW 6*. London: Blackie; 1988:329–365.
75. Turner CF, Van Hook A. The effect of ultrasonic irradiation on the formation of colloidal sulphur and ice. *J Colloid Science*. 1950;5:315–316.
76. Hickling R. Nucleation of freezing by cavity collapse and its relation to cavitation damage. *Nature*. 1965;206:915–917.
77. Tjuradi P. *The Effect of Fractionated and Nonfractionated Corn Syrups on Sucrose Crystallization*. Madison, WI: University of Wisconsin-Madison; 1993. MS thesis.
78. Shastry AV. *Batch Crystallization of Sucrose from High Concentration Syrups*. Madison, WI: University of Wisconsin-Madison; 1991. MS thesis.
79. Herrington TM, Branfield AC Physico-chemical studies on sugar glasses. I. Rates of crystallization. *J Food Technol*. 1984;19:409–425.
80. van Scoik KG, Carstensen JT. Nucleation phenomena in amorphous sucrose systems. *Int J Pharm* 1990;58:185–196.
81. Shamblin SL, Huang EY, Zografi G. The effects of co-lyophilized polymeric additives on the glass transition temperature and crystallization of amorphous sucrose. *J Thermal Anal*. 1996;47:1567–1579.
82. Blanshard JMV, Muhr AH, Gough A. Crystallization from concentrated sucrose solutions. In: Levine H, Slade L, eds. *Water Relationships in Food*. New York: Plenum Press; 1981:639–655.
83. Blanshard JMV, Franks F Ice crystallization and its control in frozen-food systems. In: Blanshard JMV, Lillford P, eds. *Food Structure and Behaviour*. New York: Academic Press; 1987:51–65.
84. Shirai Y, Nakanishi K, Matsuno R, Kamikubo T. Effects of polymers on secondary nucleation of ice crystals. *J Food Sci*. 1985;50:401–406.
85. Hogenbirk G. The influence of milk fat on the crystallization properties of cocoa butter and cocoa butter alternatives. *Manufacturing Confectioner*. 1990;70:133–140.
86. Dimick PS. Compositional impact on crystallization of cocoa butter. In: Widlak N, ed. *Proceedings of the International Conference on Physical Properties of Fats, Oil and Emulsifiers with Applications to Foods*. Champaign, IL: American Oil Chemist's Society Press; 2000:138–161.
87. Savage CM, Dimick PS. Influence of phospholipids during crystallization of hard and soft cocoa butters. *Proceedings of the Pennsylvania Manufacturing Confectioners' Association (PMCA) Production Conference*. Center Valley, PA: PMCA; 1995:54–59.
88. Herrera ML, de Leon Gatti M, Hartel RW. A kinetic analysis of crystallization of a milk fat model system. *Food Res Int*. 1999;32:289–298.
89. Koyano T, Hachiya I, Sato K. Fat polymorphism and crystal seeding effects on fat bloom stability of dark chocolate. *Food Structure*. 1990;9:231–240.

CHAPTER 6

Crystal Growth

Once nuclei have formed in a supersaturated system, they grow at a rate dependent on conditions in the surrounding environment. Crystals continue to grow as long as a supersaturated condition remains in the system and the molecules have sufficient mobility to move to the crystal interface and orient into the lattice. Growth continues until all available supersaturation is consumed and the system reaches the phase equilibrium defined by the phase diagram. In order for crystals to grow by molecular deposition into the lattice structure, a number of processes must occur. Any of these steps may be the rate-limiting step for growth of a particular species, but usually only one or two are responsible for the rate of a crystal's growth. The steps necessary for growth by molecular deposition are summarized in Exhibit 6–1.

The first step in crystal growth is migration (or mass transfer) of molecules of the crystallizing species to the surface of the growing crystal. This movement may occur by diffusion or convective (fluid motion) transport or by some combination of the two. Regardless of the extent of convective mass transfer (fluid flow) in the bulk solution, there is still a stagnant fluid boundary layer near the surface of the crystal through which diffusion of molecules must occur. In some systems, such as the reducing sugars (lactose, etc), a chemical reaction must occur to maintain equilibrium in the solution phase. For example, as α-lactose crystallizes from solution, formation of β-lactose in the liquid phase must take place through the mutarotation reaction. Under certain circumstances, the rate of this mutarotation reaction can limit the rate of crystal growth (see Figure 4–5). If the crystallizing species takes a form in solution other than the form necessary for incorporation into the crystal lattice, then a transformation must occur before crystal growth can occur. For example, the waters of hydration around a sugar molecule must desolvate prior to incorporation of those sugar molecules into the lattice. Also, if clusters of the crystallizing molecules form in the supersaturated solution, these may need to rearrange so that the molecules can fit into the lattice. As the molecules of the crystallizing species diffuse toward the growing crystal surface, all other molecules must diffuse away to maintain a driving force for crystallization. For example, the water molecules in a hydrated system must diffuse away prior to lattice incorporation of the solute. In ice crystal growth, it is the larger solute molecules (sugars, proteins, stabilizers, etc) that must diffuse away to allow growth of the ice crystal since water molecules are readily available in the vicinity of the crystal interface.

Exhibit 6–1 Summary of Potential Steps Required for Crystal Growth by Molecular Incorporation

- Mutarotation of crystallizing species to correct anomeric form
- Diffusion of molecules of crystallizing species from bulk solution to the crystal interface
- Removal of some or all waters of hydration associated with molecules of crystallizing species
- Counter-diffusion of noncrystallizing species away from crystal surface
- Orientation of molecules of crystallizing species into growth units at the crystal-liquid interface
- Surface diffusion of growth units to an appropriate kink site for incorporation into the lattice
- Removal of heat generated by phase change

Once the crystallizing species reaches the surface of the crystal, it is necessary for the molecules to orient, either singly or in the form of some multimolecular growth unit, into an arrangement that allows incorporation into the lattice. This orientation occurs in a layer of molecules near the existing lattice structure, which is thought to be only namometers in thickness.[1] This so-called adsorption layer is the region where liquid molecules become oriented in the same way as molecules in the lattice structure. Figure 6–1 presents a schematic representation of this adsorbed layer. Molecules here are thought to be semi-ordered as they begin the process of incorporating into the rigid lattice structure. Once orientation of molecular units has taken place in this adsorption layer, these units must migrate along the surface until they find a suitable site for incorporation into the lattice. This process of surface diffusion continues until the actual incorporation of units into the lattice occurs at a favorable lattice site. The choice of lattice incorporation site depends on the driving force or rate of growth. If a crystal grows slowly, growth units incorporate at the most energetically favorable site, whereas rapid growth results in a more disorganized arrangement of the growth units and the crystal lattice formed is less perfect. Once molecules have incorporated into the lattice, the latent heat of fusion associated with the phase change must be removed from the vicinity of the crystal surface. In solution systems, this primarily takes place through the liquid phase at a rate dependent on the combined mechanisms of conduction and convection. When crystallization takes place on a cold surface, as in solid-layer melt crystallization systems, heat transfer is primarily through the growing crystal itself.

Again, any of these mechanisms may influence or limit crystal growth. In fact, more than one of these mechanisms may be important and contribute varying rate-controlling aspects as conditions change. For example, growth of sucrose crystals has been found to be controlled by two steps[2]: bulk diffusion and surface incorporation (surface diffusion). At high temperatures, bulk diffusion is the primary mechanism that controls crystal growth, whereas surface incorporation is the main step that limits growth at low temperatures. It should be recognized, however, that each of the steps listed above must occur under all conditions. When a specific step is said to control growth rate, this means that this step is the one that occurs most slowly. In growth of ice crystals in food systems, it is a combination of counter-diffusion of large molecules and the rate of heat removal that controls growth rates.[3,4] Conditions that maximize counter-diffusion and heat transfer result in the fastest ice crystal growth rates.

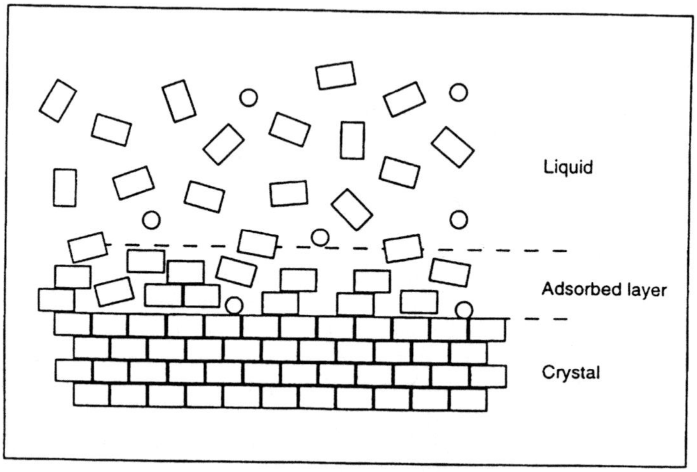

Figure 6–1 Schematic Representation of the Adsorbed Layer at the Surface of a Growing Crystal. Rectangles represent growth units of crystallizing species; circles represent adsorbing impurity.

There are other mechanisms by which crystals can grow larger, although these processes do not always lead to loss of supersaturation and may even occur when phase volume is at its equilibrium value. If many small crystals exist, they can agglomerate or aggregate together to produce larger crystals. This is primarily important in slurry systems where agitation brings crystals into contact under conditions where they stick together and form agglomerates. This mechanism may be extremely important in crystallization of fats where both molecular processes and agglomeration may occur at the same time. Another means by which crystals grow larger is the thermodynamic ripening process called Ostwald Ripening.[1] This process is driven by the difference in solubility or melting point of very small crystals (see Chapter 4). Under certain circumstances, very small (usually $< 5\ \mu m$) crystals may actually be undersaturated in a solution and dissolve at the same time that larger crystals in the surrounding environment are supersaturated and grow larger at the expense of the small ones.[5] This type of size increase occurs with no change in the overall phase volume of crystalline material, with only a redistribution of crystalline material from small crystals (which disappear) to large ones (which grow). This is the primary mechanism for recrystallization in foods,[3] and it will be discussed in greater detail in Chapter 8.

Crystal growth rates typically are given as either mass deposition rate (or mass flux) or a linear growth rate. Mass flux gives the mass of material crystallized per unit surface area per unit time (kg/m^2-sec). The total surface area of the crystal distribution must be known to calculate this value. A linear growth rate is defined as a change in characteristic dimension with time (μm/sec). However, the magnitude of this rate depends on the size dimension chosen to represent the crystal. For spherical particles, the change in diameter with time is often used to quantify linear growth rate. The choice of units used to characterize growth depends on the measurement technique and the ultimate use of those data. In general, linear growth rate is most often reported in the literature, although other units also can be found.

In solution systems, the extent of crystal growth is based on the supersaturation. Assuming no new nuclei are formed, crystal growth continues until the system reaches the equilibrium point defined by the phase diagram (the solubility point at the temperature of growth). That is, a solution starts at some elevated supersaturation (initial concentration above saturation) and as crystal growth progresses, the supersaturation decreases until equilibrium phase volume ($S = 0$) is attained. This can be seen on a typical desupersaturation curve for batch crystallization (Figure 6–2), which shows the decrease in solution concentration (supersaturation) as crystal growth progresses. At first, there is a lag period where nucleation must take place. Once nucleation occurs, the decrease in supersaturation is due to a combination of continued nucleation and subsequent growth of nuclei. Growth is initially rapid, depending on the conditions, and gradually slows down as the solubility point is approached, as shown in Figure 6–2. The relative contribution of nucleation and growth to the slope of the desupersaturation curve depends on the conditions during crystallization. The amount (mass) of crystals formed can be found from a mass balance on the initial solution and the final crystallized slurry (see Chapter 4).

In melt systems, crystal growth continues until the system attains a thermal equilibrium or all of the material has crystallized. Slow freezing of pure water, for example, occurs at the freezing point temperature until all the water has frozen. Freezing of foods is more complex since the freezing point temperature generally decreases as freezing progresses due to the freeze concentration of small-molecular-weight solutes. In frozen foods, ice crystallization ceases when the thermal driving force (difference between actual temperature and freezing point) goes to zero or when the mobility of water molecules is low enough that crystallization cannot occur. If freezing occurs very slowly, the system follows the freezing point depression curve (or slightly below it) as freezing progresses, as seen for a simple binary system in Figure 6–3. The amount of ice formed increases as temperature decreases until eventually freezing ceases when the product temperature reaches the freezer temperature (Figure 6–3B). If, however, freezing occurs rapidly, the system can follow any arbitrary path on the phase diagram (Figure 6–3A). Very rapid freezing causes the system to cool well below the freezing point depression curve. If cooling is to a very low temperature (generally below –40°C), the system will remain as small ice crystals imbedded in a glassy matrix and will remain as such until temperature is increased. If freezing stops at a temperature above the glass transition temperature, ice will continue to form at a rate dependent on the temperature difference above T_g.

CRYSTAL GROWTH MECHANISMS

Primarily, mechanisms for crystal growth can be divided into three categories: mass transfer (or diffusion) processes, surface incorporation mechanisms, and heat transfer.

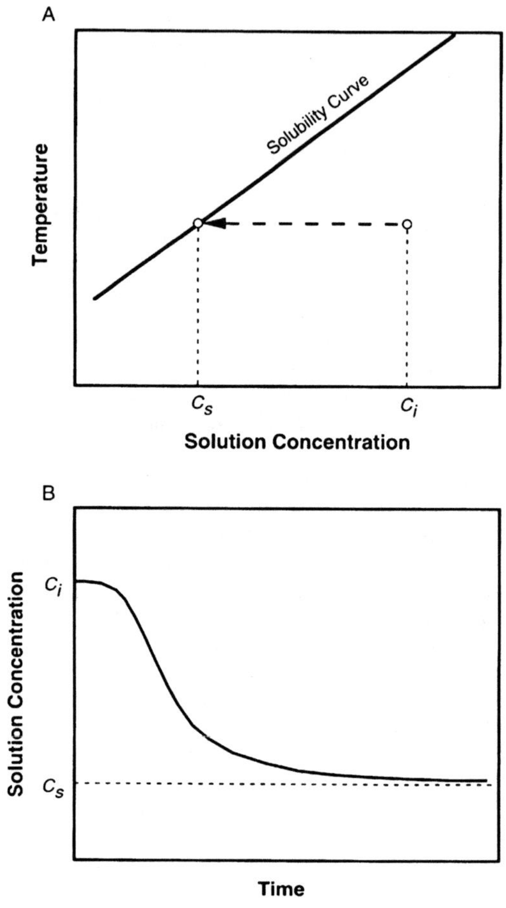

Figure 6–2 Desupersaturation of Solute during Isothermal Crystallization: (A) Progression on Phase Diagram; (B) Desupersaturation Curve Showing Progression from Initial Concentration (C_i) to Solubility Concentration (C_s) during Crystal Growth.

Mass Transfer-Based Growth

When growth is limited by mass transfer of molecules through solution to the crystal surface, mass transfer theory can be used to predict growth rates.[1,6–8] The size of the hydrodynamic boundary layer through which mass transport occurs is dependent on

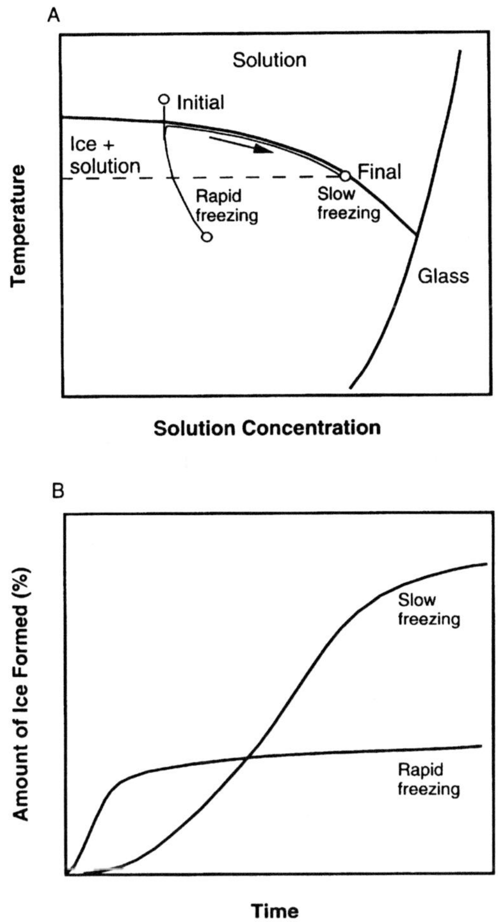

Figure 6–3 Freezing of Ice in a Binary System: (A) Slow versus Rapid Freezing Progression on a Phase Diagram; (B) Increase in Ice Phase Volume during Slow versus Rapid Freezing.

the rate of convective mass transfer in the solution and the inherent diffusivity of the material. The convective transport rate depends on temperature, fluid motion (agitation), and physical properties (ie, viscosity, density, etc) of the solution.

In crystal growth from solution, there is a concentration gradient between the bulk solution and the surface of the crystal that drives the mass transfer process (Figure 6–4).

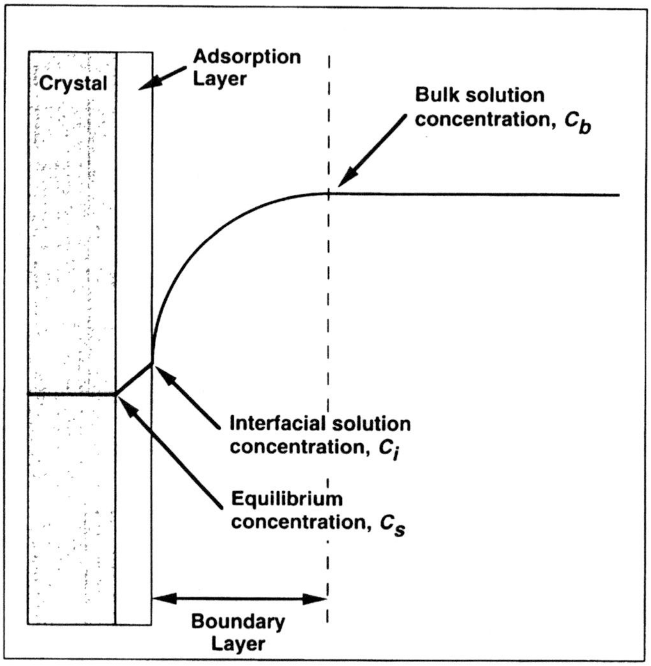

Figure 6–4 Concentration Gradient at the Surface of a Growing Crystal.

At a growing crystal surface, a boundary layer exists where the concentration varies from that in the bulk solution, C_b, to that at the crystal surface, which is often taken to be the solubility concentration, C_s. At the surface of the adsorption layer, the concentration is taken as the interfacial concentration, C_i.

Diffusion theory predicts the rate of mass transfer across the hydrodynamic boundary layer based on the concentration driving force between the bulk and interfacial concentrations. This is given as

$$\dot{m} = k_d(C_b - C_i) \tag{6.1}$$

where $\dot{m}$ is the mass flux of solute crossing the boundary layer (kg/m^2-sec), and k_d is a mass transfer coefficient (m^2/sec). The linear growth rate (in μm/min) can be expressed as

$$G = k_d k_1 (C_b - C_i)$$

(6.2)

where k_1 is a conversion constant from crystal mass to crystal size. For spherical crystals

$$k_1 = 2/\rho_c$$

(6.3)

where ρ_c is the density of the crystalline material.

The mass transfer coefficient, k_d, combines the effects of convective transport and diffusion. It is usually found from engineering correlation equations that relate flow conditions and physical properties to these transport phenomena. A general form of this equation for spheres is given as

$$Sh = 2 + dRe^b Sc^c$$

(6.4)

where Sh is the Sherwood number ($k_d L/D_v$), Re is the Reynolds number ($u L \rho/\mu$), Sc is the Schmidt number ($\mu/\rho D_v$), L is a characteristic dimension for the crystal, D_v is molecular diffusivity, u is the relative velocity of fluid across the crystal surface, ρ is the solution density, and μ is solution viscosity. The parameters b, c, and d are dependent on system geometry, fluid properties, and flow conditions (whether laminar or turbulent). Mass transfer correlations are summarized in several engineering references, such as Perry and Chilton.[9] For stirred suspensions, correlations developed by Levins and Glastonbury[10] apply. In the case where the density differences between particle and solution are small, the following equation applies

$$\frac{k_d d_p}{D_v} = 2 + 0.47\left(\frac{d_p^{4/3}\varepsilon^{1/3}}{\nu}\right)^{0\,62}\left(\frac{D_s}{D_T}\right)^{0\,17}\left(\frac{\nu}{D_v}\right)^{0\,36}$$

(6.5)

where d_p is particle diameter, ε is energy dissipation rate per unit mass, ν is kinematic viscosity, D_s is stirrer diameter, and D_T is tank diameter. For the situation where particle density is different from that of solution, the following equation is recommended:

$$\frac{k_d d_p}{D_v} = 2 + 0.44\left(\frac{v d_p}{\nu}\right)^{0\,5}\left(\frac{\nu}{D_v}\right)^{0\,38}$$

(6.6)

where v is the slip velocity between crystal and fluid.

When crystals grow in stagnant solutions, as often occurs when crystals are growing within a food product, there is no convective mass transfer and diffusion limits growth rate. In this case, Equation 6.4 simplifies to give

$$Sh = 2 = \frac{k_d L}{D_v} \tag{6.7}$$

and the diffusivity can be used to calculate growth rates. Again, this equation applies for spherical particles only. When a crystal grows, however, there is a release in latent heat and the desupersaturation in the vicinity around the growing crystal may lead to conditions where natural convection currents promote mass transfer. In this case, Equation 6.8 can be used.[11]

$$Sh = 2.0 + 0.6\left[\frac{L^3 \Delta\rho g}{\nu^2 \rho}\right]^{0.25}\left(\frac{\nu}{D_v}\right)^{0.33} \tag{6.8}$$

where $\Delta\rho$ is the difference in density, g is gravitational constant, and υ is kinematic viscosity.

As the fluid velocity across the crystal surface increases, generally due to agitation, convective mass transfer becomes increasingly important and the rate of mass transfer increases. Eventually, at sufficiently high agitation rates, mass transfer no longer limits growth rates and some other mechanism becomes rate limiting. Thus, in systems where diffusion has a major impact on growth processes, the growth rate increases initially as agitation rate increases but then levels off at some asymptotic value characteristic of surface incorporation mechanisms. This is one way to tell the influence of mass transfer on crystal growth.

Mass transfer theory predicts that growth rate increases linearly with increasing supersaturation (or concentration gradient) at low supersaturations. That is, increasing supersaturation leads to correspondingly higher growth rate according to the linear relationship in Equation 6.2. However, a linear relation between growth rate and supersaturation does not necessarily indicate diffusion-limited growth as other growth mechanisms also predict linear dependence. At higher supersaturations, growth rate decreases again in many systems due to mass transfer limitations (increases in D_v).

Surface Incorporation (or Adsorption Layer) Growth Theories

After the molecules have reached the interface, they must organize into growth units and find an appropriate incorporation site.[1] A simplified diagram of the crystal lattice structure and incorporation sites is shown in Figure 6–5. When a molecule (or growth unit) arrives at the crystal interface, it must diffuse along the surface until it finds a site for incorporation into the lattice structure. Depending on the growth conditions, different incorporation sites might be chosen. When growth occurs very slowly, the growth units generally find the most energetically favorable site for incorporation. In general, these sites are those where the attractive forces are the greatest, such as those sites where the growth units are bonded on several sides (sites a, b, and c in Figure 6–5). As growth rate increases due to higher driving forces, the growth units fit into less energetically favorable sites and the crystal typically grows more imperfectly. When rapid growth conditions exist, favorable lattice sites (eg, site c in Figure 6–5) may be passed over for less energetically favorable sites, leaving vacancies in the lattice structure. Also, impurities may not diffuse away rapidly enough and may become incorporated (or occluded) within the lattice structure. Thus, slow growth generally leads to more perfect crystals, while fast growth usually leads to crystals with many more imperfections.

When no surface imperfections exist, as for a completely smooth crystal surface, there are no favorable incorporation sites and a two-dimensional nucleation process must take place. A new center must be formed on the smooth surface in order for further growth to occur. Two-dimensional nucleation is thought to be similar to heterogeneous nucleation, where the existing crystal lattice structure provides the impetus for organizing molecules from the liquid.[1,6–8,12] Once a two-dimensional nucleus is formed on the surface, it is no longer smooth, and further incorporation of growth units can occur on the newly formed kink sites. Crystals then grow as surface layers get filled out either by two-dimensional nucleation or by kink site incorporation.

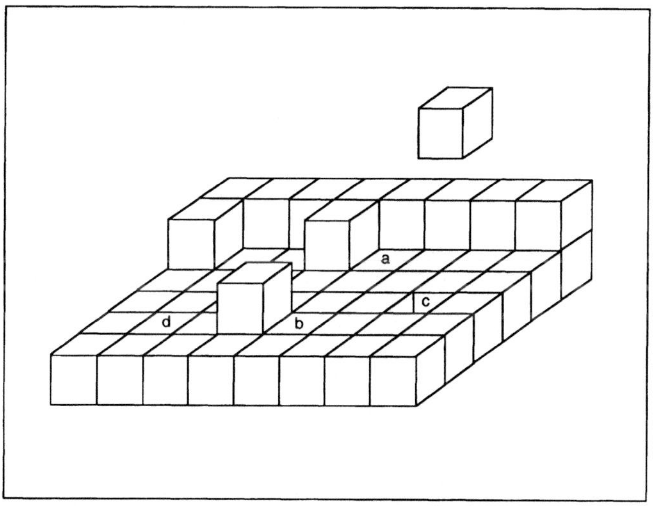

Figure 6–5 Schematic Model of Crystal Lattice and Incorporation Sites Based on Simple Cubic Units: a, b, c, and d Represent Different Potential Sites for Molecular Incorporation into the Crystal Lattice.

Surface integration theories may be categorized according to the relative rates of two-dimensional nucleation and subsequent kink site incorporation.[1] In the "birth and spread" model, nuclei form on top of uncompleted steps while the steps below are completing the layers. There are two extreme cases that can be considered: (1) spread of growth units is limited and all sites must be filled by two-dimensional nucleation, or (2) spread of growth units is instantaneous as soon as a two-dimensional nuclei is formed. In the former case, surfaces grow at the rate corresponding to the rate of the two-dimensional nucleation step, whereas in the latter case, crystals grow rapidly as each layer is filled in by kink site incorporation. According to these theories, crystals should grow quite slowly at low supersaturations since the driving force for two-dimensional nucleation is low at these conditions.

A model of crystal growth, based on regeneration of surface imperfections (spiral dislocations), was developed by Burton et al.[13] In this theory, often called the BCF (Burton, Cabrera, Frank) model of crystal growth, regenerating spiral dislocations provide continual kink sites for incorporation of growth units into the lattice. The rate of growth is then limited by surface diffusion of the growth units to the regenerating spiral dislocation. According to the BCF model, growth rate can be expressed in terms of supersaturation as

$$G = A\sigma^2 \tanh(B/\sigma) \tag{6.9}$$

where σ is supersaturation defined as $(C - C_s)/C_s$ and A and B are parameters that depend on temperature, step height and spacing, and the activity of the spiral dislocation.[1,13]

At low supersaturations, the BCF model predicts that growth is proportional to the square of supersaturation, whereas at higher supersaturations, growth increases linearly with supersaturation. This switch from second-order to first-order growth as supersaturation increases has been used by some to identify the mechanism for crystal growth based on the BCF model. For example, it is thought that the surface incorporation component of sucrose crystal growth is based on the BCF model of spiral dislocations.[2] Also, there is some evidence that growth of lipid crystals from organic solvents follows the BCF model, based in part on the observation of growth spirals on the surface of growing crystals.[14] In addition, at low levels of supersaturation, a parabolic growth curve is observed, whereas at higher supersaturations, growth becomes linear with supersaturation,[15] as shown schematically in Figure 6–6.

Heat Transfer-Limited Growth

The rate of heat transfer away from a crystal during growth also may limit the rate at which a crystal grows. This is an important step for systems that have high heat of fusion, as in the freezing of ice[3] and growth of other melt systems.[1]

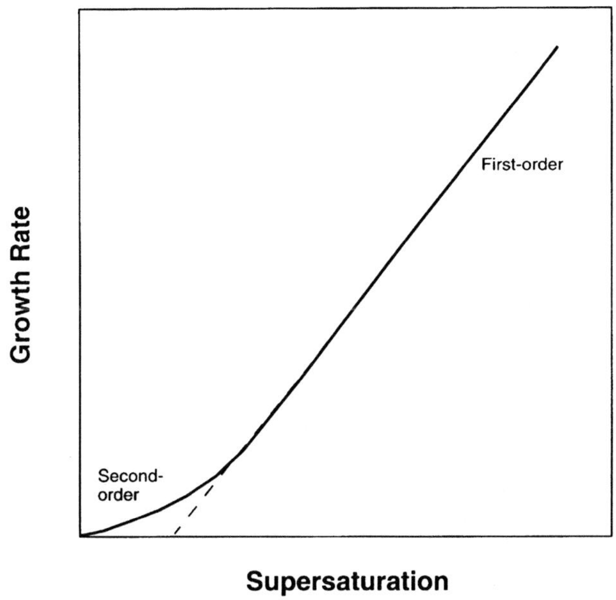

Figure 6–6 Schematic Representation of Rate Curve for Growth Based on Molecular Incorporation into Spiral Dislocations.

When heat is generated by growth at the crystal interface, it is dissipated by either convection into the solution or conduction through the crystal layer. For convective heat transfer, the rate of heat transfer, $\dot{q}$, can be expressed as

$$\dot{q} = hA(T_i - T_b) \tag{6.10}$$

where h is a convective heat transfer coefficient, A is the surface area of the crystal interface, T_i is the temperature at the crystal interface, and T_b is bulk temperature. The convective heat transfer coefficient, h, can be found from correlation equations that relate fluid velocity, geometric parameters, and physical properties to the ability of heat to be dissipated by fluid motion. Equation 6.10 is similar in form to Equation 6.1, which describes the rate of mass transfer to a growing crystal.

In melt crystallization technologies where crystals are grown on solid surfaces, heat is dissipated through the crystal layer itself.[16] The rate of heat transfer is then controlled by conduction within the solid layer.

$$\dot{q} = kA\frac{dT}{dx}$$

(6.11)

where k is thermal conductivity of the crystal layer and dT/dx is the temperature gradient applied by the cooling source to the crystal interface.

Combined Growth Models

Most often, empirical expressions are used to correlate growth rates with supersaturation when crystal growth is governed by more than one mechanism. In fact, a two-step model accounting for mass transfer and surface incorporation steps is often used to describe crystal growth.[1] An empirical expression of the form

$$G = k_G(C_b - C_s)^m$$

(6.12)

can be used, where k_G is a combined rate constant and m is a combined exponential order of growth. Here, k_G is a function of the external processing conditions such as temperature, agitation rate, and the presence of additives or impurities. The value of Equation 6.12 comes in the predictive capabilities of such a mathematical model, since no mechanistic information can be obtained.

For melt crystallization, an empirical growth model can be written as

$$G = B(T_f - T)^n$$

(6.13)

where B is a growth coefficient and n is a variable parameter depending on growth mechanisms.[1] Many systems, including ice, exhibit n values greater than one since other mechanisms besides convective heat transfer also influence growth. The value of n depends on the specific crystal face under study, the relative rates of heat and mass transfer, and the rate of surface incorporation. B also depends on the relative rates of heat and mass transfer in the liquid as well as on the physical and thermal properties of the system under study. More sophisticated growth models for melt systems, based on specific growth mechanisms, have been derived by Kirwan and Pigford.[17]

Numerous studies[4,18] on ice crystal growth have shown that the exponent, n, in Equation 6.13 is approximately 1.5 to 1.6. Omran and King[4] determined these relationships for ice crystal growth in sucrose solutions and fruit juices.

For natural lipids that crystallize when temperature is lowered, Equation 6.13 is difficult or incorrect to apply since no single melting point can be determined. Sometimes the peak melting temperature (as measured by differential scanning calorimeter [DSC], for example) may be used to define the melting point for use with Equation 6.13. However, this approach does not always provide a satisfactory reference point for thermodynamic driving force, especially for complex lipids. In addition, there may be a solubility effect in natural lipids, where high-melting components are solubilized in the lower-melting components. Thus, the driving force for crystallization of natural fats is some combination of solubility and melting point effects. For these reasons, very little kinetic data, based on thermodynamic driving force, exist for crystal growth rate of complex lipid mixtures.

RELATED GROWTH PHENOMENA

Growth of Individual Faces

The different faces of a crystal grow at different rates, due to differences in surface energy. Molecules arranged in an orderly lattice structure have different physical aspects of the molecule exposed to the solution at the different faces (see Chapter 2). Thus, each face has slightly different surface energetics, and this leads to different growth rates. Growth rates of individual faces have been measured, for example, by following the change in size and shape of a growing crystal on a microscope stage. The change in linear dimension of each face with time is used to calculate a facial growth rate.

For example, the growth rates of different faces of lactose crystals have been measured by Visser,[19] and Smythe[20] discusses the effects of relative growth rates of different faces on sucrose crystal shape. Some results from Visser[19] on lactose crystal growth rates are tabulated in Table 6–1. There are substantial differences in growth rates of the different lactose crystal faces. A model of a lactose crystal with the faces defined in Table 6–1 is shown in Figure 6–7. Typically, it is the smaller faces that have the highest growth rates, and the largest faces grow at the slowest rates (compare Table 6–1 with Figure 6–7). This makes sense from a physical standpoint since the faster-growing faces eventually grow out and become smaller, leaving the slower-growing faces as the larger planes. Thus, the

Table 6–1 Growth Rates of Different Crystal Faces of Pharmaceutical-Grade α-Lactose Monohydrate

	*Growth Rate (μm/h) of Individual Faces**			
Origin	*(010)*	*(110)*	*(100)*	*$(1\bar{1}0)$*
Brand 1; Lot 1	3.1	3.2	1.3	0.3
Recrystallized†	0.5	1.8	1.7	0.3
Brand 1; Lot 2	9.4	5.6	2.3	1.0
Brand 2; Lot 1	7.2	4.2	2.4	1.1
Brand 2; Lot 2	7.8	1.5	1.9	0.9
Brand 3	6.0	4.7	2.6	0.8

* See Figure 6–7 for definition of faces according to Miller index.
† Recrystallized samples were purified by a second recrystallization step.

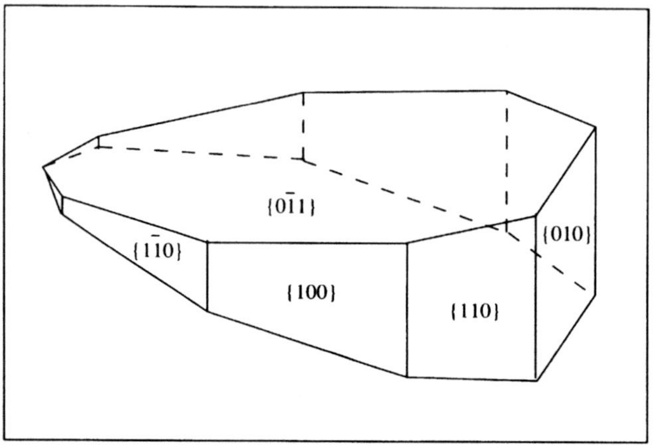

Figure 6–7 Lactose Crystal Showing Different Faces.

morphology or shape of a crystal depends on the relative growth rates of the competing faces, as shown schematically in Figure 6–8. Factors that influence these relative growth kinetics, and thereby affect crystal shape, include supersaturation, type of solvent, and the presence of additives or impurities. In some fields (eg, pharmaceuticals), control of crystal shape through addition of specific impurities is important to production of a desired crystal form.

In the food industry, a wide range of constituents can preferentially influence growth rates of certain faces and potentially affect crystal shape. In Table 6–1, differences in growth rates of different faces were observed for several different sources of pharmaceutical grade lactose.[19] Even the sample that was recrystallized (filter crystals, dissolve and recrystallize) had different facial growth rates. These differences were attributed to a minor impurity that strongly adsorbed to the lactose crystal surface.[19] In some cases, the activity of an impurity is sufficient to alter crystal habit. For example, the prismatic form of sucrose can be changed to a needle-shaped crystal by addition of certain polysaccharides. The presence of raffinose or corn syrup during sucrose crystal growth also results in formation of needle-shaped sucrose crystals under certain circumstances.

Growth Rate Dispersion

In many systems, crystals of uniform size and shape growing in the same environment do not grow at the same rate. Rather, each crystal grows at some inherent rate, and a distribution of growth rates is observed for the distribution of crystals. This distribution of growth rates is called growth rate dispersion (GRD). If a uniform (size, shape, appearance,

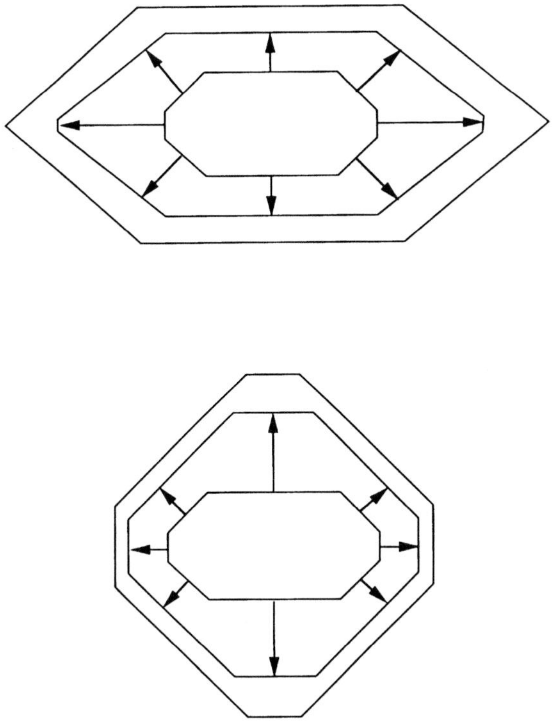

Figure 6–8 Change in Crystal Shape Related to Growth Rates of Different Faces. Lengths of arrows represent magnitude of growth rate.

etc) crystal population is placed in a batch crystallizer and allowed to grow, the distribution becomes increasingly wider as the faster-growing crystals become larger at a faster rate than the slower-growing crystals. This is seen schematically in Figure 6–9. In this case, the number of crystals present in the distribution does not change. Only the shape of the size distribution changes.

Many systems exhibit GRD,[21] including sugars, salts, and organic acids. GRD has been studied in batch and continuous crystallizers and by observations of individual crystals growing under controlled conditions on a microscope stage.[22] The widening of a distribution of crystal sizes seen during batch crystallization, as shown in Figure 6–9, has been analyzed in terms of the population balance approach[21] to quantify the extent of GRD under different operating conditions. In continuous crystallizers, the complex shape

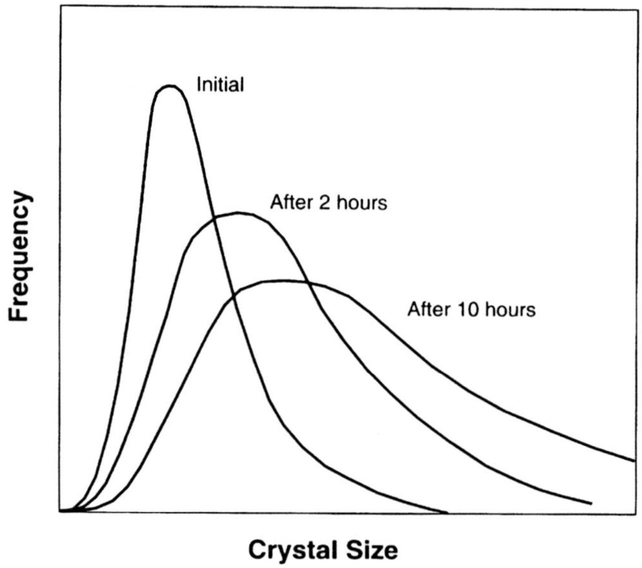

Figure 6–9 Schematic Representation of Change in Size Distribution during Growth Due to Growth Rate Dispersion.

of the size distribution at the outlet also can be analyzed in terms of GRD to quantify this phenomenon. However, the most definitive studies of GRD involve growth of multiple crystals under controlled conditions on a microscope stage. Using a specially designed cell,[23–25] the growth of individual crystals can be followed under controlled conditions of supersaturation, temperature, and hydrodynamic conditions. In this case, sufficient numbers of crystals must be observed to quantify a distribution of growth rates. In some cases,[24] there are many small crystals that do not appear to grow at all over a 1-hour time period, whereas other crystals grow rapidly under identical conditions (Figure 6–10). This leads to a distribution in growth rate for a population of growing crystals (Figure 6–11). Lactose crystals show this dramatic difference in crystal growth rates for otherwise apparently similar crystals.[26] It is important to note that no outward indication of a crystal (appearance or shape) has been correlated with the rate of growth. Also, GRD has been observed for sucrose crystals that were newly formed (by secondary nucleation under controlled conditions) and for large (400 μm) crystals of purified sucrose,[26] so size is not a factor. In fact, plots of growth rate versus size for a distribution of growing crystals shows no correlation, suggesting that GRD is not a size-related phenomenon.

The mechanisms for GRD have not been completely defined, although it is generally thought that only systems where surface mechanisms play a role in the growth process exhibit GRD. One potential mechanism is differences in dislocation densities on the

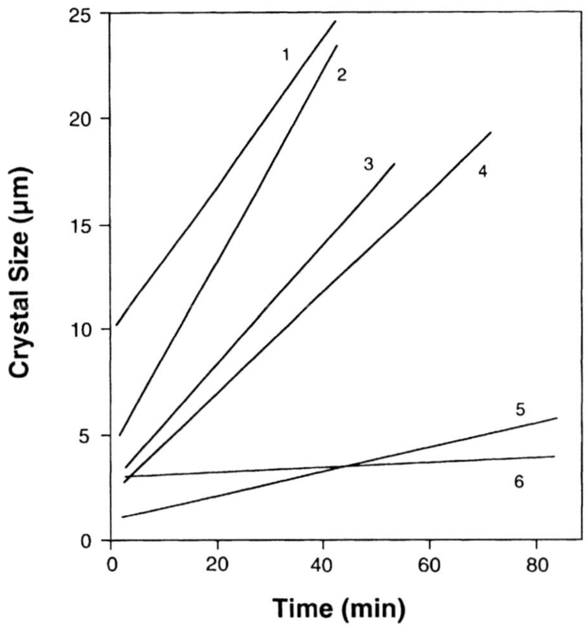

Figure 6–10 Change in Size with Time for Contact Nuclei Created by Sliding a Sucrose Crystal across a Glass Plate in a Supersaturated Solution. Numbers indicate different crystals: crystals 1 and 2 are fast growing; crystals 5 and 6 are slow growing.

growing surface that cause these differences in growth rates among crystals. Those crystals with many dislocations grow more rapidly than crystals with fewer dislocations.[27] Valcic[28] correlated growth of sucrose crystals with dislocations on the surface and showed that the BCF model of crystal growth based on activity of screw dislocations (the surface diffusion model of BCF) gave good agreement with the experimental results. Others have studied the effects of internal strain on crystal growth. In one study,[27] crystals were grown under controlled conditions, where growing crystals experienced an externally applied torsion. Crystals under no applied strain grew the fastest, and increasing levels of externally applied strain caused growth rate to decrease substantially. These results suggest that the manner in which crystals are formed plays an important role in determining how fast they grow. In a population of crystals that exhibit GRD, it is likely that the crystals created with high internal stresses make up the slow-growing population, whereas more perfect and uniform crystals grow faster. Three other possible mechanisms that may lead to GRD include (1) differences in surface microstructure due to physical effects (eg, roughening),

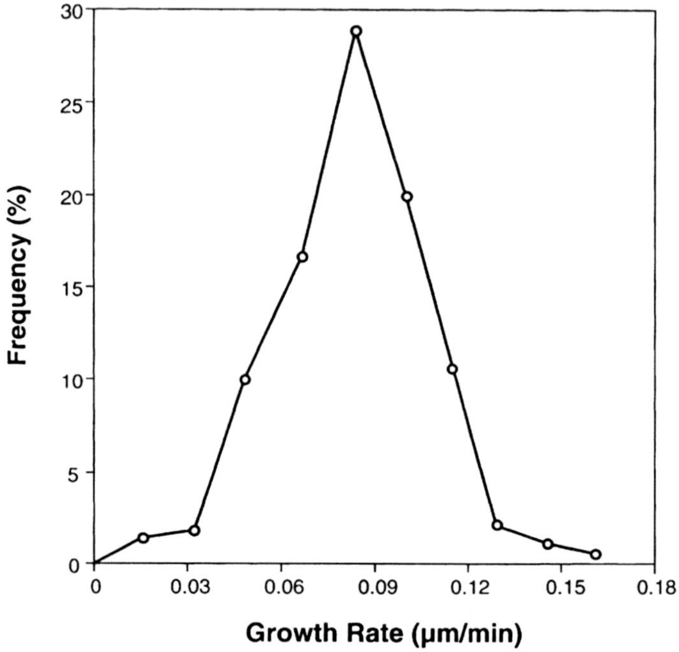

Figure 6–11 Growth Rate Distribution for Secondary Nuclei of Lactose Grown at 30°C and Supersaturation (S) of 1.52.

(2) differences in surface coverage of some foreign impurity, and (3) incorporation of a foreign impurity into the lattice to different extent.

GRD has a definite impact in certain food processing operations. In any refining operation, crystals growing in a suspension may be influenced by GRD. For example, in refining of sucrose crystals, a population of seed crystals is added to a supersaturated solution in a batch crystallizer. These seeds are grown to product size by maintaining rapid growth conditions at supersaturations that are just below the metastable zone boundary for forced crystallization (see Chapter 5). If there was no GRD, all of the seeds would grow into uniform, product-sized crystals. However, due to the influence of GRD, the growing crystals change in size distribution according to Figure 6–9. That is, the product size distribution has a wider range of sizes than the initial seeds due to GRD. Since the goal of sugar refining is to produce the narrowest distribution of sizes around the mean size, it is obvious that GRD limits the efficiency of this operation. Similar effects occur during refining of other sugars.

Dissolution/Melting

Dissolution or melting also may be important in controlling crystallization in some food processes. Manufacture of many confections, for example, requires that granulated sugar be dissolved in sufficient water to ensure that no crystal seeds enter the process equipment. Any crystals present after cooking grow during subsequent processing and potentially have negative effect on product quality. Specialized dissolving technologies have been developed to promote this step in the most efficient manner. Often, dissolution is accomplished by heating under pressure to minimize the amount of water required to dissolve the desired amount of sugar crystals. Dissolution or melting also may be important during recrystallization or ripening of existing distributions of crystals. During recrystallization of ice cream, for example, the smaller ice crystals melt away while the larger crystals grow. The rate of this melting step may impact the recrystallization process, although there has not been any experimental verification of this hypothesis.

Typically, dissolution occurs much more rapidly than growth. During crystal growth, there is always an organization step where the molecules must give up some freedom of motion in order to fit into the lattice structure. During dissolution, molecules break away from the lattice structure based on the driving force for dissolution or melting and diffuse into the liquid environment. In addition, molecules do not dissolve from the lattice

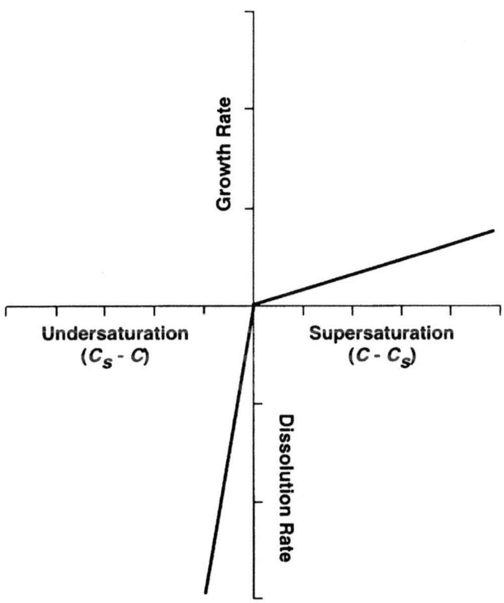

Figure 6–12 Relative Difference between Dissolution and Growth Rates, Shown Schematically.

in uniform layers, but pitting and etching of the surface provide greater surface area for dissolution than is typically available during growth. These factors result in dissolution rates that are two to five times faster than growth rates at equivalent driving forces.[1] Figure 6–12 schematically shows this difference between dissolution and growth, where for equivalent super- or undersaturation, dissolution rates are significantly greater than growth rates. Van Hook[29] showed that the rate of dissolution of sucrose was about 1.5 to 1.6 times faster than the rate of growth at equivalent undersaturations.

Although dissolution is primarily a diffusion-controlled process, there is still some surface or bulk crystal effect that impacts dissolution rates. Just as there is a growth rate dispersion, a dissolution rate dispersion (DRD) also exists. That is, different crystals dissolve at different rates even under controlled conditions.[30,31] Figure 6–13 shows data on growth and dissolution rate dispersion for sucrose crystals. The growth rates and dissolution rates of individual crystals were measured under controlled conditions on a microscope stage and correlated to either supersaturation (growth) or undersaturation (dissolution). It is clear that both types of dispersion (growth and dissolution) occur, but

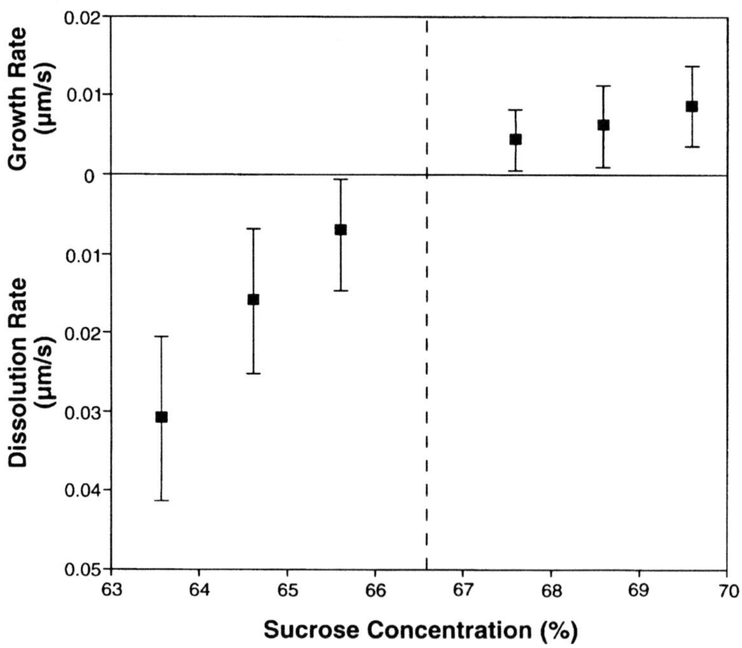

Figure 6–13 Dissolution and Growth Rate Dispersion for Sucrose Crystals at 20°C. Width of the error bars is a measure of dissolution or growth rate dispersion.

the magnitude of DRD is somewhat surprising if diffusion is the rate-limiting mechanism for dissolution. Apparently, there is a substantial effect of the surface or bulk crystal on the rate of dissolution, and this varies among crystals. It also is interesting to note that the extent of DRD is greater than that of GRD for equivalent driving force. That is, the difference between the fastest and slowest dissolving crystals is greater than the difference between fastest and slowest growing crystals.

FACTORS AFFECTING CRYSTAL GROWTH

Supersaturation

Supersaturation or supercooling is the driving force for crystallization; thus, it plays a dominant role in determining the rate of crystal growth. As discussed previously, higher driving forces result in faster crystal growth, but only up to a point. In many systems, as concentration is increased or temperature is decreased, depending on the system, the growth rate reaches a maximum value based on a balance between the thermodynamic driving force for crystallization and the diffusional limitation to molecular mobility. At very high supersaturations and/or low temperatures, molecular mobility is sufficiently reduced that crystal growth rates decrease rapidly. At the point where the glass transition is reached, growth rates effectively drop to zero. Thus, growth rate changes with supersaturation, for many food systems, in the manner shown in Figure 6–14. The exact point of the maximum growth rate depends on the nature of the system and has not been quantitatively measured for many systems.

To find this maximum in growth rate, Ben-Yoseph[32] modeled sucrose crystal growth rates according to a mass transfer limited model of growth. Equation 6.14 represents crystal growth when diffusion of molecules across the boundary layer limits the rate of growth.[33]

$$G = \frac{4D_v}{\rho_s}\left(\frac{1}{L}+\frac{1}{2\delta}\right)(C - C_t) \tag{6.14}$$

where, ρ_s is crystal density and δ is thickness of a surface sublayer. In this model, the diffusivity of molecules across the boundary layer is the rate-limiting step that controls the rate of growth. For sucrose in water, the self-diffusivity at high temperatures and concentrations has been modeled by Ben-Yoseph[32] based on previous data of Zhmyria[34] and the principle that the diffusivity must go nearly to zero when the glass transition is reached. A surface layer thickness of about 50 μm was calculated from experimental data for sucrose growth under stagnant conditions.[32] The growth rate of sucrose crystals under diffusion-limited conditions can be predicted by Equation 6.14, as shown in Figure 6–15 for a temperature of 80°C. At low supersaturation, growth rate increases with increasing concentration as expected based on increased driving force. However, as concentration increases, diffusivity decreases dramatically until, at some point, the decrease in diffusivity

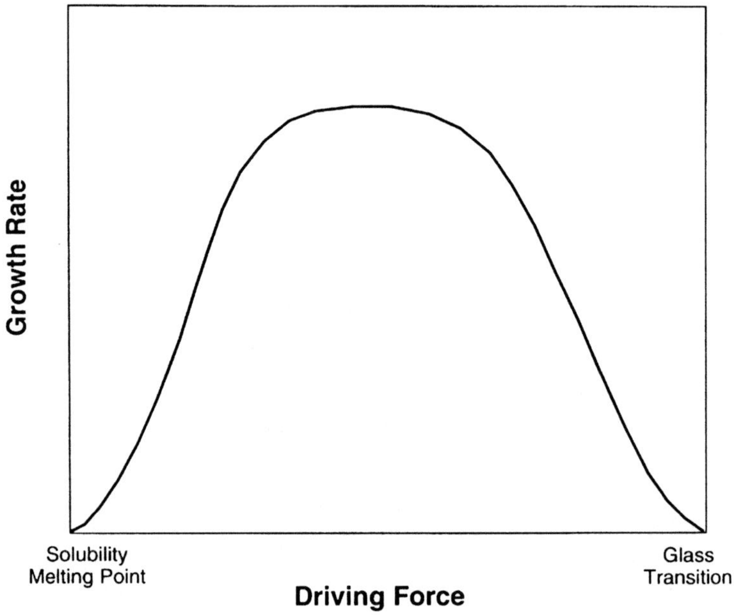

Figure 6–14 Generalized Dependence of Crystal Growth Rate on Driving Force for Many Food Systems, Including Sugars, Organic Acids, and Ice.

in Equation 6.14 becomes more important than the increase in supersaturation. At this point, growth rate begins to decrease.

In many systems, the behavior of growth rate, according to Figure 6–15, follows the same general trend as seen for nucleation rate (Figure 5–4). At some point of increasing driving force, growth rate reaches a maximum and then begins to decrease as molecular mobility limits the rate of growth. This behavior has been observed in many food systems, including sugars, organic acids, starches, and ice. In lipid crystallization, however, growth behavior is modified by polymorphic transformations. In general, the more stable polymorphs grow more rapidly than less stable polymorphs.[35] More stable polymorphs have higher driving force (greater supercooling) and thus grow faster.[36]

Temperature

The effect of temperature is intimately linked to the driving force for crystallization. In melt systems (water, etc), the temperature difference between the actual temperature and the melting point provides the driving force for growth or dissolution. Thus, lowering

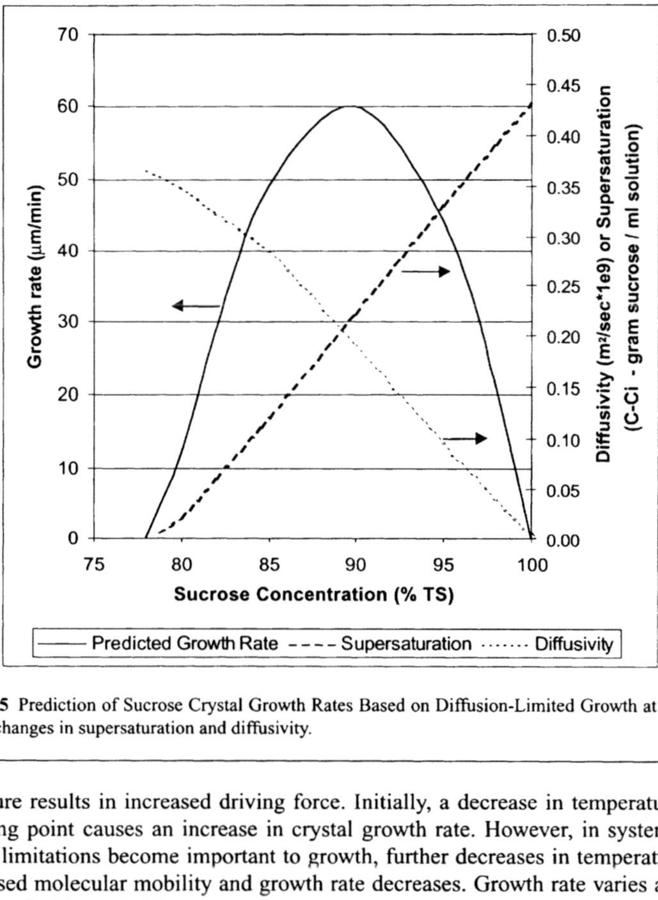

Figure 6–15 Prediction of Sucrose Crystal Growth Rates Based on Diffusion-Limited Growth at 80°C. Also shown are changes in supersaturation and diffusivity.

temperature results in increased driving force. Initially, a decrease in temperature below the melting point causes an increase in crystal growth rate. However, in systems where diffusion limitations become important to growth, further decreases in temperature result in decreased molecular mobility and growth rate decreases. Growth rate varies according to the curve in Figure 6–14.

In solution systems, especially where solubility is a strong function of temperature, supersaturation changes as temperature is changed. Thus, we can treat two cases: one where we deal with a constant concentration and change temperature, and the other where we assume concentration changes to maintain constant supersaturation as temperature changes. The first case is important commercially, whereas the latter case is more important from a fundamental standpoint. Since supersaturation and temperature are inseparable in this case, it is important to look at both situations. For example, sucrose crystal growth rates increase with both increasing temperature and supersaturation[37,38] as shown in Figure 6–16.

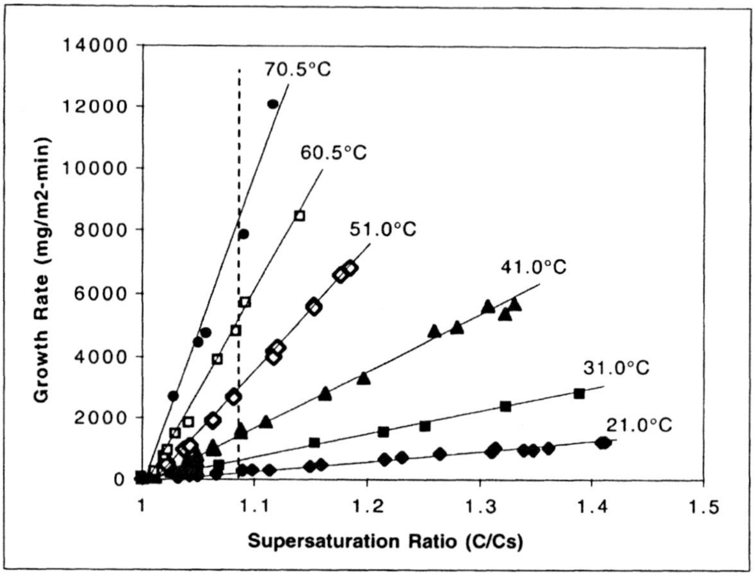

Figure 6–16 Dependence of Sucrose Crystal Growth Rates on Supersaturation and Temperature.

As temperature increases in solutions of constant supersaturation, the crystal growth rate increases based on the positive activation energy for growth, as shown by the vertical line in Figure 6–16. The rate of increase in growth rate with increasing temperature, at constant supersaturation, can be analyzed using the Arrhenius expression:

$$k_G = A_g \exp\left\{-\frac{E_G}{RT}\right\} \tag{6.15}$$

where A_g is a frequency factor, E_G is the activation energy for growth, R is the gas constant, and T is temperature. Activation energy can be found from the exponential relationship between k_G (or growth rate at constant supersaturation) and inverse temperature, as shown for sucrose in Figure 6–17. For sucrose, the activation energy representing the change in growth rate with temperature depends on the temperature range studied. At temperatures below about 40°C, the activation energy is 16 to 18 kcal/mol, whereas at temperatures above 40°C, the activation energy is 6 to 8 kcal/mol.[38,39] The difference in activation energies in these temperature ranges indicates the change in rate-limiting mechanism for growth as temperature increases. Based on these results, sucrose crystal growth is limited by combined mechanisms of diffusion in the bulk solution and a surface incorporation step.[2,40] At high temperatures, it is diffusion that provides greater limitation of growth, whereas at lower temperatures (below 40°C), growth is more limited by the surface incorporation step. Note that each of the steps required for growth, as outlined in Exhibit 6–1, occurs at all conditions. However, it is diffusion at temperatures above 40°C and surface incorporation at temperatures below 40°C that limit the rate of growth for sucrose crystals.

When a solution of given concentration is cooled below the solubility temperature (for that concentration), supersaturation increases at a rate dependent on the change in solubility with temperature. At the same time, the growth kinetics, as described by the Arrhenius expression, are slowing down. The relative magnitude of these competitive effects depends on the system, although it is usually the increasing supersaturation that is more important; this results in a growth rate that initially increases as temperature decreases from the solubility point. However, as temperature decreases to the point where molecular mobility becomes limiting, the growth rate begins to decrease. Once the temperature reaches the glass transition point for that concentration, the rate of growth again is reduced to zero, as shown in Figure 6–14. The maximum of the growth rate curve for a given food system is a complex function of many parameters, and it has not been possible to determine this point except through experimentation for each particular food product.

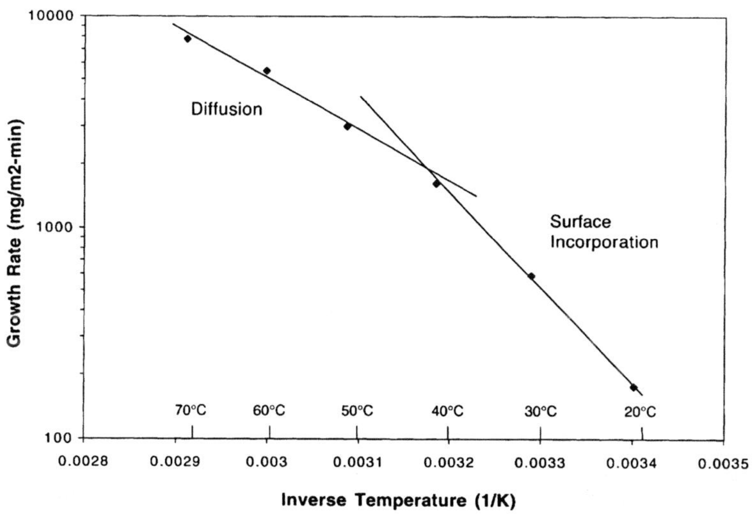

Figure 6–17 Arrhenius Analysis for Sucrose Showing Growth Rates (at Constant Supersaturation of S = 1.09) at Different Temperature. Regions where growth is primarily limited by either diffusion or surface incorporation are indicated.

From a processing standpoint, crystal growth occurs most rapidly at this temperature where the supersaturation increase and mobility limitations are balanced. Operation at any other temperature results in slower growth.

In some systems, such as the reducing sugars (lactose, fructose, etc), the molecules undergo a chemical reaction (mutarotation) in solution to maintain equilibrium between forms. The rate of this reaction is temperature dependent, and under certain circumstances (high crystal surface area and/or low temperature), the rate of mutarotation may become a rate-limiting step in crystal growth.[41,42] For example, the mutarotation reaction between α- and β-lactose in solution becomes increasingly slower as temperature is lowered.[43] At low temperatures (eg, near the freezing point of water), mutarotation rates are sufficiently slow that lactose crystal growth rates may be influenced by the rate of mutarotation. At these temperatures, α-lactose monohydrate crystallizes from solution. As an α-lactose molecule leaves the solution to become incorporated into the lattice, a β-lactose molecule must mutarotate to maintain equilibrium in solution, as seen in the two-step mechanism of Figure 4–5. If this mutarotation rate is slower than the rate at which α-lactose crystallizes (mass deposition rate), then the mutarotation rate limits growth. This can be seen in lactose systems by monitoring the optical rotation (to distinguish between the relative amounts of α- and β-lactose in solution) during crystallization. Figure 6–18 shows how optical rotation decreases during the fastest growth of lactose, evidenced by the fastest rate of desupersaturation. The surface area of growing crystals also is important in this case since it is the mass deposition rate of crystallization that determines how fast α-lactose molecules are removed from solution.

Viscosity

For the most part, the effects of viscosity are accounted for through the effects of temperature and concentration, as described previously. In fact, it is widely misconstrued that bulk viscosity is the limiting factor influencing crystal growth. Rather, it is the mobility of the molecules that influences crystallization (either the crystallizing species itself or impurities that impact growth of that species), and this mobility is not always related to bulk viscosity. Budiaman and Fennema[44] showed that various hydrocolloids reduced the rate of growth of an ice crystal front, but that for equivalent viscosity, the extent of growth inhibition was dependent on the type of hydrocolloid. Thus, it was the specific effect of each hydrocolloid (perhaps from counter-diffusion away from the growing crystal surface) that inhibited ice crystal growth and not simply the bulk viscosity. Similar results were found for growth of sucrose crystals. Smythe[39] found a general trend for the relationship between viscosity and growth rate when the effects of various additives on growth rate were compared. However, there were several examples of substances (ie, dextran and calcium chloride [$CaCl_2$]) that did not follow the general trend. That is, the growth rate could not be attributed to the viscosity of the solution.

At very low addition levels, gelling agents have a dramatic effect on bulk viscosity, but do not influence the mobility of the water molecules. Nevertheless, ice crystal growth is inhibited in a gel, but probably due to the mechanical interference on growth of ice

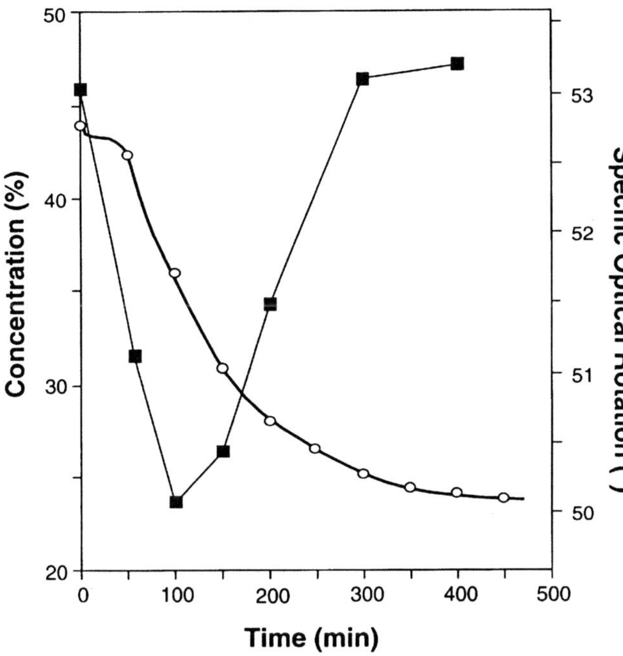

Figure 6–18 Lactose Crystal Growth Rate (Open Circles) Showing Limitation of Growth by Mutarotation as Evidenced by a Decrease in Optical Rotation (Solid Squares). Crystallization at 40°C, with initial lactose content of 44%, in a batch crystallizer with 2.5 g of seed crystals added.

crystals.[45] Thus, the bulk viscosity of a solution (a macroscopic property) is not necessarily a good indicator of crystallization rate.

Agitation Rate

When mass transfer of molecules either to or from the growing crystal surface affects growth, changes in agitation rate can have a significant effect on growth rates by molecular deposition. However, when mass transfer is not a main factor in determining crystal growth rates, agitation will have little effect on growth rate. Agitation enhances the rate of bulk mass transfer, for both crystallizing species toward the surface and noncrystallizing species away from the surface. Thus, crystal growth rate increases with increasing agitation for these systems. However, this increase in growth rate continues only to the point where mass transfer no longer limits growth. Typically, growth rate increases with agitation rate, as shown in Figure 6–19. Van Hook[46] and Smythe[38] found such behavior for the effect of agitation on growth rate of sucrose. Similar behavior is seen in systems where heat transfer is a contributing factor to growth rate. In this case, growth rate increases with agitation rate until heat transfer is no longer a limiting factor.

Source of Nuclei

The manner by which nuclei were formed may impact the rate at which they grow, since growth rate is often dependent on the extent and activity of surface dislocations to provide kink sites for molecular incorporation. In addition, the internal stresses on the molecules in the lattice due to imperfections have been shown to influence crystal growth rates significantly. Nuclei that are formed by heterogeneous nucleation most likely have different dislocation density distributions and internal stresses than nuclei formed by a contact with a stirrer, for example. This may be a concern in suspension crystallizers where secondary nucleation is the dominant form of nuclei generation.[21] It has been observed in some suspension crystallization systems that a large population of extremely slow-

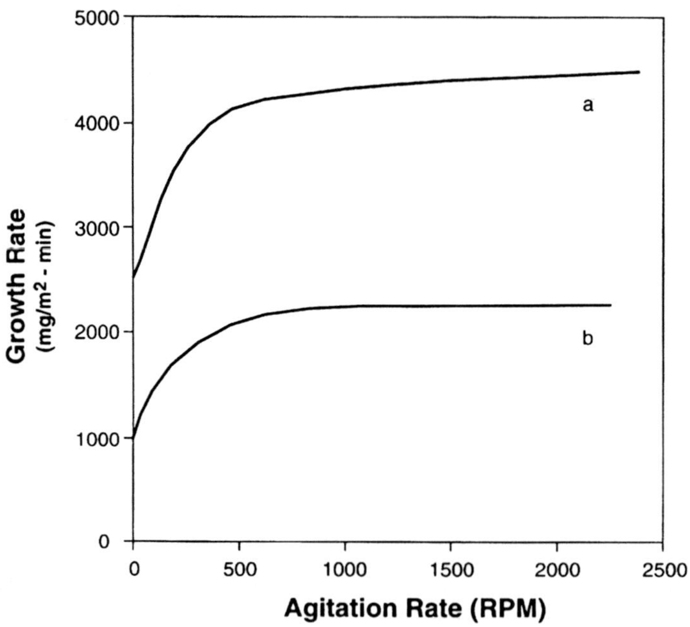

Figure 6–19 Effects of Agitation on Growth Rate of Sucrose Crystals. Line (a): 40°C; ΔC = 70 g/100 g H_2O; line (b): 50°C; ΔC = 17 g/100 g H_2O.

growing crystals exists. These nuclei do not grow fast enough to be part of the product-sized distribution, but may have an adverse effect on the separation step. There has been some speculation that these slow-growing crystals are those that have been formed with excessive internal stresses so that growth is inhibited.

Van Hook[47] studied the growth rates of "good" and "poor" sugar crystals. The "poor" crystal, which was cloudy and had irregular-shaped facets, grew at a faster rate than the crystal of "good" quality. The dependence of growth rate on supersaturation also was different between the two crystals. The growth rate of the "poor" crystal increased linearly with supersaturation, whereas the "good" crystal had a second-order dependence of growth rate on supersaturation at low supersaturations and a first-order dependence at higher supersaturations. Apparently the different kink sites or dislocations available for crystal growth resulted in the different behavior for the two crystals.

In some food processes, seed crystals are added to avoid the necessity of nucleating a population of crystals. In this way, it is sometimes possible to control the crystallinity of a product to a greater extent. Seeding of grained caramels or fudge with fondant or powdered sugar is an example of this. In some cases, these seed crystals are obtained by grinding or milling product-sized crystals to the proper size. However, grinding causes significant damage to crystal structure and typically causes some loss in crystallinity. For example, milling of sugars results in formation of a layer of amorphous sugar at the surface where high temperatures have caused the crystal to melt; recooling is sufficiently fast so that an amorphous layer is formed. Although there is no real evidence that these effects cause any difficulties in crystal growth in seeded systems, the food processor must take care to choose seed material that ensures proper crystallization.

Crystal Mass Density

The number of crystals in a unit volume competing for available supersaturation may have an effect on the rate of growth of each crystal. The close proximity of these crystals causes the diffusion fields of individual crystals to overlap. Under conditions of importance in most foods, where supersaturation decreases naturally in a batch crystallization system, growth rates of individual crystals decrease when the crystals are crowded together.[48] This may only be important in products where many crystals are crowded into a very small volume. For example, sugar crystals in a fondant are separated by small distances (only a few micrometers), and there may be some effect of crowding. In fact, it has been suggested that growth of sugar crystals in a fondant is limited by how fast the diffusion field around a growing crystal can be replenished by diffusion of sugar molecules into the region. No experimental verification of this hypothesis has been published.

External Fields

Electric, magnetic, and especially ultrasonic fields often are claimed to enhance crystallization processes under certain conditions.[12,49] In sugars, growth rates were enhanced by up to 70% in a constant electric field,[50] with a linear increase in growth rate as electric field strength increased. Although the mechanism for this process is unclear, it was suggested that the mechanism involved interfacial phenomena (at the crystal surface) rather than mass transfer. Similar growth enhancement has been observed for several crystallizing systems (including sugars) when exposed to ultrasonic fields.[49] Again, the mechanism for this effect has not been clearly identified, and van Hook[12] suggests that any effect of ultrasonic energy on crystallization is most likely due to a nucleation effect and not to growth. If there is a specific effect of ultrasonic energy, it is likely that enhanced mass transfer rates are the cause of any enhancement in growth rates.

Additives and Impurities

Crystallization in impure systems is of considerable interest from both theoretical and practical standpoints. In practice, most crystallization occurs from systems containing more than one component. This is certainly true in complex food products, but it also is true even in relatively pure systems. For example, the concentrated syrups in sugar refining contain mostly sucrose, but they also contain numerous other components (salts, other sugars, various dextrins, etc). These other components may have significant impact on the crystallization processes and the characteristics of the final product. Most impurities negatively affect crystal growth rates, although the extent of inhibition depends on the mechanism by which the impurity acts. In addition, impurities often affect crystal morphology since the impurity may have greater impact on certain faces and change the relative growth rates of the different crystal faces. Crystal habit modification via selective impurities may be an important mechanism for controlling crystal shape in the food industry, although a better understanding of the relationships between crystal shape and product texture will be needed to make efforts in this area worthwhile.

In general, there are three main mechanisms by which impurities affect crystal growth. The first mechanism involves a limitation to the mass transfer of molecules either toward or away from the growing crystal surface. Many molecules, particularly large and bulky ones, inhibit mass transfer coefficients and diffusivity of the crystallizing species. If mass transfer becomes a rate-limiting step by addition of an impurity, then the growth rate will be decreased.

The second mechanism of growth inhibition by impurities involves adsorption of the impurity at the growing crystal surface. Once adsorbed, the impurity hinders further incorporation of molecules of the crystallizing species since the impurity molecule has to be displaced before surface incorporation can occur. Compounds with high affinity to the growing crystal surface inhibit crystal growth to a significant extent. An example of this type of inhibition is the reduction in growth rate of sucrose crystals on addition of corn syrup or invert sugar to the solution. The glucose end-groups in corn syrup and invert sugar adsorb to the sucrose lattice structure and inhibit crystal growth since energy is needed for a sucrose molecule to displace the glucose unit and incorporate into the lattice.

The third mechanism involves actual incorporation of the impurity into the crystal lattice, which then disrupts further addition to the lattice structure of the crystallizing species and inhibits growth. These "tailor-made" impurities actually fit into the lattice structure through a portion of their molecular structure, but then leave a molecular moiety that physically impedes further incorporation of the crystallizing species. For example, raffinose is an effective inhibitor of sucrose crystal growth due to the similarity in structure. Raffinose is a trisaccharide containing a sucrose disaccharide attached to a galactose molecule. The sucrose aspect of raffinose fits well into the crystal lattice, but then the galactose remains above the surface, inhibiting further incorporation of sucrose. Since different crystal faces have different molecular arrangements, such tailor-made impurities often alter crystal habit. In this case, raffinose causes sucrose crystals to appear needle-shaped rather than prismatic since raffinose selectively inhibits certain faces more than others.

When comparing the inhibition of crystal growth rate by addition of an impurity, any effects on equilibrium conditions must be taken into account. As discussed in Chapter 4, addition of impurities or additives often influences either solubility or melting point, which can affect the driving force for crystallization. In order to truly compare the effect of a specific impurity on crystal growth, systems with constant supersaturation must be studied. If addition of a certain level of an impurity depresses solubility, then the concentration of the crystallizing species must be increased to maintain constant supersaturation.

Sugars

Crystallization rates of sugars can be affected dramatically by addition of many common food ingredients, as summarized by Hartel and Shastry.[22] For example, growth rates of sucrose crystals, compared at constant supersaturation, in the presence of either invert sugar or raffinose is shown in Figure 6–20. The monosaccharides, glucose and fructose, in invert sugar adsorb to the sucrose crystal surface and inhibit incorporation of the sucrose molecules.[39] Raffinose, on the other hand, is actually incorporated into the sucrose crystal lattice and acts as a tailor-made impurity.[51] From the relative effects of these two impurities, it is clear that the tailor-made impurity has much greater inhibition of sucrose crystal growth.

The presence of lactose and sucrose in a mixture, as might occur in caramel or the continuous phase of ice cream, influences crystallization of each sugar. Nickerson and Patel[52] studied crystallization of sucrose/lactose mixtures at ratios from 100:0 to 50:50. Rates of desupersaturation for each individual sugar were dependent on solution ratio. At ratios of 100:0 and 90:10 of sucrose to lactose, only sucrose crystallized from solution, with the rate of crystallization decreasing with increasing lactose content. At ratios of 80:20 and 70:30, both sugars crystallized from solution, whereas at ratios of 60:40 and 50:50 it was only lactose that crystallized. The rate of lactose crystallization increased as the level of sucrose decreased, indicative of the higher lactose supersaturation. At the same time, the rate of sucrose crystallization decreased as lactose content increased due to the decrease in supersaturation of sucrose. In addition, lactose crystals changed shape in the presence of sucrose, becoming shorter and broader rather than the pyramid shape typically found for lactose crystals. When both sucrose and lactose crystals formed at the same time, it was virtually impossible to distinguish between them based on microscope images.

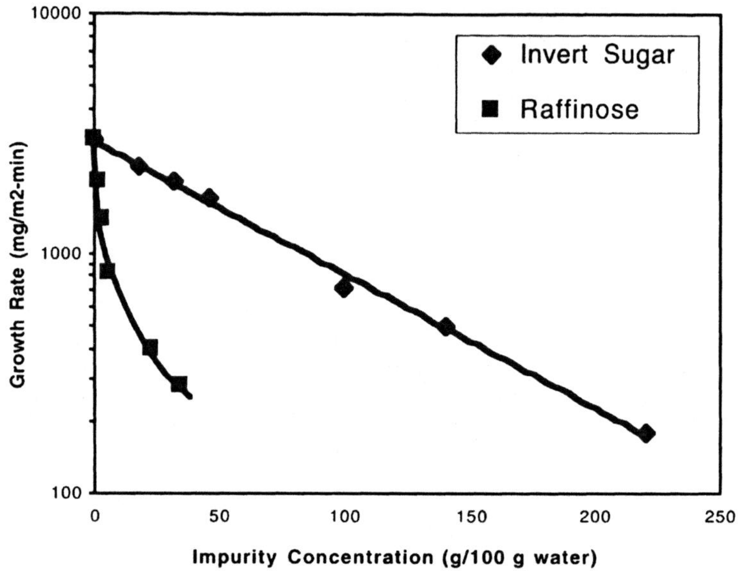

Figure 6–20 Effects of Invert Sugar and Raffinose on Sucrose Crystal Growth Rates at 41°C and Constant Supersaturation of ΔC = 40 g Sucrose/g Water.

In many food products, corn syrup is added to control sucrose crystallization. For example, corn syrup is added to fondants and caramels, in part, to control sucrose crystallization. Also, hard candies contain corn syrup for the same reason, except in this case, crystallization from the amorphous phase is inhibited. Tjuradi and Hartel[53] evaluated the effects of corn syrup saccharides on sucrose crystal growth from solution. The different saccharide components of corn syrup were fractionated according to molecular weight, and each fraction added back to a sucrose solution to give the same supersaturation ratio (S = 1.46; based on concentration units expressed in g sucrose/100 g water). The growth rate of a seed distribution was followed by image analysis of the growing crystal distribution. When added on an equivalent weight basis of addition (5, 10, or 20% of the sugar solids by weight), the corn syrup components with lower molecular weight (lower degree of polymerization) were more effective at reducing sucrose crystal growth rate. However, when the results were expressed in terms of average molar addition levels of each fraction, there was no difference in inhibition due to chain length of the saccharides (Figure 6–21). Higher addition levels (on an equivalent molar basis) of all components inhibited sucrose crystal growth to approximately the same extent. This suggests that glucose polymers from corn syrup inhibit sucrose crystal growth by adsorbing at the surface and that the energy of adsorption is the same regardless of the chain length of the glucose polymer. Perhaps

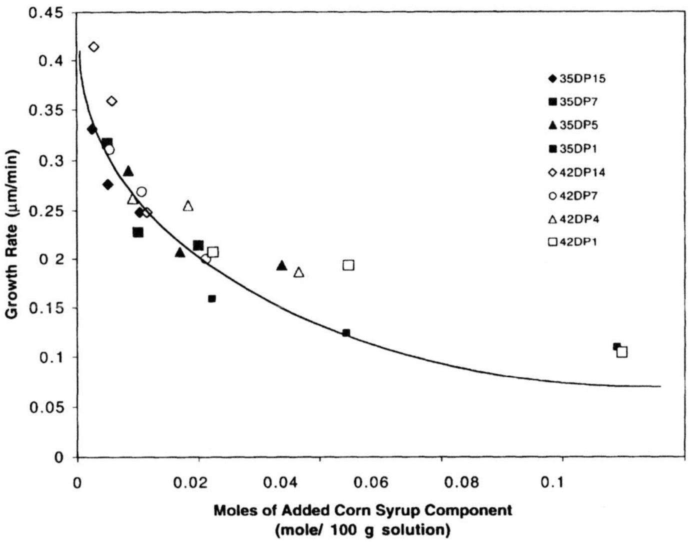

Figure 6–21 Effect of Corn Syrup Saccharides with Different Average Degree of Polymerization (DP) on Sucrose Crystal Growth from Solution. Samples labeled as dextrose equivalent (DE) of initial corn syrup and average DP (eg, 42DP14 indicates a saccharide fraction with average DP of 14 produced from a 42 DE corn syrup).

---

there is only one active glucose end unit, regardless of the degree of polymerization, that adsorbs to the surface. Thus, the energy required (and degree of growth inhibition) is not dependent on molecular size in this case. Since the results of Tjuradi and Hartel[53] do not agree with common opinion (where longer-chain saccharides are thought to provide greater growth inhibition), further studies are required to document that this proposed mechanism is indeed correct.

However, this raises the point of studying inhibition based on constant concentration or constant supersaturation in the solution. Industrially, it may be more important to maintain constant concentration or total solids of the syrup, whereas the experiments above were done with constant supersaturation (concentration altered). For example, take a solution with concentration of 600 g sucrose/100 g water (80% total solids) that is crystallizing at 30°C. Pure sucrose has a solubility concentration of about 215 g sucrose/100 g water (68% total solids) at this temperature and the pure solution would be supersaturated by $\Delta C = 385$ g sucrose/100 g water. If 30 g corn syrup solids/100 g water (11.5% corn syrup) is added and initial sucrose content is decreased by the same amount (to maintain the same initial dissolved solids content), the solubility of sucrose in this mixture decreases to 131 g sucrose/100 g water, according to the data presented in Table 4–1. In this case, the supersaturation is $\Delta C = 570 - 131 = 439$ g sucrose/100 g water. Thus, addition of corn syrup in this case causes a substantial increase in supersaturation, yet there is a lower growth rate (Figure 6–21). This clearly shows that corn syrup components indeed have a separate effect on growth rate, besides their effect on supersaturation.

Higher-molecular weight substances (proteins, dextrans, gums, etc) also can have an impact on growth rates of sugars.[39,54] These large molecules can potentially inhibit mass transfer of sugars to the growing crystal interface. However, there also may be some inhibition of growth rate by adsorption of these molecules to the growing crystal surface. It is this preferential adsorption to specific faces of a sucrose crystal by polysaccharides and dextrans that causes change in shape from prismatic to needle-like.[55,56]

Many other components in foods may influence crystal growth rates of sugars.[22] Typically, addition of salts decreases the rate of sugar crystal growth,[57] although some salts apparently have little effect. The effect of salts are particularly important during growth of lactose crystals, where natural salts (calcium chloride [$CaCl_2$, potassium chloride [KCl], sodium phosphate [$NaPO_4$] etc) are present in the whey. According to Jelen and Coulter,[58] these salts generally caused an increase in lactose crystal growth rate, although the effect of these salts on solubility of lactose was not taken into account. Surfactants may potentially influence crystal growth through interactions at the growing crystal surface. However, generally surfactants have been found to have little effect on sugar crystal growth rates.[54,59] Organic compounds (urea, sorbic acid, amino acids, etc), such as may be found naturally in the raw materials, also may impact sugar crystal growth rates.[39] These compounds typically have about the same extent of inhibition as invert sugar on sucrose, and they probably act in the same manner (surface adsorption). Of particular importance to lactose refining is the presence of riboflavin in whey. Riboflavin affects both growth rate and morphology,[54,60] and causes a distinct yellowish color in the refined product. The riboflavin adsorbs preferentially on certain faces and changes the crystals to a plate-like shape. Levels of 50 to 100 ppm are sufficient to decrease the growth rate of lactose crystals.

Inhibition of crystallization also occurs due to additives and impurities when these are present in sugar glasses. In the glassy state, corn syrup–sucrose mixtures are quite stable (resistant to crystallization) due to the low molecular mobility. However, once the glass transition temperature is exceeded, any crystals present in the amorphous material can grow. In the same way, if the glass picks up moisture from the surroundings, the moisture content may cause a decrease in the glass transition temperature and provide sufficient mobility for crystals to grow. The presence of corn syrup and other saccharides has been shown to reduce the rate of crystallization of amorphous sucrose.[61–65] However, it is not clear whether this effect comes from an inhibition of crystal growth or nucleation. Under certain conditions (eg, freeze drying of sugar mixtures), small crystallites are present within the sugar glassy structure.[66] If this is the case, then additive molecules present within the glass may influence crystal growth.[67] However, our understanding of crystallization from amorphous matrices is not sufficient to determine the relative effects of impurities on nucleation versus crystal growth.

Ice

Even the simplest frozen foods contain numerous other ingredients that may influence ice crystal growth. In fact, a wide variety of components has been shown to influence ice crystal growth rates, including sugars, salts, proteins, organic acids, and gums (stabilizers), as summarized in Fennema et al.[3]

Ice crystal growth generally is limited by (1) counter-diffusion of solutes away from the growing crystal surface and (2) removal of the heat energy released by crystallization. Mass transfer of water molecules to the growing ice interface is usually considered to be rapid,[3] with the exception of extremely low-temperature conditions where all molecular mobility is substantially reduced (below the glass transition). Ice crystal growth is typically slower in foods or formulations than in pure water due to the inhibitory effect of the other ingredients.[3] Dissolved solutes (sugars, salts, organic acids, etc) must diffuse away from the growing ice crystal interface in order to allow incorporation of water molecules into the lattice. The rate at which the dissolved solutes diffuse away from the surface is often the rate-limiting step that determines growth rate of ice, particularly when the concentration of dissolved solutes is high.[68] Heat transfer away from the growing ice interface can occur either through the solution or the growing crystal itself. Heat transfer rates are determined by thermal conductivity (conduction) and fluid motion (convection). In well-stirred systems, heat transfer through the solution may be quite rapid, whereas in stagnant systems, heat transfer may become the rate-limiting step to ice crystal growth.

Gums and stabilizers often are added to foods to help control ice crystallization. For example, gelatin, carrageenan, or locust bean gum are common additives in ice cream to help control the growth of ice crystals during storage due to recrystallization. However, the mechanisms by which these stabilizers work remain unclear. The effects of several hydrocolloids (gelatin, pectin, alginates, methyl cellulose, etc) on the inhibition of ice crystals growing in a sucrose solution have been studied due to their importance in frozen dessert formulations and other frozen foods.[44,45] The presence of sucrose caused a substantial decrease in ice crystal growth rate, as compared with pure water. The addition of soluble hydrocolloids to the sugar solution caused a further retardation of ice crystal growth, although not nearly to the same extent as the effect of sucrose. Shirai et al[69] suggested that polymers did not influence ice crystal growth due to an effect on mass transfer, but rather the effect of polymers was due to heat transfer limitations. Rapatz and Luyet[70] documented the decreased growth rate of an ice crystal by using optical microscopy, although the gelatin also caused a change in ice crystal shape. Hydrocolloids that formed a gel had more significant effect on both growth rate and shape of ice crystals.[45,71] The gel structure provided a mechanical inhibition to ice crystal growth even though the mobility of the water molecules was not altered.

Anti-freeze proteins have been shown to have a significant effect on ice crystal growth rates.[72] At very low levels (ppm), these proteins can essentially stop the growth of existing ice crystals below some critical temperature driving force.[73] Harrison et al[73] found this critical temperature to be –0.5°C for the concentration and type of anti-freeze protein studied. Once this critical driving force was surpassed, ice crystal growth rate in the presence of the anti-freeze protein was actually more rapid than in pure water. However, at higher supercoolings (lower temperatures) of greater than $\Delta T = 2°C$, the anti-freeze proteins again caused a reduction in ice crystal growth rate compared with that in pure water. The mechanism for this strong inhibition of ice growth rate is a direct adsorption of the protein on to the ice crystal surface with a high energy of affinity. The action of anti-freeze proteins also causes a change in morphology of ice crystals. Harrison et al[73] found that anti-freeze proteins generally caused more needle-like growth as compared with dendritic growth under the same conditions in the absence of the anti-freeze proteins.

When concentrations of solutes are high and temperatures are low, diffusional limitations cause ice crystal growth rates to decrease. Charoenrein et al[74] show the growth rate of ice crystals in 60% sugar solution at supercoolings between 4 and 21°C. Under the conditions of this experiment, growth rate (measured as a linear propagation velocity) increased between supercoolings up to about 10°C. At higher supercoolings, the growth rate decreased to nearly zero when supercooling was over 20°C. At this low temperature, the solution was approaching, although still above, the T_g value, and kinetic rates were decreased substantially. When temperature goes below T_g, growth rates decrease to near zero and changes in ice crystal size are very slow to occur. This is the basis behind rapid freezing of samples for scanning electron microscopy (SEM) analysis. However, for pure water or even gelatin solutions, ice crystal growth rates increased even down to very low temperatures (between –50 and –80°C) with no indication that growth rates would decrease due to diffusional limitations.[3] Since T_g of pure water is about –135°C,[75] temperatures of –50 to –80°C must not be close enough to the glass transition zone to provide diffusional limitation to ice crystal growth in pure water.

Lipids

Due to the similarity in chemical structure and the wide range of molecular structures present in natural lipids, crystallization of lipids is quite a complex process. Undoubtedly, the presence of different triacylglycerols in complex lipids causes a reduction in growth rate as the molecules attempt to organize into a crystal lattice. In relatively simple natural lipids, like cocoa butter, where there are only a few main triglyceride species, the minor components act as impurities to growth of the major triglycerides. For more complex natural fats, like milk fat, there is mixed crystal formation as triglycerides with similar molecular structures crystallize together. However, very little quantitative data are available to document the effect of the various components of natural fats on crystal growth rates. This lack of quantitative data is in part due to the difficulty of studying individual lipid crystals; it also is in part due to the difficulty of determining the true driving force for crystallization of lipids. In addition, impurities also are likely to impact the rate of polymorphic transitions,[76] making the study of impurities on lipid crystal growth extremely complex.

Smith et al[77] investigated the effects of lauric acid, monolaurin, and dilaurin on crystal growth rates of trilaurin (β-polymorph) using a moving-front microscopy technique. The movement of a growing trilaurin crystal front on a microscope slide with controlled temperature gradient was used as a measurement of growth rate. The presence of the various impurities, as well as the magnitude of the temperature gradient, influenced crystal growth rates and crystal morphology. The addition of 2% lauric acid and monolaurin slightly increased the growth rate of trilaurin; however, the addition of dilaurin had a significant retarding effect on the growth of trilaurin. The 1-3 dilaurin was especially effective at reducing trilaurin growth rate, by about 60 to 70% under the conditions of this study. These impurities also had different effects on morphology of the growing crystal. The results were interpreted according to the hypothesis that the impurities fit more or less easily into the trilaurin crystal lattice. In a situation such as this, the moving front of the growing trilaurin crystal causes a concentration gradient of the impurity in the melt such that some of the impurity is excluded from the crystal and must move away from the moving interface, while other molecules are incorporated into the trilaurin lattice structure. This causes both internal and external effects on trilaurin growth. The possibility of changing polymorphic stability in this situation also must be considered. Even in this relatively simple model system, the situation is quite complex with several effects occurring simultaneously. In real lipid systems, the complex effects are compounded by the variety of both naturally occurring and added components.

SUMMARY

Once crystals have formed in a product or process, they will continue to grow until the thermodynamic equilibrium between the fluid and crystalline states has been attained. This thermodynamic equilibrium is given by the phase diagram, from which the end-point conditions can be determined as well as the total amount of crystalline phase volume.

It is important to understand the factors that influence crystal growth rates in foods so that the nature of the crystalline structure (size, number, shape, polymorph, etc) can be controlled. It is this crystalline structure that influences texture and quality of many food products, and it allows for the most efficient separations in refining operations.

Many parameters can influence the rate of crystal growth. As is the case for nucleation, the process conditions (rates of heat and mass transfer) and the formulation (additives or impurities) are the primary parameters that influence crystal growth. Process parameters such as temperature of crystallization, rate of heat transfer, and agitation rates have significant effect on crystal growth rates. Also, the presence of various components/ingredients or impurities in a food formulation can have a dramatic impact on crystal growth rates.

Our understanding of the general principles and mechanisms of crystal growth is quite advanced, although the application of these principles to foods is not always complete. For example, our understanding of the synergistic effects of the various ingredients in complex formulations is extremely limited. Also, the more extreme processing conditions found during processing (agitation, shear, high concentrations, etc) are often not duplicated in laboratory experiments. More fundamental experiments designed to determine rate-limiting effects on crystal growth at conditions of importance in food processing should be performed. In this way, the size and shape of crystals can be controlled through proper choice of the parameters that influence growth.

REFERENCES

1. Mullin JW. *Crystallization*. 2nd ed. London: Butterworth; 1993.
2. van Hook A. Growth of sucrose crystals: a review. *Sugar Technol Rev*. 1981;8(1):41.
3. Fennema OR, Powrie WD, Marth EH. *Low-Temperature Preservation of Foods and Living Matter*. New York: Marcel Dekker; 1973.
4. Omran AM, King CJ. Kinetics of ice crystallization in sugar solutions and fruit juices. *Am Inst Chem Eng J*. 1974; 20(4):795–803.
5. Hartel RW. Mechanisms and kinetics of recrystallization in ice cream. In: Reid DS, ed. *The Properties of Water in Foods: ISOPOW 6*. London: Blackie; 1998:287–319.
6. Dirksen JA, Ring TA. Fundamentals of crystallization: kinetic effects on particle size distributions and morphology. *Chem Eng Sci*. 1991;46(10):2389–2427.
7. Doremus RH. *Rates of Phase Transformations*. Orlando, FL. Academic Press; 1985.
8. Strickland-Constable RF. *Kinetics and Mechanisms of Crystallization*. New York: Academic Press; 1968.
9. Perry RH, Chilton CH. *Chemical Engineers' Handbook*. New York: McGraw-Hill; 1973.
10. Levins DM, Glastonbury JR. Particle-liquid hydrodynamics and mass transfer in a stirred vessel, Part II—mass transfer. *Trans Inst Chem Eng*. 1972; 50:132–146.
11. Treybal RE. *Mass-Transfer Operations*. 3rd ed. New York: McGraw Hill; 1980.
12. van Hook A. *Crystallization: Theory and Practice*. New York: Reinhold; 1961.
13. Burton WK, Cabrera N, Frank FC. The growth of crystals and the equilibrium structure of their surfaces. *Philos Trans R Soc Lond*. 1951;A243:299.
14. Boistelle R. Fundamentals of nucleation and crystal growth. In: Garti N, Sato K, eds. *Crystallization and Polymorphism of Fats and Fatty Acids*. New York: Marcel Dekker; 1988:189–226.
15. Sato K. Crystallization of fats and fatty acids. In: Garti N, Sato K, eds. *Crystallization and Polymorphism of Fats and Fatty Acids*. New York: Marcel Dekker; 1988:227–266.
16. Ulrich J. Melt crystallization. In: Myerson AS, ed. *Handbook of Industrial Crystallization*. Stoneham, MA: Butterworth-Heinemann; 1993:151–164.
17. Kirwan DJ, Pigford RL. Crystallization kinetics of pure and binary melts. *Am Inst Chem Eng*. 1969;15:442–449.
18. Margolis G, Sherwood TK, Thibaut Brian PL, Sarofim AF. The performance of a continuous well stirred ice crystallizer. *Industrial Engineering Chemistry Fundamentals*. 1971;10(3):439–452.
19. Visser RA. A natural crystal growth retarder in lactose. *J Dairy Sci*. 1980;34:255.
20. Smythe BM. Sucrose crystal growth: III. The relative growth rates of faces and their effect on sucrose crystal shape. *Aust J Chem*. 1967;20:1115–1131.
21. Randolph AD, Larson MS. *Theory of Particulate Processes* 2nd ed. New York: Academic Press; 1988.
22. Hartel RW, Shastry AV. Sugar crystallization in food products. *CRC Crit Rev Food Sci Nutr*. 1991;30:49–112.
23. Shanks BH, Berglund KA. Contact nucleation from aqueous sucrose solutions. *Am Inst Chem Eng J*. 1985;31(1):152.
24. Liang BM, Hartel RW, Berglund KA. Contact nucleation in sucrose solutions. *Chem Eng Sci*. 1987;42(11):2723.
25. Liang BM, Hartel RW, Berglund KA. Growth rate dispersion in seeded batch sucrose crystallization. *Am Inst Chem Eng J*. 1987;33(12):2077.
26. Shi Y, Hartel RW, Liang BM. Formation and growth phenomena of lactose nuclei under contact nucleation conditions. *J Dairy Sci*. 1989;72(11):2903.
27. Bhat HL, Sherwood JN, Shripathi T. The influence of stress, strain and fracture of crystals on the crystal growth process. *Chem Eng Sci*. 1987;42(4):609–618.
28. Valcic AV. The influence of dislocations on the growth rate of saccharose crystals. *J Crystal Growth*. 1975;30:129–136.
29. van Hook A. Observations on the dissolution of sucrose crystals. *Int Sugar J*. 1988;90:65–67.
30. Fabian J, *Untersuching zur Auflosungskinetik in der Losungkristallisation* [in German]. Bremen: Universitat Bremen, Germany; 1996. PhD dissertation.

31. Fabian J, Hartel RW, Ulrich J. Growth and dissolution rate dispersion of sucrose crystals. In: Myerson AS, Green DA, Meenan P, eds. *Crystal Growth of Organic Materials*. Washington, DC: American Chemical Society; 1996:216–219. American Chemical Society Conference Proceedings Series.
32. Ben-Yoseph E. *Computer Modeling of Sucrose Crystallization during Drying of Thin Sucrose Films*. Madison, WI: University of Wisconsin–Madison; 1999. PhD dissertation.
33. Wey JS. Basic crystallization processes in silver halide precipitation. In: Wilcox WR, ed. *Preparation and Properties of Solid State Materials*. New York: Marcel Dekker; 1981:6.
34. Zhmyria LP. Concentration and temperature ratios of coefficients of mutual diffusion. *Izd Krasnodarskogo ita Pishchevo I Promyshlennosti*. 1972;2:125–128.
35. Timms RE. Crystallization of fats. In: Hamilton RJ, ed. *Developments in Oils and Fats*. London: Blackie; 1995: 204–223.
36. van Putte KPAM, Bakker BH. Crystallization kinetics of palm oil. *J Am Oil Chem Soc*. 1987;64:1138.
37. Smythe BM. Sucrose crystal growth. *Sugar Technol Rev*. 1971;1:191–231.
38. Smythe BM. Sucrose crystal growth rate: I. Rate of crystal growth in pure solutions. *Aust J Chem*. 1967;20:1087–1096.
39. Smythe BM. Sucrose crystal growth rate: II. Rate of crystal growth in the presence of impurities. *Aust J Chem*. 1967;20:1097–1114.
40. Maurandi V. Correlations between the growth kinetics of sucrose crystals and other crystals. *Zuckerindustrie*. 1984; 109(3):233.
41. Twieg WC, Nickerson TA. Kinetics of lactose crystallization. *J Dairy Sci*. 1968;51(11):1720.
42. Brinkman GE. New ideas for the utilization of lactose—principles of lactose manufacture. *J Soc Dairy Technol*. 1976;29(2):101.
43. Haase G, Nickerson TA. Kinetic reactions of alpha and beta lactose. I. Mutarotation. *J Dairy Sci*. 1966;49:127.
44. Budiaman ER, Fennema O. Linear rate of water crystallization as influenced by viscosity of hydrocolloid suspensions. *J Dairy Sci*. 1987;70(2): 547–554.
45. Muhr AH, Blanshard JMV. Effect of polysaccharide stabilizers on the rate of growth of ice. *J Food Technol*. 1986; 21:683–710.
46. van Hook A. *Kinetics of sucrose crystallization Industrial Engineering Chemistry*. 1945;37(8):782.
47. van Hook A. Growth rate curves of sucrose crystals. *Zuckerindustrie*. 1984;109(7):638.
48. Zagrodski S. Influence of the amount of sugar crystals present in a solution on the crystallization rate. *Zucker*. 1968;21:658.
49. Kasputin AP. *The Effects of Ultrasound on the Kinetics of Crystallization*. Moscow: Institute Crystallography; 1962.
50. Gonzalez PV, Ramirez M. Crecimento de la sacarosa en presencia de un campo electrico [in Spanish]. *Revue Cubana Fisica*. 1983;3(3):21.
51. Vaccari G, Mantovani G, Sgualdino G, Aquilano D, Rubbo M. The raffinose effect on sucrose morphology and kinetics. *Sugar Technol Rev*. 1986;13(3):133.
52. Nickerson TA, Patel KN. Crystallization in solutions supersaturated with sucrose and lactose. *J Food Sci*. 1972; 37:693–697.
53. Tjuradi P, Hartel RW. Effects of corn syrup oligosaccharides on sucrose crystallization. *J Food Sci*. 1995;60(6): 1353–1356.
54. Michaels AS, van Kreveld A. Influences of additives on growth rates in lactose crystals. *Neth Milk Dairy J*. 1966; 20(3):163.
55. Kelly FHC, Mak FK. Sucrose crystal elongation. *Zeitschriff Zuckerindustrie*. 1977;27(3):138–144.
56. Leonard GJ, Richards GN. Polysaccharides as causal agents in production of elongated sucrose crystals from cane juice. *Int Sugar J*. 1969;71(849):263–267.
57. Mak FK. Ionic impurities and habit modification in sucrose crystals. *Zuckerindustrie*. 1978;103(11):947.
58. Jelen P, Coulter ST. Effects of certain salts and other whey substances on the growth of lactose crystals. *J Food Sci*. 1973;38(7):1186.
59. van Hook A. Events in sugar crystallization. *Zuckerindustrie*. 1988;113(7):591
60. Leviton A. Adsorption of riboflavin by lactose. *Industrial Engineering Chemistry*. 1943;35(5):489.

61. Herrington TM, Branfield AC. Physico-chemical studies on sugar glasses. I. Rates of crystallization. *J Food Technol.* 1984;19:409–425.

62. van Scoik KG. *Nucleation and Crystallization Phenomena in Amorphous Sucrose Systems.* Madison, WI: University of Wisconsin; 1987. PhD dissertation,

63. Roos YH, Karel M. Phase transitions of mixtures of amorphous polysaccharides and sugars. *Biotech Prog.* 1991; 6:159–163.

64. Saleki-Gerhardt A, Zografi G. Nonisothermal and isothermal crystallization of sucrose from the amorphous state. *Pharm Res.* 1994;11:166–173.

65. Gabarra P, Hartel RW. Corn syrup solids and their saccharide fractions affect crystallization of amorphous sucrose. *J Food Sci.* 1998;63(3):523–528.

66. Mathlouthi M. Amorphous sugar. In: Mathlouthi M, Reiser P, eds. *Sucrose: Properties and Applications.* Glasgow: Blackie; 1995:75–100.

67. Gabarra P. *The Effects of Corn Syrup Solids and Their Oligosaccharide Fractions on the Glass Transition and Crystallization of Amorphous Sucrose.* Madison, WI: University of Wisconsin–Madison; 1996. MS thesis.

68. Blanshard JMV, Muhr AH, Gough A. Crystallization from concentrated sucrose solutions. In: Levine H, Slade L, eds. *Water Relationships in Foods.* New York: Plenum Press; 1991:639–655.

69. Shirai Y, Nakanishi K, Matsuno R, Kamikubo T. On the kinetics of ice crystallization in batch crystallizers. *Am Inst Chem Eng J.* 1985;31(4):676–682.

70. Rapatz G, Luyet B. Patterns of ice formation and rates of growth in gelatin solutions. *Biodynamica.* 1972;11:117–123.

71. Holt C. The effect of polymers on ice crystal growth. In: Bald WB, ed. *Food Freezing: Today and Tomorrow.* London: Springer-Verlag; 1991:81–86.

72. Blanshard JMV, Franks F. Ice crystallization and its control in frozen-food systems. In: Blanshard JMV, Lillford P, eds. *Food Structure and Behaviour.* London: Academic Press; 1987:51–65.

73. Harrison K, Hallett J, Burcham TS, Feeney RE, Kerr WL, Yeh Y. Ice growth in supercooled solutions of antifreeze glycoproteins. *Nature.* 1987;328(6127):241–243.

74. Charoenrein S, Goddard M, Reid DS. Effect of solute on the nucleation and propagation of ice. In: Levine H, Slade L, eds. *Water Relations in Foods* New York: Plenum Press; 1991:191–198.

75. Roos YH. *Phase Transitions in Foods.* San Diego: Academic Press; 1995.

76. Garti N. Effects of surfactants on crystallization and polymorphic transformation of fats and fatty acids. In: Garti N, Sato K, eds. *Crystallization and Polymorphism of Fats and Fatty Acids.* New York: Marcel Dekker; 1988:267–303

77. Smith PR, Cebula DJ, Povey MJW. The effect of lauric-based molecules on trilaurin crystallization. *J Am Oil Chem Soc.* 1994;71(12):1367–1372.

Chapter 7

Controlling Crystallization

In order to truly control crystallization in foods, a sound understanding of the principles underlying crystallization is necessary. However, even that is no guarantee of success as the process of crystallization is often extremely complex, with many factors that can influence the final results. In addition, the food manufacturer does not always have access to some of the tools available for controlling crystallization due to product limitations. For example, use of certain ingredients that could modify crystallization may not be allowed under the code of federal regulations (standard of identity) for a product. Alternatively, use of a desired ingredient for controlling crystallization may produce an undesired taste or texture in the product. Thus, controlling crystallization in foods may be quite challenging at times.

In foods, several circumstances for controlling crystallization can be distinguished (Table 7–1). In the first case, control of the correct number, size (and distribution), shape, and polymorph of crystals is required to provide the desired processing characteristics, quality (texture, flavor, etc), appearance, and shelf stability of the product. Here, the kinetics of nucleation and growth must be controlled—through proper choice of formulation and processing conditions—to give the desired crystalline structure. Controlling crystallization in sugar products (sugar-frosted cereals, fondants, panned candies, caramels, etc), frozen foods (ice cream, frozen desserts, and other frozen products), and lipid-based products (tempering of chocolate, butter, margarine, shortening, etc) are examples of products where a certain crystal size distribution gives the desired attributes. In many of these products, numerous small crystals are desired to give the proper texture and structure to the product. However, in other processes, fewer but larger crystals are desired with proper size distribution for efficient separation. Refining of sugars, fractionation of fats, and freeze concentration are examples of processes where control of crystallization is necessary to provide efficient separation. In both types of products and processes, control of the crystal size distribution is important to a high-quality product.

In other products, controlling crystallization means preventing crystallization from occurring even though the product is thermodynamically prone to crystallize. In some products, a thermodynamic driving force for crystallization exists, but crystallization is undesired because its appearance would cause a defect in the product. Some examples of products in this category include hard candies, some ungrained caramels, milk and whey powders, honey, and ice cream. In these products, crystallization of the sugars leads to an

Table 7–1 Characteristics of Food Products Where Controlling Crystallization Is Important

Product	*Ingredient*	*Desired Characteristic*
	Control Number, Size, Shape, and Polymorph	
Caramel	Sucrose or lactose	Smooth and short texture
Fondant	Sucrose	Smooth texture
Panned confections	Sucrose	Brittle texture (hard panned) Soft texture (soft panned)
Cereal coatings	Sucrose	Appearance
Refined sugars	Sucrose, lactose, etc	Size distribution for separation
Refined salt	Sodium chloride	Size distribution for separation
Organic acids	Citric acid, etc.	Size distribution for separation
Bread, other baked product	Starch	Texture
Frozen foods	Ice	Thawed quality
Frozen desserts	Ice	Smooth texture
Freeze concentration	Ice	Size distribution for separation
Chocolate	Cocoa butter	Texture (snap) and gloss Shelf stability to blooming
Butter	Milk fat	Hardness, spreadability
Margarine	Vegetable fats	Hardness, spreadability
Peanut butter	Vegetable fats	Texture, spreadability
Fat fractionation	Palm oil, tallow, milk fat	Size distribution for separation
	Control To Prevent Crystallization	
Hard candy	Sugars, polyols	Rapid dissolution, dispersion of color and flavor
Honey	Glucose	Undesired appearance, texture
Powders	Sucrose, lactose, etc	Flow properties
Ice cream	Lactose	Prevent sandiness defect
Cheese	Calcium lactate	Undesired appearance
Wine	Potassium tartrate	Undesired appearance

undesirable texture, flavor loss, and/or poor appearance. The formation of calcium lactate crystals in some cheeses also is undesirable as it leads to an undesirable appearance.[1–3]

One final aspect of controlling crystallization in foods occurs during storage. Subtle changes in the nature of the crystalline material during storage can sometimes lead to product defects. Since no food products are in equilibrium, changes in crystalline characteristics are likely to occur as the system approaches a different energy minimum (ie, surface energy or internal energy). For example, cocoa butter crystals in chocolate approach a lower energy state by converting to a more stable polymorph during storage. Also, ice crystals in ice cream and frozen foods approach a lower surface energy state through a thermodynamic ripening process. Crystallization of starches after manufacture of bread and some extruded products also may fit into this category. These changes during storage are often called recrystallization, a topic which will be covered in greater detail in Chapter 8.

CONTROL TO PRODUCE DESIRED CRYSTALLINE STRUCTURE

Control for Product Quality

In many food products, the desired crystalline structure provides a characteristic texture to the product. The crystals influence texture by disrupting the continuous phase in which they are embedded. Ice cream or frozen desserts are good examples of this, where the ice crystal structure provides a certain characteristic smoothness and mouthfeel desired in the product. Sugar crystals in a grained caramel also demonstrate the effect of crystallization on product texture. In these products, many small crystals of uniform shape and size are desired. In the case of grained caramel, the presence of sugar crystals results in a "short" texture, where the caramel matrix is broken by the dispersed phase. If the crystals are too large, they can be detected in the mouth, and the product is regarded as coarse. Crystal shape also may be important in some cases, although the effect of different shapes on product texture is not fully understood.

Probably the most important step in the crystallization process for controlling the crystal size distribution is nucleation. By controlling nucleation, the desired crystalline microstructure can be attained. An overlay of the crystal growth and nucleation kinetic curves (using primary heterogeneous nucleation as the mechanism of crystal formation) is shown in Figure 7–1. The point at maximum nucleation rate also corresponds to rapid crystal growth, somewhere in the middle of the crystallization zone between the equilibrium point (either solubility or melting point) and the glass transition point. At conditions both above and below this maximum, fewer nuclei are formed and they grow more slowly. The point on this rate curve where nuclei are formed during manufacture of the product is critical to the nature of the size distribution generated.

If nucleation occurs under conditions in the process that are at or near the peak of the nucleation rate curve, then the maximum number of crystals will be formed. To obtain many small crystals in the product, nucleation must occur at process conditions (temperature and concentrations) correlated with this maximum rate. If nucleation takes place at some condition off the peak in the nucleation rate curve (ie, temperature or supersaturation too high or too low), fewer crystals are formed and these ultimately grow to a larger size. Another important condition for producing many small crystals of uniform size is that nucleation takes place all at the same time. That is, process conditions must be set such that the entire mass of product attains the desired temperature and concentration at the same time. If the duration of nucleation lasts for an extended period of time, then a wider range of crystal sizes will be produced. The crystals that nucleate first have the opportunity to grow to larger size than those that form later in the process. Thus, process conditions that produce the maximum number of nuclei in the shortest nucleation time will result in a crystal size distribution with small mean size and a narrow distribution of sizes. In some processes, flow conditions are not ideal so that a portion of the material attains

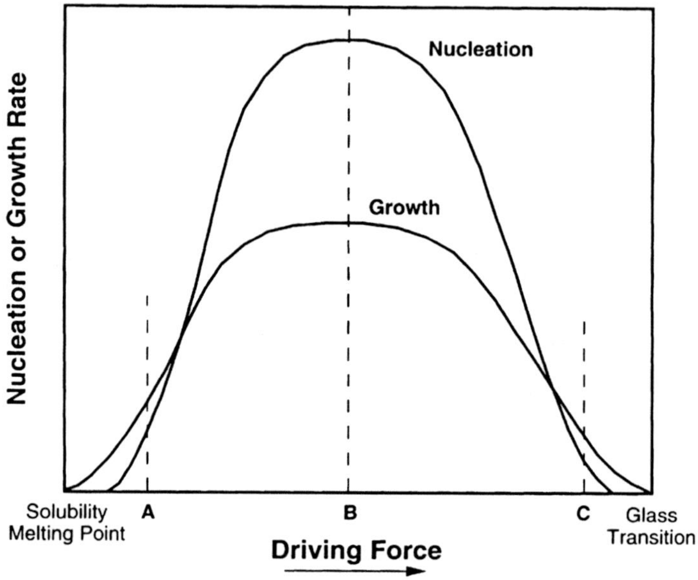

Figure 7–1 Schematic Diagram of Nucleation and Growth Rate Curves.

supersaturated conditions well before the bulk of the mass does (as in the colder material closer to cooling surfaces). In this case, the crystals that form in the regions of nonideal conditions can grow to significantly larger size than the bulk of the crystals and result in a size distribution that is skewed toward larger sizes (a log-normal type distribution). If sufficient numbers of particles are much larger than the mean, the product may be perceived as coarse or grainy. In extreme cases, the consumer may perceive very large crystals to be particulates of undesirable material (eg, glass).

A comparison of size distributions attained when nucleation occurs at different points on the rate curves is shown in Figure 7–2. When nucleation takes place at point B of Figure 7–1, the maximum of the nucleation rate curve, a narrow crystal size distribution with many small crystals results (bottom half of Figure 7–2). This is ideal for products that contain crystals, but where a smooth texture is still desired. Nucleation at point A in Figure 7–1, a condition with lower driving force, results in fewer crystals being formed, and they can all grow to larger size (for the same crystal phase volume), as seen in the top half of Figure 7–2. The same distribution is obtained if nucleation occurs at point C in Figure 7–1, although the time required for completion of crystallization is much longer due to the slower growth rate. These products have a distinctly coarser crystalline structure and do not have the same desired quality.

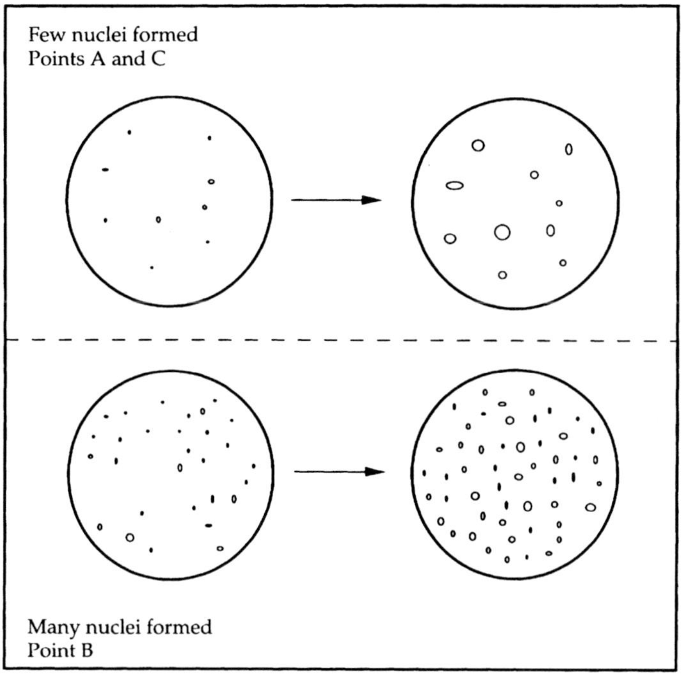

Figure 7–2 Schematic Diagram of Crystal Size Distributions Generated by Nucleation at Different Points on the Crystallization Rate Curves. Points A, B, and C correspond to points on Figure 7–1.

If nucleation occurs over a broad time span in a process, the crystal size distribution is generally skewed to larger sizes, as shown in Figure 7–3. The mean size of the distribution may be sufficiently low, but the presence of significant numbers of larger crystals may cause this product to be perceived as grainy or coarse. In general, the wider the range of time over which nucleation occurs, the wider the resultant size distribution.

Another way to think about controlling the relative rates of nucleation and growth is in terms of a driving force that must be dissipated by formation of crystalline structure. If the bulk of desupersaturation occurs in nucleation, the resulting average crystal size will be small. However, if only a small portion of the supersaturation is relieved during nucleation, and most of it is taken up in crystal growth, the crystals will be few in number and large in size. This is shown schematically in Figure 7–4. If nucleation occurs rapidly and supersaturation decreases rapidly, as shown by the curve that starts at $(C_i)_1$ in Figure 7–4, many crystals will be formed and the mean size will remain small. In contrast, if nucleation occurs slowly, the curve that starts with $(C_i)_2$ in Figure 7–4, there will be fewer

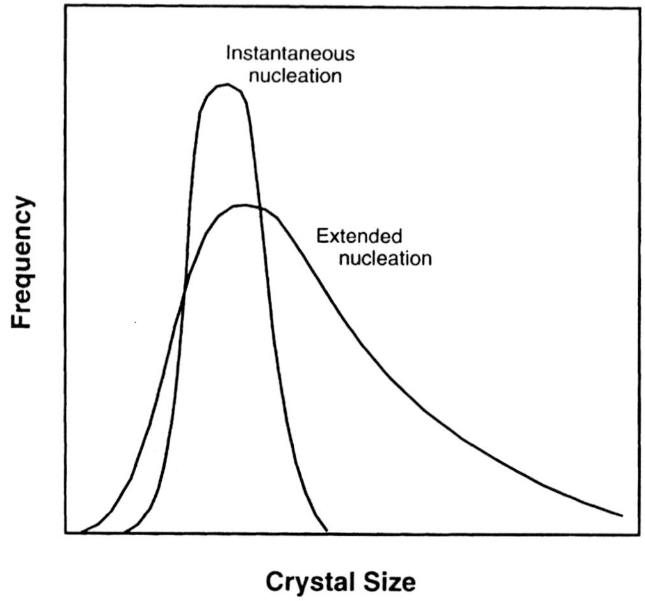

Figure 7–3 Schematic Depiction of Differences in Crystal Size Distribution When Nucleation Is Instantaneous or Occurs over an Extended Time Period.

crystals formed and these will grow to larger size. The nature of the desupersaturation curve in Figure 7–4 depends on many factors, as discussed in Chapter 5. These factors include initial supersaturation (which is shown in Figure 7–4), extent of agitation, and the presence of impurities that either promote or inhibit nucleation. Thus, each of these parameters plays an important role in controlling crystallization for generation of the desired crystal size distribution (CSD).

For lipid products, controlling crystallization also means controlling the polymorph formed during processing. Less stable polymorphs have nucleation rate curves (see Figure 5–14) that give maxima at lower temperatures. The most stable polymorphs may not nucleate very rapidly at any temperature. Thus, it may be necessary to nucleate a lower-stability polymorph to get sufficient numbers of crystals and then raise the temperature slightly to promote rapid transformation to the more stable, desired polymorph. This is exactly what happens during tempering of chocolate.

Control for Separation

In some cases, controlling crystallization to provide the most efficient separation is the desired goal. In this case, crystallization conditions that produce large mean size and

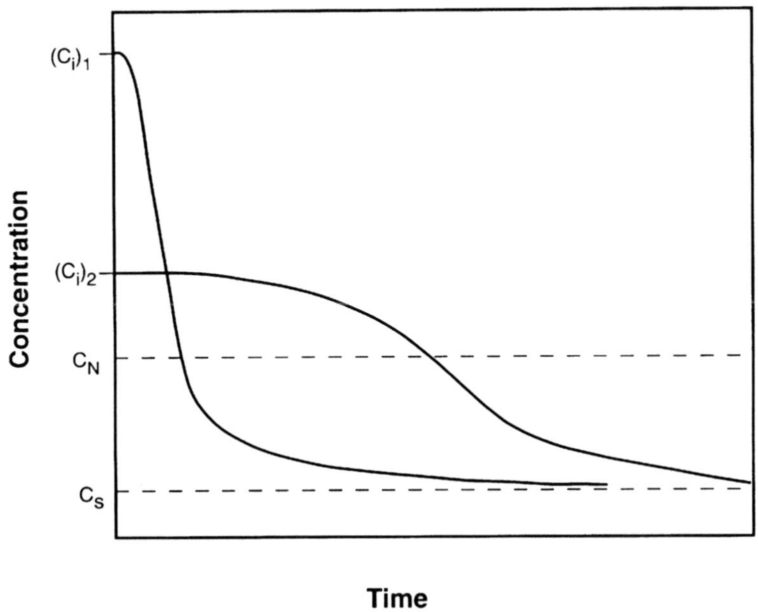

Figure 7–4 Different Rates of Desupersaturation for Solutions with Different Starting Concentrations (C_i). Concentrations above some critical point (C_N) result in nucleation whereas concentrations below C_N are insufficient to induce nucleation. C_S is solubility concentration.

---

a narrow distribution of sizes generally result in most efficient separation by filtration. Conditions that result in distributions equivalent to point A in Figure 7–1 with nucleation occurring over a narrow time period generally result in the crystal size distributions that give the desired separation efficiency.

Fat Fractionation

Fractionation is important for processing such fats as tallow, palm oil, palm kernel oil, and milk fat to produce components with desired physical and chemical composition. Typically, the high-melting triacylglycerol components are separated from the low-melting triacylglycerols by selective crystallization controlled by temperature changes. In this case, nucleation is controlled by cooling to the desired temperature to produce the desired nuclei (few in number) and then allowing them to grow to sufficiently large size for efficient separation. With lipids, it is important to recognize the different nucleation rates of the different polymorphs. Typically, less stable polymorphs nucleate at lower temperatures, and this nucleation is more rapid than for the more stable polymorphs (Chapter 5). In some cases, fats are fractionated with the assistance of a solvent. For example, acetone may be added to help dissolve the fat, promote crystallization, and provide more efficient separation (reduced viscosity during filtration). However, potential solvent residual in the product limits the use of this technology in the food industry.

There have been numerous suggestions for processing conditions to control lipid crystallization during fat fractionation. Generally, the process is as follows. The molten fat is cooled to a temperature just below the melting point of the desired polymorph, but not so low that massive spontaneous nucleation occurs. At this point, sufficient agitation is required to promote nucleation. The formation of a crop of nuclei can be observed by a slight increase in temperature. After sufficient nuclei have been formed, the temperature is lowered slowly to promote crystal growth without further generation of nuclei. Agitation should be kept to a minimum to prevent secondary nucleation. As temperature decreases, the mass of crystalline material (yield) increases. However, in general, the lower the temperature (higher yield), the less efficient the separation between high- and low-melting components of the fat.

Freeze Concentration

Freeze concentration of aqueous solutions is often the best concentration technique for providing optimal product quality of heat-sensitive foods. Low-temperature operation means that no thermal degradation reactions occur, and it is widely recognized that the flavor quality of freeze concentrated fruit juices, for example, is superior to the flavor of evaporated products.[4] In products containing alcohol, freeze concentration is one potential technique for concentrating the aqueous phase without removal of alcohol. Ice beers are made in this way. Many different freeze concentration technologies have been developed over the years, although the method generally applied to food freeze concentration involves a two-step process for production of ice crystals suitable for separation.[5] Rather than limiting the production of ice, which crystallizes readily and is often difficult to control, the commercial freeze concentration unit involves generation of many small nuclei in a scraped surface heat exchanger. These small crystals are then adiabatically ripened to large size in a separate growth tank.[6,7] The resulting crystals, with mean size of 200 to 300 μm, are efficiently separated in a wash column with very little residual solids remaining with the melted ice.

Sugar Refining

In sugar refining, control of nucleation is critical to manufacture of a product with the proper size distribution.[8] The proper number of initial nuclei (or seeds) is required to ensure that larger crystals are produced that are efficiently separated. In many cases, the nucleation step is bypassed by seeding with the appropriate crystalline material. In refining sucrose, the seed crystals most commonly used are powdered sugar produced by grinding product-sized crystals to the appropriate small size. In crystalline glucose production, a portion of the previous batch is retained in the crystallizer vessel to act as seeds, or a footing, for the next batch. In contrast, refining of lactose typically does not use seeding to generate the desired distribution, but rather the concentrated lactose solution is allowed to nucleate as it cools in the crystallizing tank. In all cases, however, the goal is to grow crystals uniformly up to product size through careful control of the growth conditions. After an initial seeding, or burst of nucleation in the case of lactose, no further nuclei formation is desired. If nucleation does occur during refining, typically due to uncontrolled secondary nucleation, a wider size distribution is the result; this may cause problems in subsequent separation. Thus, careful control of nucleation (or seeding) and growth is necessary to produce the desired crystal size distribution for optimal separation. However, even careful control of nucleation may not be enough to produce a narrow crystal size distribution since growth rate dispersion causes inherent widening of the distribution. The extent of widening due to growth rate dispersion is dependent on the nature of the seed crystals and the growth conditions in the crystallizer.

CONTROL TO PREVENT CRYSTALLIZATION

In some products, crystallization is undesired even though, thermodynamically, there is a driving force (supersaturation or subcooling) for crystallization to occur. Products in the glassy or amorphous state, like hard candies and dried sugar (fruit) powders, fit into this category. Also, the unfrozen phase of ice cream, which is supersaturated in lactose, may crystallize during storage, causing a sandy defect in the ice cream. Other products that may be supersaturated in sugars, but where crystallization is undesired, are caramel and similar ice cream toppings. In caramel, the aqueous phase contains sucrose and lactose, both of which may be supersaturated during storage at room temperature. Crystallization of lactose in caramels or ice cream toppings made with the addition of whey powders or skim milk solids is a problem to many confectionery manufacturers. These products may be satisfactory directly after processing and packaging, but may develop sufficient crystalline structure during storage to be considered grainy. The shelf life of these products is thus limited by the onset of crystallization and growth to a detectable size. Typically, reformulation is necessary, with reduced milk solids and/or sucrose levels (and increased corn syrup), to ensure stability of the product to graining for the desired shelf life.

Controlling conditions to prevent crystallization in products such as caramel generally requires one of two approaches. Either the system must have sufficiently low molecular mobility (high viscosity) so that the molecules cannot form into crystals, or the crystallizing species must be inhibited by addition of other molecules. Caramels provide an excellent example. Prevention of crystallization in some ungrained caramels is accomplished by increased viscosity. These products are solid-like in nature and may be found on a stick, for example. On a state diagram, these products are located in the vicinity of the glass transition curve (point A on Figure 7–5). These products are stable to crystallization for long periods of time as long as the product remains in this highly viscous state. The lower the viscosity of the product, and the farther into the crystallization zone this product extends, the shorter the time before crystallization occurs. Shelf life of these products, due to crystallization, is related directly to the position on the state diagram and the kinetics of crystallization at the storage conditions. Crystallization in other ungrained caramels may be prevented by addition of substantial amounts of ingredients like corn syrup. These caramels may be fluid and even runny at elevated temperatures. Typically, these

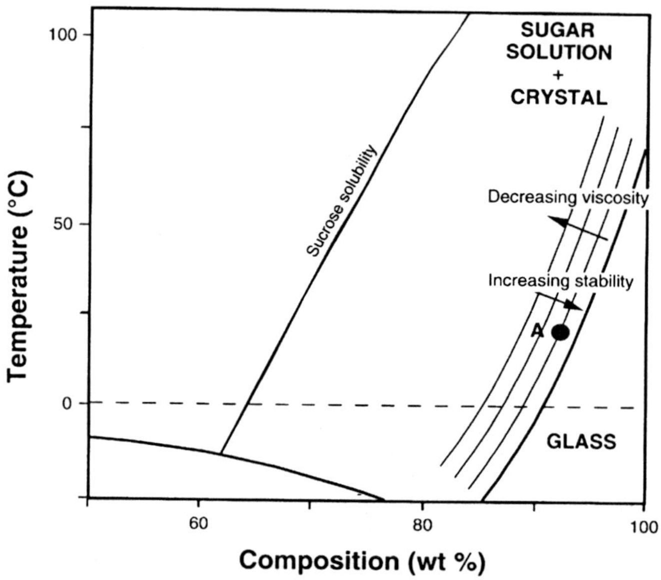

Figure 7–5 Stability of Caramel against Crystallization as Related to Viscosity on the State Diagram.

products are located within the crystallization zone of the state diagram but crystallization is inhibited through proper formulation. The kinetics for onset of nucleation are delayed substantially by addition of ingredients such as corn syrup and other sugars, as discussed in Chapter 5.

Crystallization of honey provides another interesting example of a product that is supersaturated, but where crystallization is undesired. Honey is supersaturated in glucose and it is the glucose content that is probably the main determinant in whether a particular honey crystallizes or not. Doner[9] notes that when the glucose/water ratio is less than 1.7, crystallization is substantially delayed, whereas honey with a glucose/water ratio greater than 2.1 usually crystallizes within weeks. However, the presence of fructose in honey has been shown to provide protection against crystallization.[10] Higher fructose contents delay onset of crystallization of honey. Even processing conditions have been shown to influence the likelihood of crystallization.[11] Crane[10] suggested that honey be pasteurized at 77°C for 5 minutes, cooled rapidly to 0°C, and stored at this temperature for at least 5 weeks before warming to room temperature. This process would produce honey that remained uncrystallized for over 2 years. Assil et al[11] studied the effects of filling temperature on crystallization of honeys and found that filling temperatures below 40°C led to a greater incidence of crystallization of that honey during storage. They suggested the effect was related to viscosity of the honey during filling. At filling temperatures above 40°C, the honey had low viscosity and flowed easily into the bottles. At temperatures that were less than 40°C, the honey was much more viscous, and bubbles were entrained in the honey during filling. They suggested that the entrained bubbles served as catalytic sites for nucleation of glucose. Assil et al[11] also suggested that moisture loss (and a subsequent increase in glucose supersaturation) during storage influenced crystallization of honey. Despite the previous work that has been done to understand the nature of glucose crystallization in honey, our understanding of all the factors that influence this process is still not complete.

CONTROL OF CHANGE IN CRYSTALLINE STRUCTURE

Subtle changes in crystalline structure occur in many foods during storage; these changes often lead to undesirable product characteristics. For example, ice crystals in ice creams and frozen foods gradually increase in size during storage, depending on the conditions, leading to products that the consumer rates as coarse and undesirable. Changes in lipid crystals—as in bloom formation in chocolate or sintering of fat crystals in spreads—or sugar crystals—as in coarsening of fondant—are other examples of this phenomenon.

The mechanisms that govern these changes are many and complex, but the overall process is often called recrystallization.[12] These changes are typically driven by a reequilibration to either a lower surface free energy or lower internal energy state without a change in the amount of crystalline mass, although in some cases the crystalline mass increases too. These changes can be controlled by understanding the effects of external conditions (eg, storage temperature) and internal conditions (due to formulation) on the mechanisms and kinetics of recrystallization. In Chapter 8, recrystallization is covered in detail.

FACTORS AFFECTING CONTROL OF CRYSTALLIZATION

There are a number of factors that influence crystallization, or prevention of crystallization, in food products. Processing conditions must be chosen for the desired effect—to either promote or prevent crystallization. These conditions depend on the interactions of the ingredients as the product is manufactured. Thus, formulation plays an important role in selection of the optimal processing conditions. Once the product has been made and packaged, any changes during distribution and storage influence shelf life. Proper storage conditions are necessary to maintain the desired state of the product and allow the longest possible shelf life.

Heat and Mass Transfer Rates

During processing, the rates of heat and mass transfer are important to provide proper control of crystallization. The rates of heating and cooling during the manufacturing process can impact crystallization as can the holding times and temperatures. If evaporation of water occurs during the process, this can impact crystallization as well. In fact, some processes require controlled evaporation (drying) to produce the desired product quality; the rate of this drying significantly influences crystallization. Also, mechanical energy input, whether intentional through forced mixing or unintentional through fluid shear forces, can promote crystallization (as discussed in Chapter 5).

The rate of cooling of a material to a desired crystallization temperature, as in production of fondant or crystallization of grained caramel and fudge, determines the extent and type of nucleation. If the mass is cooled uniformly and rapidly to the desired processing temperature, all of the mass will crystallize at the same condition and at the same time (Figure 7–4). This leads to the maximal production of nuclei and the narrowest distribution of crystal sizes. On the other hand, if cooling is slow or nonuniform, crystallization also will be nonuniform. In regions that cool more rapidly, nucleation takes place first; these crystals continue to grow as the rest of the mass crystallizes. The nuclei that form first grow larger than the other crystals and may result in a coarse texture. It is, therefore, important to ensure that processes are designed so that the entire mass cools rapidly and uniformly to the desired crystallization temperature.

If crystallization is not desired, cooling must be sufficiently rapid that the material passes through the crystallization zone on the state diagram before nucleation occurs. In this case, heat transfer must occur more rapidly than the time scale for onset of nucleation (induction time) in the product.

The temperature at which crystallization occurs also is important to control for the desired CSD and product texture. As discussed in Chapter 5, there is an optimal temperature for many systems where nucleation rate is at its maximum. This point of maximal nucleation is based on the balance between thermodynamic driving force (supersaturation) and mobility effects (glass transition). At this optimal temperature, the maximum number of crystals will be formed, resulting in formation of a narrow distribution with small mean size. This is generally the desired characteristic for products with smooth texture (caramels, ice cream, frozen foods, etc). For lipids, processing temperatures must be chosen to maximize formation of the desired polymorphic form. Since rapid nucleation does not occur for the most stable polymorphs, temperature and time conditions that allow rapid nucleation and proper polymorphic formation are necessary. This time-temperature protocol is called tempering, and it is a very important step in production, for example, of quality chocolates.

On the other hand, if larger crystals are desired for efficient separation, then nucleation must occur at a higher temperature (or lower supersaturation) so that fewer nuclei are formed. It still is important that the nuclei form at the same time so that they all will grow to about the same product size for efficient filtration. If crystallization is undesired in a product, cooling must be sufficiently rapid so that the mass passes through the crystallization zone to a point on the state diagram where induction times for nucleation are very long. The driving force at this point depends on formulation and other processing requirements. For example, forming of hard candies from an amorphous mass requires sufficient mobility that the desired shape of the candy can be attained during processing, but that nucleation does not occur. This intermediate temperature for processing is very critical to production of a stable product, but depends on the type of formulation used.

In products where massive nucleation is desired, it is important to provide sufficient agitation to promote maximal nucleation. As discussed in Chapter 5, mechanical energy input enhances the nucleation process. During processing of fondant, for example, the sugar mass is cooled to the desired crystallization temperature and then worked extensively in a beater to promote massive crystallization. This results in production of many small crystals and produces a smooth fondant. If nucleation must be avoided (as in hard candy), agitation during processing must be kept to an absolute minimum. Sometimes even the shear forces that exist as a supersaturated fluid is passed through valves or pipe fittings may be sufficient to initiate nucleation.

Some processes involve simultaneous crystallization and drying. Controlling crystallization in the sugar coating of a cereal product, for example, requires a careful balance between crystallization and drying since, in a sense, they are competitive processes. A similar balance is required during crystallization of sugar in thin films around a panned confectionery piece (jaw breaker, sugar-coated peanut, etc). In these processes, a concentrated sugar solution is sprayed or applied to the piece (whether a candy piece or a cereal base) under conditions where sugar crystals are formed. The film is then dried to produce a hard brittle shell in panned items or a sugar coating with desired appearance on a cereal piece. Crystallization causes a decrease in concentration in the solution phase, whereas drying causes concentration to increase as water is removed. Thus, drying influences the rate of crystallization and can significantly impact the product characteristics. Controlling crystallization in this case requires an understanding of the relative rates of drying and crystallization.[13]

Product Formulation

Choice of ingredients in a product formulation is based on many factors, but rarely is the formulation designed solely to control crystallization. In some cases, formulation may be governed by standards of identity or some other desired product characteristic (eg, all natural). However, the impact of formulation on crystallization kinetics must be taken into account when developing a process for a product. For example, the use of corn (or glucose) syrups in hard candy may be based on economic decisions (in the United States), but corn syrup also inhibits graining of the sugar mass during processing and storage.

One of the main factors affecting crystallization in solution systems is the water content, or total solids content. Water content is a determining factor for supersaturation and the subsequent crystallization rate kinetics as temperature is changed. Higher solids content generally indicates higher supersaturation, although this is dependent on the type of ingredients that are used.

As discussed in Chapters 5 and 6, addition of various soluble components in a formulation typically inhibits crystallization, although the effects on supersaturation or thermodynamic driving force also must be considered. The choice of ingredients in a formulation also may influence the glass transition zone. These effects are shown in Figure 7–6 for mixtures of lactose and sucrose. Addition of lactose and sucrose in solution causes a decrease in solubility concentrations of both components, as shown by the dashed

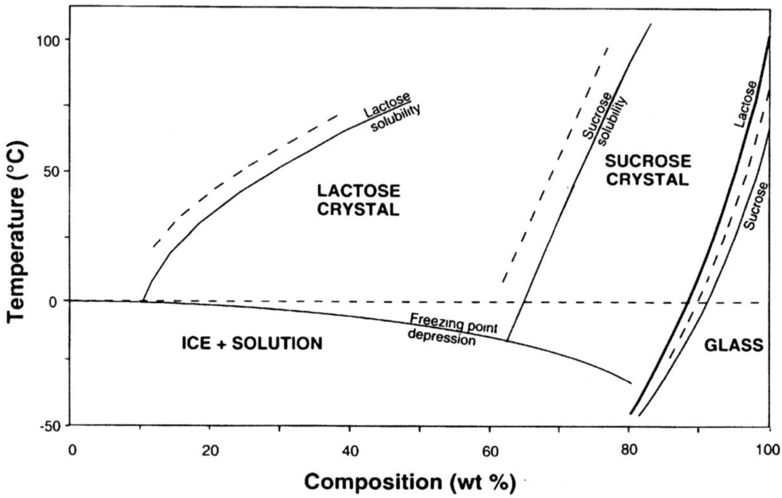

Figure 7–6 Effect of Mixtures of Lactose and Sucrose on the State Diagram. Dashed lines represent mixtures of sucrose and lactose.

lines in Figure 7–6. Since both lactose and sucrose have the same molecular weight, both components have the same effect on freezing point depression. However, lactose has substantially higher glass transition temperature (about 102°C) than sucrose (65 to 70°C). Mixtures of the two components have a glass transition temperature somewhere between these two curves, depending on the relative addition levels of each. The lines shown in Figure 7–6 represent approximate compositions for mixtures of lactose and sucrose found in typical caramels and in the unfrozen phase of ice cream under storage conditions. Note also that Figure 7–6 represents a simplification of the ternary phase diagram of a sucrose-lactose-water system.

The exact position of phase equilibrium curves and glass transition curves for complex food products is often not known. However, optimal control of crystallization processes in foods during processing and storage requires as much detail about this diagram as possible.

Post-Processing Effects

Even though a product has been manufactured and placed in a package, it is still possible that further changes in crystalline structure can occur. It is widely recognized in some products that crystallization is not complete for a period of time after packaging, perhaps even up to several weeks. The texture of some confectionery products, for example, is different at the point of packaging as compared with the texture of products sold to the consumer. These changes in crystalline structure may come about for several reasons.

First, it is possible that crystallization has not been completed during manufacture of the product. That is, the maximum phase volume of crystalline material, or full desupersaturation of the solution, may not be attained prior to packaging. The balance between time of processing and completion of crystallization must be weighed carefully in determining process conditions. Often, high product throughput rates result in products that are not fully crystallized during processing. However, as long as crystallization continues under reasonably controlled conditions, product quality will not suffer. In some frozen foods, particularly those that are frozen rapidly, it is possible that insufficient ice is formed in the process to relieve the temperature driving force for ice formation. If the product is frozen rapidly and stored at sufficiently low temperatures, only a little ice may form. Also, subsequent growth of ice is inhibited by the limited molecular mobility. On the state diagram, the product will be at a point well below the freezing point depression curve, indicating incomplete crystallization (point A on Figure 7–7). Warming of the product, either in subsequent processing steps or during storage, results in continued ice formation until eventually the system equilibrates in ice phase volume as the freezing point depression curve is reached at the new temperature (point B on Figure 7–7). Growth of the existing ice nuclei may cause product damage as the larger ice crystals begin to break cellular structure and cause damage to the texture. An example of incomplete freezing would be freezing of a food product, such as ice cream mix, by immersion in liquid nitrogen. In this case, the product is frozen so rapidly that there is not enough time to form an equilibrium amount of ice. That is, rapid freezing results in less ice than would be expected based on the phase diagram. A product with small ice crystals surrounded by a glassy matrix is produced, and this physical state remains as long as temperature is maintained below T_g' (the temperature denoting the maximally freeze concentrated solution). Preparation of samples for cryo-scanning electron microscopy (cryo-SEM) is another example of this process. However, when such products are warmed, the higher mobility of water molecules at the higher temperature allows continuation of ice crystallization.

Other examples of products where crystalline changes take place after processing are butters, margarines, and spreads. When butter comes off the processing line, it is still sufficiently fluid or pliable that it can be formed into the desired shape. After packaging, the butter is cooled to refrigeration temperature where further crystallization takes place. Lowering the temperature allows more of the butter to crystallize, which results in a significant hardening of the butter and gives the product some of its desired attributes. In this case, the additional heat of crystallization must be removed as the product approaches a phase equilibrium. If this heat is not removed effectively, melting and recrystallization can occur within the food product, causing degradation of texture. In general, any changes in CSD during packaging and storage must be taken into account for development of the desired overall product quality.

Other changes in crystalline structure take place during storage due to a thermodynamic ripening effect. These processes will be discussed in more detail in Chapter 8. Briefly, crystals in a food product may not be at their lowest thermodynamic equilibrium

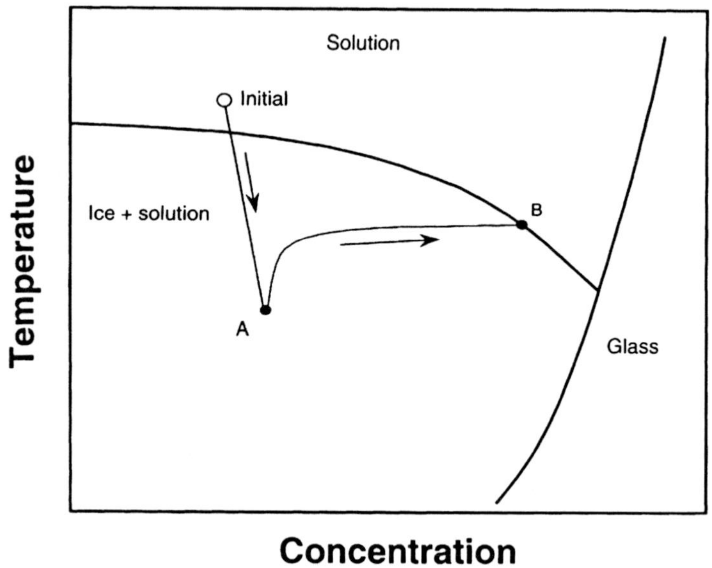

Figure 7–7 Freezing Rapidly (to Point A) Results in Less Ice Than Can Be Formed. Subsequent warming allows crystal growth to occur until phase volume equilibrium is attained (point B).

even though the phase equilibrium may have been attained. That is, changes in internal energy or surface free energy can still take place during storage that can cause significant changes in CSD or crystalline structure. These changes may affect the quality of the final product. Growth of ice crystals in ice creams and fat bloom in chocolates are two examples of this thermodynamic recrystallization process.

Changes in crystalline characteristics can take place in foods due to improper storage conditions. As temperature fluctuates, the equilibrium conditions within the product also change. For example, in frozen foods at equilibrium phase volume, the amount of ice present is defined by the freezing point depression curve. Thus, at any storage temperature, there is an equilibrium phase volume of ice that is related to the concentration of the unfrozen phase. When temperature changes, the ice must either melt or grow to maintain this equilibrium in phase volume (as shown for ice cream in Figure 4–21). In some circumstances, these changes can lead to dramatic loss of product quality. An extreme example occurs when ice cream is left out at room temperature for long periods of time. The outer layers melt due to the thermal exposure (heat shock); if the product is refrozen by simply placing it back in the freezer, ice crystallization is no longer controlled. The resulting product contains large ice crystals formed under uncontrolled conditions and with negative effects on product quality. A similar problem occurs when a chocolate bar melts during exposure to high temperature and then is allowed to cool again. The resulting crystallization is uncontrolled and unstable polymorphs form, which then transform to the more stable state and lead to bloom formation.

Another change that may occur during storage is equilibration in moisture content. All products have an equilibrium moisture content dependent on the relative humidity of the air around it. This equilibrium may be influenced by choice of packaging material, but there also may be significant moisture migration within different components of the same food product. This may arise from moisture equilibration between two physical regions of a complex food item (eg, between the sauce and crust of a frozen pizza) or through equilibration with the storage air conditions. In either case, moisture content changes may have significant impact on the crystalline structure of a food. As a food dries out, the concentration of the remaining aqueous phase increases with a subsequent increase in supersaturation. The system then moves toward a new equilibrium between crystals and liquid to maintain the solubility concentration in solution. If the product picks up moisture from the air, the reverse process occurs, where the crystals dissolve to maintain the equilibrium situation in solution.

Significant changes in CSD may occur during storage due to changes in both moisture content and temperature. This is especially true when temperatures or moisture contents fluctuate, and the system moves back and forth along an equilibrium curve (either the freezing point depression curve or the solubility curve). These fluctuations allow the crystals to reduce the surface free energy, where smaller crystals disappear and larger crystals grow. The result is an increasingly coarse CSD that can eventually lead to the end of shelf life in some products.

CONTROLLING CRYSTALLIZATION: SEVERAL EXAMPLES

There are numerous examples of how crystallization is controlled in food processing. Controlling crystallization of sugars, ice, lipids, starches, salts, and proteins in foods plays an important role in determining the quality attributes of these products. Product attributes of importance include texture and rheology as related to the crystalline (or noncrystalline) microstructure, flavor release (crystalline microstructure or glassy state), and appearance (frosted cereals or bloomed chocolates). The following examples provide an overview of how the crystallization principles discussed in this book apply to specific applications.

Fondant Manufacture

Fondant sugar is a confectionery ingredient used for producing frostings, icings, and creams for chocolate-covered pieces and a variety of other products. Fondant also is used as a source of seed crystals for manufacture of other products. For example, the crystalline microstructure in caramel and fudge can be developed by addition of the proper amount of fondant sugar. Fondant is a highly crystalline product, with about 50 to 60% of the mass in the form of crystalline sugar contained in a viscous concentrated sugar syrup matrix. Water content is typically between 8 and 12%, with all of the water contained in the syrup phase.

The texture of fondant is strongly dependent on the CSD, so control of crystallization is extremely critical. If the sugar crystals are too large, the fondant has a coarse texture and is unacceptable for use in other products. In fact, the threshold size for detection of sugar crystals in a fondant is somewhere between 10 and 20 μm.[14] A high-quality fondant will have the vast majority of sugar crystals smaller than 10 μm. Thus, controlling crystallization to ensure that most of the crystals are smaller than this size is very important.

The study of sugar crystals in fondant can be accomplished by light microscopy or light scattering. Optical microscopy involves dispersion of the sugar crystals in a suitable organic solvent to allow light transmittance through a thin film of sample. Solvent to fondant ratio may be about 50:50. Use of crossed polarizers enhances the observation of very small crystals. A light microscope with high resolution is needed to see the very smallest sugar crystals (less than 2 to 3 μm). Alternatively, the fondant can be dispersed in solvent to prepare a dilute suspension for measurement of particle size by using a light-scattering technique. These techniques were described in Chapter 3.

There are numerous technologies for production of fondant; however, they all involve basically the same steps. An example of a continuous fondant manufacturing process is shown schematically in Figure 7–8. To make fondant, sucrose and corn syrup are mixed at the appropriate concentrations. Then the syrup is heated to ensure that any sucrose crystals have been dissolved. In some equipment, heating and dissolving occur under pressure so that all available water goes to dissolving sucrose crystals. If there are any sucrose crystals present in the syrup, they will grow readily during processing and produce a coarse-textured fondant. Thus, the dissolution step is very important if crystalline sucrose is used in the formulation. Then the syrup is concentrated rapidly in an evaporator. Some systems use a microfilm evaporator that provides rapid (on the order of seconds) concentration. The concentration at this point is usually between 80 and 90% total solids. This hot concentrate must be cooled rapidly to the desired temperature for controlled crystallization. In some equipment, it is done on a rotary cooling drum that provides a high heat transfer surface area and results in cooling to 40 to 50°C in a matter of minutes. The rate of concentration and the rate of cooling are very important to controlling crystallization. If these rates are too slow, crystallization of sugar may occur in an uncontrolled manner during the process.

The temperature to which the fondant is cooled should be close to the maximum point on the nucleation rate curve (Figure 7–1). At this point, extensive agitation of the sugar mass is used to initiate massive crystallization of the sucrose. This may be accomplished in a pin mixer or an extruder that also pumps the crystallizing mass through the system. Residence time within this mixer or beater is on the order of minutes. In this case, the large supersaturation (*S* was estimated to be about 1.67[15]) is removed in a very short time interval through massive nucleation and very little supersaturation is available for growth (the steep curve in Figure 7–4). Thus, there are numerous crystals (estimated at 360×10^9 crystals per mL), the crystal size distribution is very narrow, and the mean size is very small (5 to 10 μm).[14] Also, there may be a slight temperature rise of several degrees in this beater due to the release of latent heat upon massive crystallization. When the mass exits the beater or extruder, the supersaturation has almost all been relieved so that only

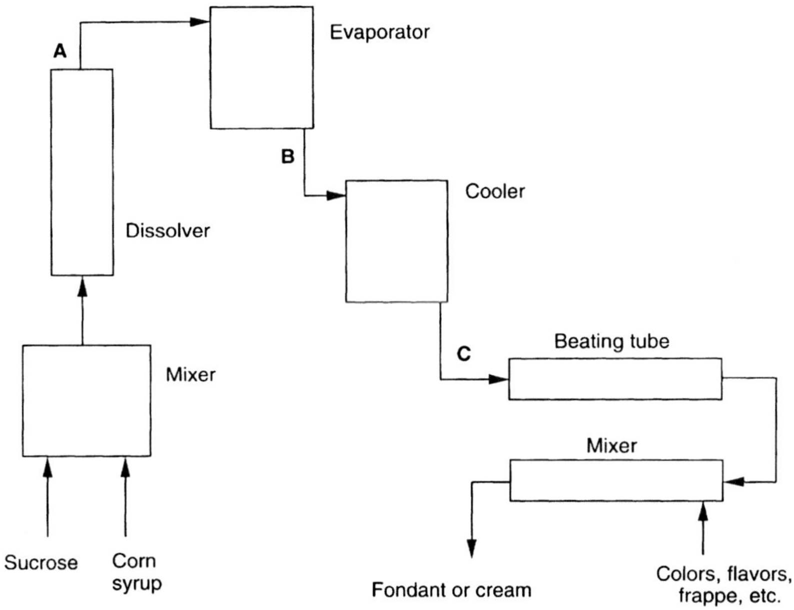

Figure 7–8 Schematic of a Fondant Manufacturing Process. Points A, B, and C correspond to points on the state diagram in Figure 7–9.

a slight amount of crystallization occurs thereafter. Sometimes, other ingredients such as colors, flavors, or frappes are added in a second beating unit prior to packaging or further processing.

The fondant manufacturing process may be followed on the state diagram, as shown schematically in Figure 7–9, to show how supersaturation is generated and relieved during the process. The initial starting point is a slightly undersaturated sugar syrup that is evaporated to produce a concentrated syrup with low moisture content (10 to 20%). This hot, concentrated syrup is cooled rapidly to the optimal beating temperature where the supersaturation is relieved within a few minutes. The product is cooled to room temperature for storage, and the system follows the solubility curve (approximately) with the associated increase in sucrose phase volume and mean crystal size. If the process has been controlled correctly, the product is a dispersion of uniformly fine sucrose crystals in a highly concentrated sucrose syrup. This structure produces a semi-solid material that can be shaped and formed into desired products, or warmed up and used as an ingredient in other products.

There are many factors that can influence crystallization in fondant. These may be divided into formulation and process factors. Choice of formulation, including the amount

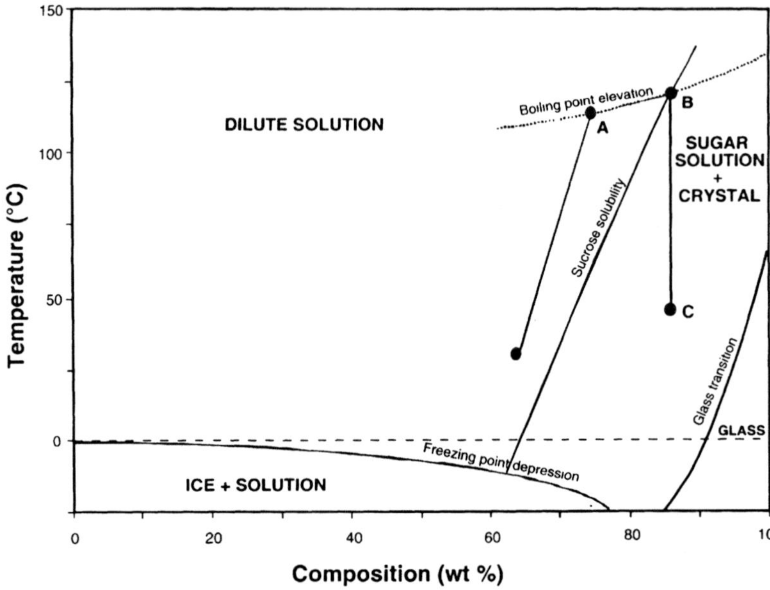

Figure 7–9 Path of the Fondant Manufacturing Process on the State Diagram. Points A, B, and C correspond to points in the fondant process shown in Figure 7–8.

of water in the final syrup and the ratio of sucrose to corn syrup, affects both the phase/state diagram and the crystallization kinetics at given process conditions. The total solids of the syrup prior to crystallization impacts the supersaturation at crystallization. The ratio of sucrose to corn syrup also affects the supersaturation since the addition of corn syrup depresses the solubility of sucrose, as discussed in Chapter 4. Corn syrup is added to fondant, in part, to control product texture. This effect is due to the decrease in crystal phase volume and increase in dissolved solids concentration in the solution phase with an increase in corn syrup content, as shown in Table 7–2.[15] The presence of corn syrup also affects the nucleation and growth rate kinetics, and it may influence the relative shapes of the kinetic curves (Figure 7–1). However, little data are available to quantify this effect. Some data showing the effects of corn syrup on nucleation and growth of sucrose have been shown previously (Figures 5–20 and 6–21, respectively), but the application of these data directly to the situation during fondant manufacture is difficult. Typically, data must be taken under the conditions directly of importance in the process of interest.

Processing factors also influence the ability to control crystallization during fondant manufacture. Of primary importance in this process are the rates of evaporation and cooling, the beating temperature and the extent of mechanical agitation during crystallization.

Table 7–2 Effect of Glucose Syrup Addition Level on Solid (Crystal) and Liquid (Syrup) Phases in Fondant with 10% Moisture at Room Temperature

Glucose Syrup (g Dry Solids/100 g Sucrose)	*Syrup Phase (%)*	*Solid Phase (%)*
0	30.0	70.0
5	32.8	67.2
10	35.6	64.4
15	38.3	61.7
20	40.8	59.2
25	43.1	56.9
30	45.3*	54.2

* Numbers do not total 100% in original data.

The presence of seed crystals, from improper processing of rework, for example, also may impact the final product distribution. The rates of evaporation and subsequent cooling are extremely important, especially for processing of fondants with high total solids. The concentration must be increased rapidly and the concentrated syrup cooled quickly so that no crystallization occurs until the sugar mass enters the beater. If either concentration or cooling is too slow, crystallization can take place prior to the beating device and at a temperature above the optimal point for maximal nucleation. The temperature at which mechanical agitation (beating) is applied is a critical variable for controlling crystallization. If this temperature is too high or too low, less than optimal nucleation will occur and there will be fewer and larger crystals present. Defining the optimal temperature, however, depends on the formulation since the specific mixture of ingredients produces a unique nucleation kinetic rate curve. Each formulation has a slightly different optimal temperature for beating, and this temperature can only be found by experimentation. Thus, finding this optimal beating temperature for a given formulation generally is based on a trial-and-error process or past experience.

Both formulation and processing conditions influence the composition of the fondant and the resulting CSD[14,16] as shown in Table 7–3. For a fondant with 10% water, formulations based on different corn syrups and beating temperatures were studied.[14] At each condition of corn syrup addition (10 and 25%) and beating temperature (40, 55, and 70°C), Table 7–3 gives the amount of crystalline and solution phases, the total solids in the solution phase, and a rough CSD. In these formulations, increasing corn syrup content caused a decrease in the amount of crystalline sucrose since the solubility concentration of sucrose is depressed in the presence of corn syrup. Also, the higher level of corn syrup results in high total solids concentration in the liquid phase because all of the corn syrup solids remain solubilized in the liquid phase along with soluble sucrose. The effect of decreasing the beating temperature was to produce smaller crystals, a result that also was seen by Desmarais and Ganz.[17] This was undoubtedly due to progression along the nucleation rate curve to a higher level of nucleation at the lower temperature. Lees[14] reported that beating at temperatures less than 30°C causes the product to become translucent and vitreous in nature. Thus, one would expect the peak in the nucleation rate curve to occur somewhere between 30 and 40°C for these formulations. Under the conditions of these formulations, the fondant made with higher corn syrup had smaller crystals than the fondant made with lower corn syrup. Several factors might influence this size distribution. First, the amount of solid phase is different between formulations; typically, higher phase volume results in larger crystals unless the nucleation rate is significantly different. This might explain the larger mean size for the fondant made with lower corn syrup (more sucrose). However, the missing piece of information to understand fully the results in Table 7–3 is the number of crystals. Again, this is related to the point on the nucleation rate curve where beating of the fondant mass occurred.

Numerous studies have investigated the effects of other ingredients on CSD in fondant. The addition of cellulose gum was shown to decrease the final crystal size in fondant[17] as did sorbitol.[18] In both cases, the final phase volume of sucrose crystals was decreased due to addition of these ingredients; thus, smaller crystals were obtained. In the case of cellulose gum, an additional inhibitory effect on sucrose crystal growth rate was likely.

The rate of agitation during beating also is important to final crystal size in fondant.[15] According to Lees,[14] distinctly larger mean size is obtained from slow beating. This is understandable in terms of the effects of agitation on nucleation. Higher agitation generally results in more rapid nucleation and formation of more crystals (which then remain smaller), and this finding is particularly true for systems with high viscosity where diffusion limitations must be overcome. Enhanced agitation allows the sugar molecules to migrate more easily and come together into a crystal lattice.

Tempering of Chocolate

Crystallization of cocoa butter must be carefully controlled to manufacture chocolate with the desired texture and quality characteristics and adequate shelf life.[19–21] During

Table 7–3 Effect of Crystallization Temperature on Crystal Size Distribution in Fondant Manufacture. Fondant made with 90% sucrose and 10% corn syrup in dry matter and 90% total solids.

Temperature	*Crystal Size Range (%)*			
(°C)	*0–10 μm*	*10–20 μm*	*20–30 μm*	*> 30 μm*
10% glucose syrup				
70	83.2*	12.2	3.6	1.1
55	89.1	9.7	1.1	1.1
40	92.3	7.4	1.1	
25% glucose syrup				
70	89.5	7.7	1.8	0.4
55	96.2	3.5	0.3	
40	99.9	0.2		

* Numbers do not add up to 100% in original data.

manufacture, chocolate is tempered through a careful temperature and time protocol that produces a large number of cocoa butter seed crystals of the stable polymorph (the β V form in this case). In the well-tempered state, the chocolate mass has the desired flow properties for subsequent processing (molding or enrobing). Once the chocolate is cooled to set into a hard product (molding or enrobing), these crystals form a structure that gives the desired surface appearance (gloss) and texture (snap). Many small crystals are desired to give the proper snap as well as a fine internal grain structure to the chocolate. In addition, the proper amount of crystallization during cooling is necessary to give the desired extent of contraction for proper mold release, where a molded chocolate bar can be released from the mold during processing. The cocoa butter crystals also must be in the correct polymorphic form to resist oil migration and the appearance of fat bloom. Due to the nature of chocolate, researchers have been unable to visualize the size and shape of cocoa butter crystals in chocolate directly. Microscopy techniques (optical, confocal, and electron microscopy) are incapable of distinguishing individual cocoa butter crystals in a chocolate mass. Thus, our understanding of chocolate structure is based on indirect evidence (X-ray spectroscopy, calorimetry, etc).

When chocolate is cooled directly to room temperature, regardless of whether or not there is agitation, typically a combination of α- and β′-polymorphs forms. These polymorphs would gradually transform to the more stable forms over a time span of minutes to hours. The resulting product would be undesirable since the structure would be grainy (large crystals), and fat migration during the polymorphic transformation (accompanied by release of latent heat) would lead to fat bloom at the surface. For this reason, chocolate must be cooled according to a well-designed protocol to ensure formation of the most stable crystalline structure. A tempering process typically involves cooling rapidly to a temperature where nucleation of an unstable polymorph (β′) occurs and then heating to promote transformation to the desired polymorph (β). The tempered chocolate then can be carefully cooled to storage temperature to produce the desired crystalline microstructure. A typical temperature-time protocol for tempering of chocolate is shown in Figure 7–10.

In the initial stages of tempering, the chocolate is cooled rapidly under agitation to a temperature of 26 to 27°C to promote formation of the then β′-polymorph. As seen in Figure 5–14, the nucleation rates of the α- and β′-polymorphs are significantly higher than that of the β-polymorph. Thus, nucleation is rapid, and many small crystals (of unstable polymorph) will be formed. Due to the decreased temperature and the onset of crystallization, the viscosity of the chocolate mass increases substantially at this point in tempering. After nucleation is completed, the temperature is raised to 31 to 32°C to promote transformation to the more stable β-polymorph and melting of the unstable polymorph. Some of the crystalline fat melts as the temperature increases until only a small (2 to 3%) percentage of the fat phase is crystalline at tempering temperature.[22,23] Viscosity decreases in this stage as temperature is increased and the mass of crystals is reduced. At this point, the tempered chocolate mass can be pumped for use in molding or enrobing operations. A well-tempered chocolate mass cools readily to produce a product with many small crystals of the stable β-polymorph and gives the desired textural characteristics and

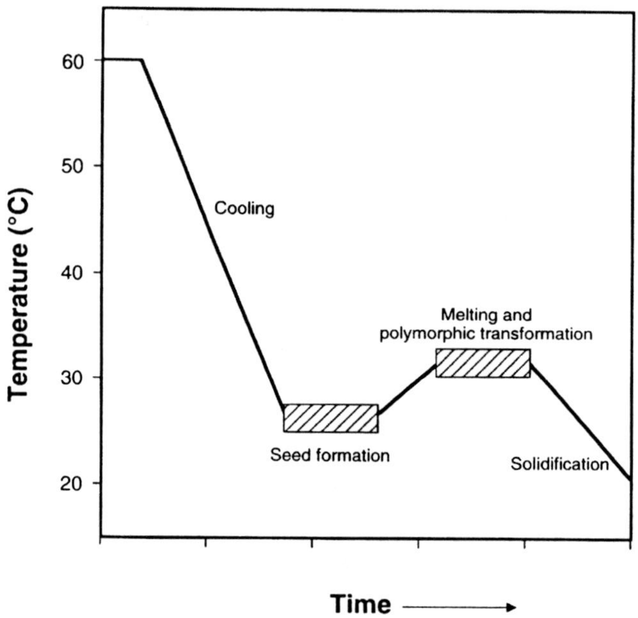

Figure 7–10 Typical Temperature-Time Treatment during Tempering of Chocolate Showing the Approximate Range of Temperatures at Each Point in the Tempering Process.

appearance. Ideally, only growth of the existing seeds, or formation of more stable phases, occurs during cooling.

If the temperature conditions during tempering are not appropriate for the type of chocolate being produced, chocolate that is not "in temper" may be the result.[21] Chocolate can be either under- or over-tempered, depending on the state of the crystalline material. Under-tempered chocolate contains too little crystalline mass. When it is cooled, the seed crystals present are insufficient to remove the supersaturation and nucleation must occur. Since nucleation favors formation of the less stable β'-polymorph, the solidified chocolate contains a mix of unstable β'-polymorphs formed during cooling along with the stable β-polymorph formed in tempering. The chocolate will be unstable, and rapid fat bloom formation is likely to occur. If the chocolate mass is over-tempered, it either has too much crystalline mass or the existing seeds are too large. Over-tempered chocolate may be more viscous than tempered chocolate, which can cause problems with pumping and filling. In addition, when over-tempered chocolate is cooled, the result is a more grainy structure that may bloom a little more rapidly over time. Thus, it is important that the chocolate processor understand the correct tempering conditions for the given chocolate. At this point, design of tempering units for chocolates is still somewhat a trial-and-error process since each different chocolate formulation requires slightly different temperature-time conditions to produce ideal temper.

Variability in chocolate formulation is one of the primary causes of variation in tempering conditions. It is widely known that tempering conditions for milk chocolate must be 1 to 2°C lower than for an equivalent dark chocolate due to the inhibition effect of the milk fat (Table 5–1) on cocoa butter crystallization. Higher levels of milk fat addition generally require lower tempering temperatures to obtain satisfactory temper. Differences in tempering requirements also arise due to differences between cocoa butters, since different cocoa butters crystallize with different nucleation and growth kinetics. Regional differences in cocoa butter production can result in substantial differences in triacylglycerol composition that affect crystallization rates.[24–26] Harder cocoa butters, with higher solid fat content (SFC) curves, typically crystallize more readily at higher temperatures than softer cocoa butters. The nontriacylglycerol lipid composition (particularly the phospholipids) also have been shown to impact crystallization kinetics of cocoa butter dramatically.[27,28]

Cooling of tempered chocolate mass (eg, in a cooling tunnel to solidify chocolate bars) also must be controlled to maintain the desired crystalline structure.[21] If a well-tempered chocolate is cooled too rapidly, for example, a portion of the chocolate mass may nucleate in the less stable β′-polymorph. This will cause unwanted problems later as the β′-polymorph converts to the more stable β-polymorph, releasing some latent heat as it does. Rapid cooling also may cause cracking of the chocolate piece, along with other undesired physical changes. In general, cooling tunnels for chocolates are designed so that the initial stages of cooling are relatively slow so that severe undercooling does not occur. The middle stages of the cooling tunnel have lower temperatures and are designed to remove the latent heat of crystallization during setting of the chocolate. Prior to the end of the cooling tunnel, the temperature must be increased slightly to ensure that condensation does not occur as the piece exits the tunnel into room air.

Butter

In making butter, crystallization of the milk fat occurs in the fat globules as the cream is conditioned prior to churning. Upon churning of butter, phase inversion and loss of water (in the form of butter milk) are controlled to obtain a water-in-oil emulsion. The milk fat crystals in the cream are dispersed into the continuous lipid phase, and they continue to grow as processing continues. Over time, these lipid crystals grow together to form a network structure that impacts the texture of the final product.[29] Through mechanical working, the network structure of these fat crystals can be interrupted and a softened butter produced. However, this network reforms during further storage, and the hardness increases once again. The composition of the lipid phase also impacts hardness of butter through an influence on crystal microstructure. It is widely recognized that butter made from milk obtained in the winter is typically harder than that made from milk obtained in the summer. This effect arises from differences in lipid composition due to different feeding strategies between the winter and summer seasons, with winter fat often having a higher SFC than summer fat. However, recent feeding trends (year-round supplementation of the cow's diet) have led to decreases in these differences between summer and winter fat, at least in the United States.[30]

Although the exact nature of crystallization in cream and butter to control butter quality is not completely understood, some general principles for controlling crystallization to give the desired product quality have been developed.[29] The first step in controlling butter texture and quality comes in the treatment of cream. The nature of the fat crystals present in cream prior to churning into butter affects the final crystalline microstructure of the butter. The nature of crystalline fat in cream has been widely studied, particularly with polarized light[31] and scanning electron microscopy.[32] These studies suggest that there are four different types of fat globules in cream, between the temperatures of 4 and 20°C, as differentiated primarily by the pattern of crystallization of milk fat.[29] Type 1 fat globules are primarily noncrystallized with mainly liquid fat, although there may be a few microcrystals present. Type 2 globules contain crystal platelets and irregular aggregates throughout the globule, with an occasional liquid fat region seen within the crystalline microstructure. Type 3 globules contain a thick crystalline shell at the periphery of the droplet with liquid fat contained within. Type 4 globules have a thick crystalline shell at the periphery with irregular crystal aggregates and a few pools of liquid fat contained within. Differences in crystallization between fat globules in cream have been attributed to a number of mechanisms including differences in heterogeneous nucleation sites between droplets, differences in molecular composition of the fat contained in each droplet, and differences in size (which influences heat transfer rates). Regardless of the source of these different globule characteristics, the relative concentration of each form in cream influences the texture of the butter produced from that cream. For example, type 4 fat globules are the predominant form of globules left after churning into butter.[29,33] Thus, a higher content of type 4 globules in the cream can significantly influence the consistency of the butter.

The nature of the crystalline shell in the fat globules of cream suggests that milk fat fractionates during crystallization within the droplet. That is, the high-melting components that form first exclude the liquid fat from the region of nuclei formation. Often, the result is a globule with high-melting fat crystallized as the shell and containing low-melting fat components in either liquid, semi-liquid, or crystalline form within the droplet. The nature of the crystalline fat in milk fat globules also influences the stability of these droplets against coalescence during storage.[31] The degree of crystallization plays a significant role in determining extent of coalescence. If most of the fat is crystallized in an emulsion, coalescence is not possible. The presence of some liquid fat within the droplets is important as is the nature of the crystals within the partially crystallized globules. Bridge formation may occur when liquid fat is present between two adjacent globules or when crystals that break through the outer membrane of one fat globule lead to coalescence with another fat globule in contact.[34] Liquid fat may appear at the surface of a fat globule where either internal or external forces are sufficient to disrupt the surface membrane layer.[29]

During churning of butter, the fat globules are sheared and compressed in such a way that liquid fat is squeezed out of the droplets.[35] In batch churning, air incorporation is thought to play a role in conversion of cream to butter.[36] The result of churning is phase inversion with separation of butter milk. Butter contains about 20% water in the form of a fine emulsion. The continuous phase is made up of crystalline fat, aggregates of milk fat globules from the cream (butter granules), and liquid fat (depending on the temperature).[37] During churning, clumping and melting of globules occur; this leads to restructuring of the fat crystals both within and outside the globules.[36] Small milk fat crystals, composed primarily of the high-melting components of milk fat, are found adsorbed to the surface of the water droplets.[29] The surface of air cells found in butter also contains crystallized fat as well as intact fat globules.[29]

It is the interaction between the different phases in butter that leads to the observed texture or physical properties (hardness, spreadability, plasticity, etc). Factors that have been found to influence the physical properties of butter include conditions of physical cream ripening, temperature, air content, and fat composition.[29] Crystallization conditions in particular influence the number of intact fat globules, the size of fat crystals, the solid-to-liquid-fat ratio, formation of mixed crystals, the degree of fractionation between high- and low-melting components of milk fat, and the relative proportion of primary (stronger) or secondary (weaker) bonds between crystals.[29] In particular, these factors influence the phenomenon of setting in butter, where a substantial increase in firmness is found during the early stages (within the first week) of storage. This phenomenon has been related to continued crystallization and recrystallization (rearrangement of crystalline microstructure) of the milk fat.[38]

Control of crystallization of milk fat in cream and butter is critical to control the physical properties (spreadability and firmness) of the final product. Although much work has gone into this area, our understanding of all the factors that influence the physical properties of butter is still lacking. This is in part due to the limitations of the techniques capable of investigating the crystalline microstructure. As these techniques are improved, our understanding of the physical phenomena that occur during production of butter also will improve.

Freezing of Ice Cream

The texture and quality of ice cream are governed, at least in part, by the nature of the ice crystalline state. The desired characteristic is a smooth, creamy texture that comes from having many small ice crystals. When ice crystals are too large, they can be detected in the mouth, and the ice cream has a coarse, grainy mouthfeel. Russell et al[39] showed the relationship between ice crystal size and sensory perception of smoothness and ice crystal detectability for a particular ice cream formulation. These relationships are shown schematically in Figure 7–11. The perception of smoothness decreased and the detectability of ice crystals increased linearly with an increase in mean ice crystal size. In general, ice crystals larger than about 50 μm in ice cream are detectable in the mouth, and too many ice crystals above this size yield a coarse product.[40] Thus, controlling

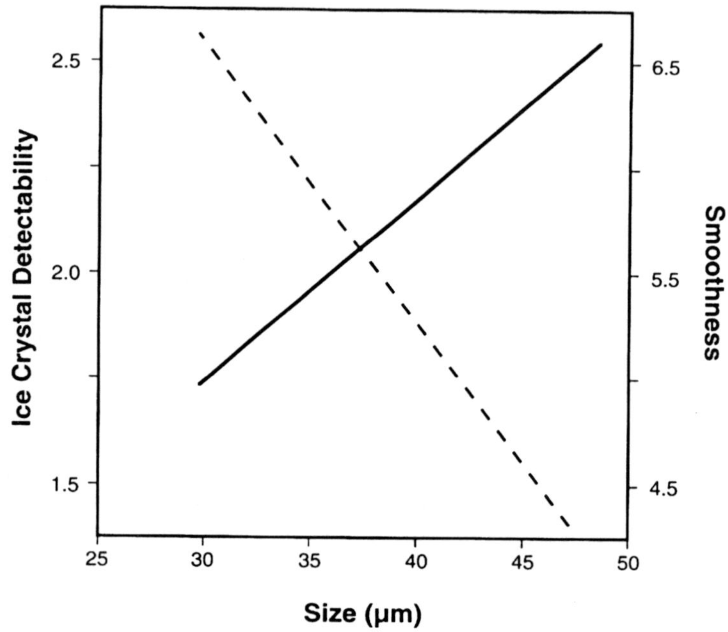

Figure 7–11 Relationship (Shown Schematically) between Mean Ice Crystal Size and Sensory Perception of Smoothness and Ice Crystal Detectability for an Ice Cream Product. Dashed line, smoothness; solid line, ice crystal detectability.

ice crystallization during manufacture of ice cream and frozen desserts is extremely important.[39,41] The initial size of the distribution also may impact storage characteristics, as discussed in Chapter 8.

In a typical fresh ice cream, it is estimated that there are about 7.8(10^6) crystals per gram.[42] Russell et al[39] reported between 1 and 25 × 10^6 ice crystals per gram of ice cream, depending on the ice phase volume in the ice cream. Their results showed that, in the range of 40 to 55% ice phase volume, the number of ice crystals was not influenced by the ice phase volume. The mean ice crystal size in hardened ice cream is typically between 35 and 45 μm, with crystals ranging in size from a few micrometers to over 100 μm. The distribution of ice crystals in a premium vanilla ice cream is shown in Figure 7–12. The ice crystal size distribution depends greatly on the ice cream formulation and the process employed to freeze the product.

To quantify a CSD in ice cream, generally either SEM or optical microscopy is used. Preparation of ice cream for SEM analysis requires careful freezing, fracturing, and drying

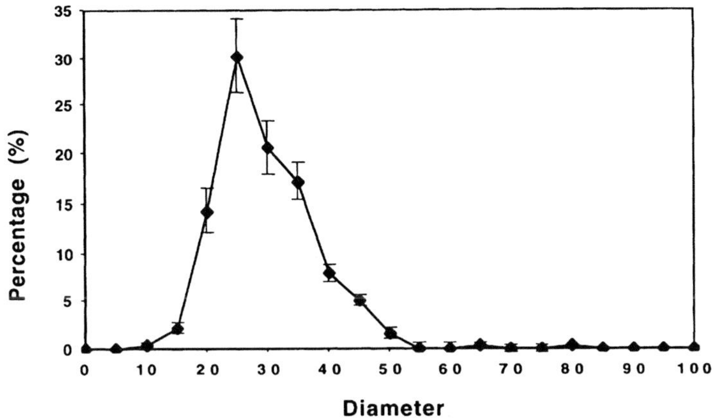

Figure 7–12 Typical Ice Crystal Size Distribution for a Premium Vanilla Ice Cream.

to produce a sample with representative structural characteristics. The use of cryo-SEM (cold-stage SEM) allows imaging of the ice cream structure at freezer temperatures so that drying is no longer required.[43] However, it is still possible to introduce artifacts, or unreal structural aspects, through improper sample preparation. Optical microscopy also has been widely used to quantify ice crystal size in ice cream. Maintaining freezer temperatures at the microscope slide is critical so that ice crystals do not melt or grow during imaging. Most optical microscope techniques involve "squashing" a sample on a microscope slide, sometimes with the assistance of an organic solvent as dispersant.[44] Under carefully controlled conditions, ice crystals can be stabilized for long periods of time (hours) so that image analysis can be performed. The advantage of optical microscopy over SEM is the relative ease of sample preparation and the high number of ice crystals that can be counted. However, SEM can provide a more realistic image of the original ice cream structure.

In most commercial processing plants, ice cream is frozen in a scraped surface heat exchanger (SSHE) with liquid ammonia as the refrigerant to provide freezing. The process (Figure 7–13) essentially involves feeding ice cream mix (at about 1 to 2°C) to the SSHE, where freezing and air incorporation occur at the same time. A portion of the water in ice cream is frozen in the SSHE and a semi-frozen product is extruded into the desired packaging. Typically, the semi-frozen ice cream exits the SSHE at about –5 to –6°C. The ice cream is then further cooled in a hardening tunnel or room, typically between –30 and –40°C, until product temperature is below –18 to –20°C. Figure 7–13 also indicates the various ice crystallization phenomena of importance during processing of ice cream.

The ice cream mix entering the SSHE experiences a driving force for ice formation at the point in the unit where the ammonia jacket begins. Liquid ammonia vaporizing under pressure at a temperature of about –30°C on the other side of the SSHE barrel wall causes substantial cooling of the mix. When the mix temperature is sufficiently reduced, nucleation occurs. Ice forms readily on metal walls, and it is likely that a layer of ice forms on the barrel surface.[45] Since the heat transfer resistance of the metal wall between vaporizing ammonia and ice cream mix is minimal, the temperature just inside the barrel wall may be as low as –26 to –28°C. This provides substantial driving force for ice crystallization, and ice forms and grows rapidly. This rapid crystallization results in release of latent heat that increases the temperature in the vicinity. The scraper blade in the SSHE cleans off the metal surface and disperses the ice slurry within the center of the chamber. As this slurry continues to grow, additional heat of fusion is evolved. Furthermore, recrystallization, or reorientation of the crystalline phase volume, occurs as the crystals formed at the cold surface are distributed into the warmer environment in the center of the freezer. In this process, the crystals scraped off the freezer wall reform into the block-shaped crystals characteristic of frozen ice cream. Recrystallization also leads to release of latent heat. The characteristic temperature profile for ice as it freezes along the length of the freezer barrel was measured by Russell et al[39] and is shown in Figure 7–14. In this case, the temperature of the ice cream decreased rapidly during initial freezing within the first 10% to 15% of the freezer length. After that, temperature decreased only

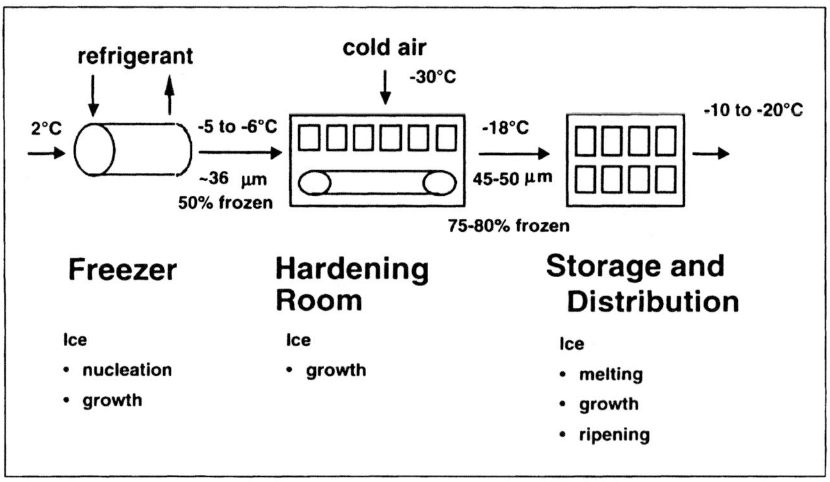

Figure 7–13 Process for Freezing of Ice Cream with Ice Crystallization Phenomena That Occur in Each Stage.

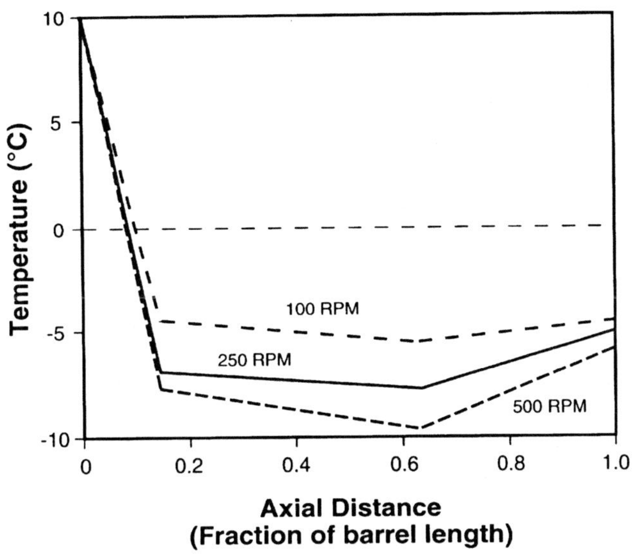

Figure 7–14 Axial Profile of Ice Cream Temperature within the Barrel of a Scraped-Surface Ice Cream Freezer.

slowly until a slight warming was observed at the end of the freezer. Russell et al[39] also measured the temperature of the ice cream after exit from the freezer to determine if there was additional release of latent heat (due to crystallization). Their results show that there was no increase in temperature, which indicates that the ice cream exiting the freezer had already reached the phase volume equilibrium defined by the freezing point depression curve. Assuming that in most commercial ice cream freezers, freezing is sufficiently slow that the freezing point depression curve is followed, the amount of ice formed at any temperature is approximately at the equilibrium phase value[39] and can be calculated from the phase diagram.

In the vicinity of the metal wall, a substantial temperature driving force for crystallization is generated. Estimating the rate of heat transfer through the metal wall gives a temperature at the freezing surface of –26 to –28°C. Assuming that the freezing point of the mix is about –2 to –4°C, a subcooling of 22 to 26°C has been generated.[41] The rate of nucleation at this subcooling is extremely high, although it is unclear if this subcooling lies at the peak of the nucleation rate curve (Figure 7–1). Experimental freezers using colder refrigerant can produce ice crystals of smaller size than the conventional process.[46,47] However, this occurs at the expense of pumpability of the ice cream mass as it exits the freezer. Mixing of ice cream mix with liquid nitrogen results in a product with very small ice crystal size (less than 20 μm), although the final size depends on the process used to warm the ice cream back up to storage temperature (Hartel, 1997, unpublished results). When ice cream is made using brine as the refrigerant, as in a home ice cream freezer, the ice crystals are significantly larger (for storage at equivalent temperature). In the case of brine as refrigerant, the subcooling may only be 4 to 8°C. Clearly, this condition gives a much lower nucleation rate than freezing with liquid ammonia. Thus, freezing with liquid ammonia (commercial freezer) produces freezing conditions closer to point B in Figure 7–1, whereas freezing with brine (home freezer) produces conditions closer to point A. The difference in ice crystal size distribution (Figure 7–2), and quality of the ice cream, based on these conditions is obvious.

Under conditions of the high subcoolings experienced at the freezer barrel wall, ice crystals grow in the shape of dendrites.[48] Thus, it has been suggested that the situation at the barrel wall in an SSHE is likely to be that depicted in Figure 7–15.[45] In a commercial ice cream freezer, the time between scrapings at a point on the freezer barrel is less than a tenth of a second (200 rpm with four blades). The blade scrapes the dendritic ice crystals off the surface and disperses them into the bulk solution. At the warmer temperature in the bulk solution, as compared with the material at the wall, the dendritic ice crystals ripen into block-shaped crystals and continue to grow. Although this mechanism for ice crystal formation and growth seems reasonable, only indirect evidence for this mechanism exists. Current research on formation of ice crystals in an SSHE using a submersible microscope probe shows that dendrites do indeed ripen into disc-shaped crystals. However, large sheets of ice appear to be removed from the cold surface and the dendrites grow out of these polycrystalline sheets (Si and Hartel, 2000, unpublished results), similar to the process suggested by Russell et al.[39] These ice sheets also are seen to ripen into pellet-shaped crystals in a slush-making machine for frozen drinks.[49] Further research is needed to verify the mechanisms of ice crystallization in the SSHE.

Processing factors that can influence ice formation in the scraped surface freezer include the rate of heat transfer removal (draw temperature and throughput rate), the average residence time of product within the freezer, the scraper speed and design, the type of dasher (open or closed), and blade sharpness. Russell et al[39] found that mean ice crystal size varied directly with average residence time. Higher average residence time resulted in larger mean size, regardless of other operating parameters (eg, dasher type). Thus, average residence time appears to be an important parameter in controlling ice crystal size distribution in ice cream. It is logical to think that higher scraper speeds will result in smaller ice crystals produced at the end of the SSHE, based on the dendritic ice mechanism. However, production of more nuclei at higher speed of scraping is offset by the increase in frictional heat as scraper speed is increased. Thus, faster scraper speeds can actually produce larger ice crystals at the exit as some of those formed in the freezer are melted by the additional heat input.[39] Scraper blades also must be maintained as sharp as possible to reduce the thickness of the ice layer built up on the freezer wall. This ice layer acts as an insulator and decreases the driving force for nucleation and growth. The result is larger ice crystals and a coarser ice cream product.

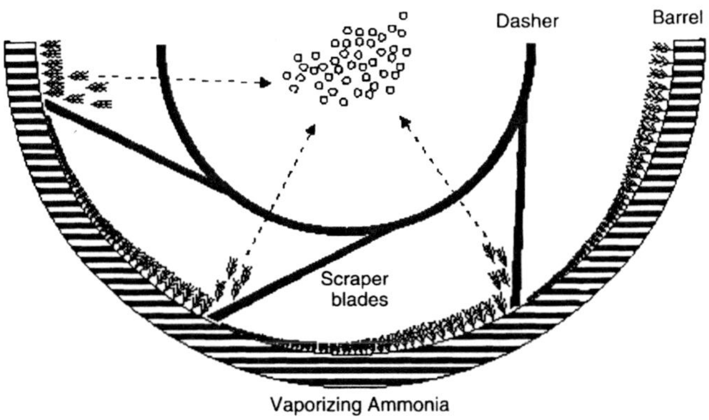

Figure 7–15 Crystallization of Ice from Ice Cream Mix at the Metal Wall of a Scraped Surface Heat Exchanger.

The ice cream formulation also influences the ice CSD. The total solids of the ice cream mix affect the amount of ice formed at any temperature, as seen in the freezing point depression curve. A mix with higher water content results in an ice cream with more ice at any given temperature. More ice typically means a harder ice cream and, if processing occurs at the same conditions for different water contents, mean ice crystal size is larger for the ice cream mix with higher water content (lower total solids),[44] as shown in Figure 7–16.

The various components of ice cream can affect the initial ice CSD in ice cream.[40,41] Sweeteners, milk solids, and stabilizers are the primary ingredients in ice cream mix that impact ice crystallization. These components may (1) affect the freezing point depression curve, (2) influence solution viscosity, and (3) provide some specific surface inhibition to ice crystal growth. In particular, stabilizers are claimed to produce smaller ice crystals than those found in unstabilized ice cream. Compounds usually used as stabilizers (gums, proteins, etc) may have some specific affinity to ice crystal surfaces and result in slower growth rates. However, it also is likely that the change in viscosity of the ice cream mix due to addition of stabilizer causes significant differences in heat and mass transfer during freezing that also influence crystallization. Thus, the effect of stabilizers is not easily understood. In fact, the interrelations between all of the ingredients of ice cream mix generally present a difficult task for sorting out specific effects of individual components.

Fractionation of Fats

It is possible to separate different triacylglycerol components of a complex natural fat based on melting point. Milk fat, palm oil, tallow, and palm kernel oil are fats that

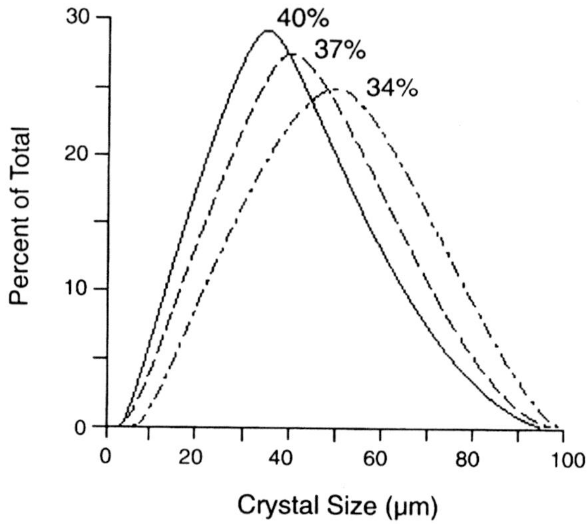

Figure 7–16 Crystal Size Distribution in Ice Creams Made with Different Initial Water Contents.

are commercially fractionated to produce components with specific characteristics. One technique for separating components of fats based on melting point is dry fractionation. The term *dry* is used to denote that no solvent is added to promote separation. In dry fractionation, the fat is melted at high temperature prior to controlled cooling to initiate crystallization. Once crystallization has been completed, the slurry is separated in some manner, usually by filtration. A typical fractionation process, shown schematically in Figure 7–17, is a batch operation where temperatures are sequentially lowered to produce a range of fractions with different physical and chemical properties.

In this application, crystallization is controlled to produce the desired size distribution for the most efficient filtration. To enhance filtration, particle size must be reasonably large (at least 100 to 200 μm) and the distribution of sizes reasonably narrow (low variance). To obtain larger size, nucleation must be limited so that only relatively few crystals are formed and these can then grow in size to produce the desired distribution. Nucleation also must be controlled so that all nuclei grow to the desired size at about the same time. Growth conditions should be uniform throughout the crystallizer. In batch systems, nucleation should occur within a narrow time range so that all crystals can grow to the same size. In continuous systems, the residence time distribution should be such that all nuclei formed during a given time span reach product size at about the same time. Crystallization temperature also may be controlled to produce the desired polymorphic form.

Traditionally, natural fats have been treated as melt systems, where the crystallization driving force was caused by cooling below the melting temperature. However, the defini-

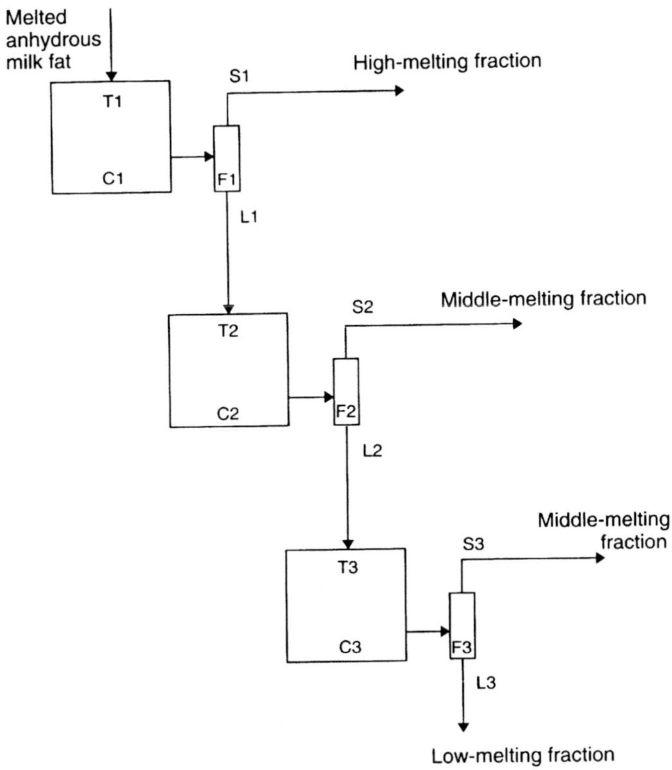

Figure 7–17 Sequential Fractionation of Fats in Multiple Stages. Crystallizers (C1, C2, and C3), operated at sequentially lower temperatures (T1, T2, and T3), feed filters (F1, F2, and F3) to produce a series of solid (S1, S2, and S3) and liquid (L1, L2, and L3) fractions.

tion of subcooling in natural lipids is ambiguous because of the wide range of temperatures over which the fat melts. For example, milk fat exhibits melting over the temperature range from –40 to 40°C. Cocoa butter, on the other hand, melts over a relatively narrow temperature range, with most of the melting taking place between 25 and 35°C. Because of this range of melting temperatures, crystallization kinetics for natural fats are rarely based on any type of thermodynamic driving force. Instead, kinetics are often correlated against operating conditions, such as temperature and agitation.[50]

In some cases, complex lipids for fractionation can be treated as a solution system, where the high-melting components are dissolved in the low-melting components, which act as a solvent. Researchers have attempted to simplify crystallization in lipid systems by studying model systems that can be treated as binary solutions. For example, Ng[51] studied crystallization of tripalmitin from a triolein solution, Dibildox-Alvarado and Toro-Vazquez[52,53] studied crystallization of tripalmitin from sesame oil, and Kloek et al[54] studied crystallization of fully hydrogenated palm oil from sunflower oil. Herrera et al[55] studied crystallization of a milk fat model system, with different ratios of high-melting to low-melting components of milk fat.

The high-melting component of milk fat, which is concentrated in the triacylglycerols that contain long-chain saturated fatty acids, can be treated as a solute that is dissolved in the liquid oil (triacylglycerols that contain unsaturated and short-chain fatty acids). The high-melting component makes up about 5 to 10% of milk fat, whereas the low-melting components make up about 70%.[56,57] The remainder of milk fat is an intermediate-melting component. By measuring an effective solubility concentration of the high-melting component in the liquid (low-melting) component, a supersaturation based on composition can be derived. In fact, this supersaturation can be related to the composition of the high-molecular-weight components (high melting). An operational phase diagram based on this concept has been developed for milk fat, as shown in Figure 7–18 (Liang and Shi, 1997, unpublished results). The saturation line represents the effective equilibrium concentration of the long-chain triacylglycerols in the liquid component of milk fat. The supersaturation of any starting lipid milk fat can be obtained by comparing the concentration of these

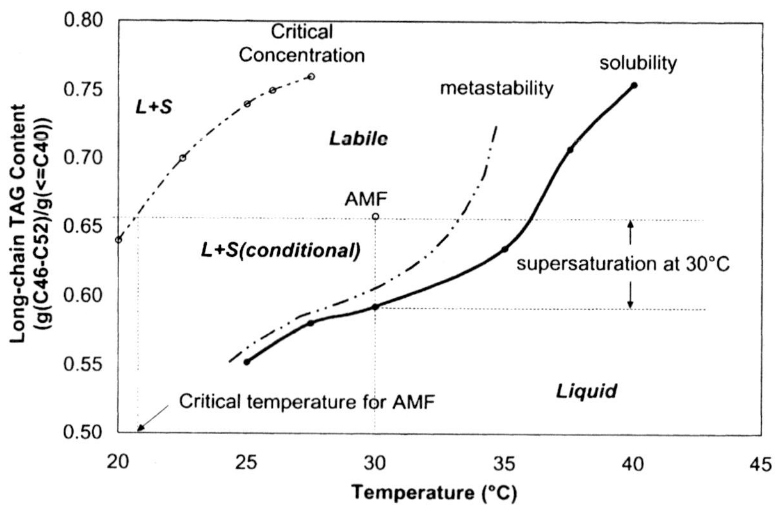

Figure 7–18 Operational Phase Diagram for High-Melting Components in the Low-Melting Components of Milk Fat Based on Triglyceride Composition (Acyl Carbon Number).

triacylglycerols with the solubility concentration, in the same manner as for a true binary system. This supersaturation then can be used to predict crystallization kinetics, as shown in Figure 7–19. As the content of these long-chain triglycerides in the original fat increases, the crystallization rate also increases.

On the operational phase diagram of Figure 7–18, note the region at very high concentration of the high-melting component and low temperature. Under these conditions, massive crystallization of different species of triacylglycerols occurs, and this simple approach to phase equilibrium breaks down. At these low temperatures, which may occur during setting of butter, for example, crystallization is dominated by mixed crystals in a complex melt system.

To control crystallization during fractionation, the composition of the starting material should be known and compared with the driving force for crystallization. In addition, the operating conditions during crystallization (cooling rate, crystallization temperature, agitation rate, etc) are important in determining the nature of the size distribution obtained and the subsequent filtration efficiency. One parameter of utmost importance is agitation rate.[58] Mixing and shear profiles in the crystallizer control rates of heat transfer and mass transfer as crystallization proceeds. Secondary nucleation also may occur under conditions of high shear agitation. At low tip speeds (low agitation rate), there is insufficient agitation and milk fat crystals readily agglomerate. At very high tip speed, substantial breakage

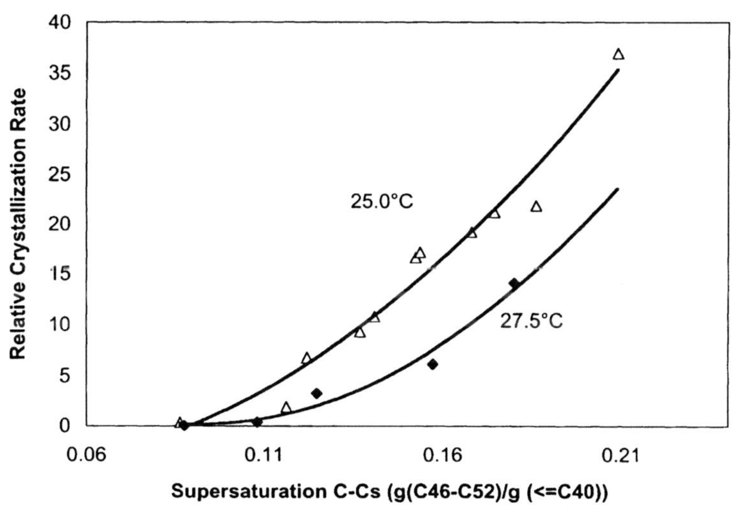

Figure 7–19 Crystallization Kinetics of High-Melting Component from Low-Melting Components of Milk Fat Based on Supersaturation Defined in Figure 7–18.

or secondary nucleation takes place, resulting in numerous small crystals. Neither size distribution filters very well, with extremely high filtration resistance (see Figure 3–1). At some intermediate agitation condition, agglomeration and breakage counteract to produce a size distribution that results in the lowest filtration resistance (most efficient filtration). Typically, the crystals are reasonably uniform in size and large enough for efficient separation.

Graining of Sugar Glasses

Hard candy is an extremely concentrated (about 2% water) sugar mass that is sufficiently viscous to be in the glassy state. The glassy state is desired for hard candies for several reasons: (1) rapid dissolution in the mouth (compared to the crystalline state), (2) uniform dispersion of flavors and colors, and (3) the ability of the amorphous state to be molded into the desired shapes. Typically, hard candies are made with sucrose and corn syrup as the main ingredients, with addition of colors and flavors, and perhaps acid to provide tartness. The amount and type of corn syrup in the formulation determine the characteristics of the product. Table 7–4 documents the differences in physical properties of hard candies made with different levels of corn syrup and sucrose.

The process for manufacture of hard candies is shown in Figure 7–20. The raw materials are mixed and heated to ensure that all sugar crystals are dissolved. This sugar syrup is concentrated rapidly by boiling at high temperature to low water content (2 to 3%). Thin-film evaporators often are used for this step since they provide rapid and efficient concentration. The speed of concentration is critical to production of a stable sugar glass. If evaporation is too slow, crystallization can occur during the concentration step resulting in a crystalline, rather than glassy, product. The concentrated sugar mass exiting the evaporator is cooled rapidly to about 75 to 80°C and then formed and cut into the desired shape. After forming, the hard candy pieces are cooled to room temperature. The path of this process on the phase/state diagram is shown in Figure 7–21. Once the sugar mass has been concentrated, it must be cooled quickly to prevent crystallization. However, the sugar mass must remain somewhat fluid and pliable so that it can be formed into the desired shape and size. This is why the intermediate temperature of 75 to 80°C is chosen. At this point, the sugar mass is sufficiently viscous that crystallization is inhibited

Table 7–4 Characteristics of Hard Candies Made with Higher Amounts of Sucrose or Corn Syrup

Higher Sucrose	*Higher Corn Syrup*
Increased sweetness	Decreased Sweetness
High flavor release	Low flavor release
Grains	Cold flow
Lower T_g*	Higher T_g
Low machinability	High Machinability

*T_g, glass transition temperature.

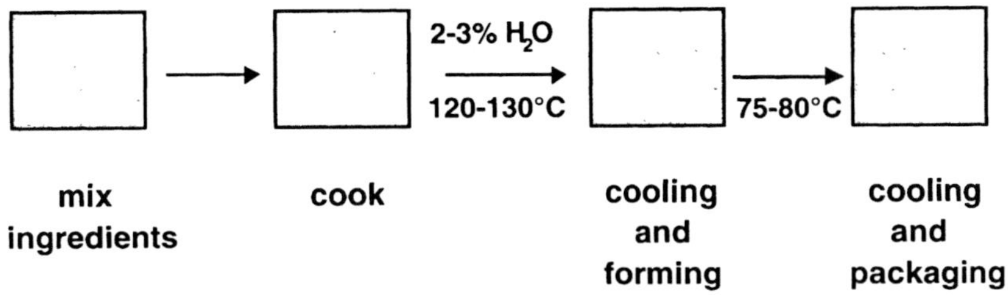

Figure 7–20 Schematic of Processing Steps for Hard Candy Production.

for long times, but it is still sufficiently fluid so that it can be formed into the desired shape. The candy must be cooled below the glass transition temperature (T_g) after forming and packaging to ensure long-term stability of the product. If the sugar mass is held too long at high temperatures (above T_g), the chance of graining (crystallization of sucrose) is increased substantially. Also, the mass exiting the evaporator should be cooled with the minimum amount of agitation to prevent accidental crystallization due to mechanical energy input.

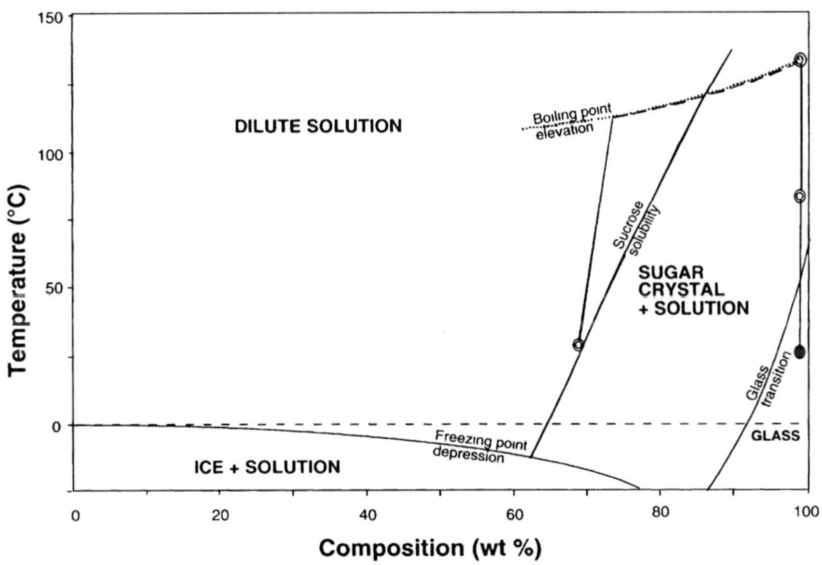

Figure 7–21 Path of Hard Candy Manufacture on the State Diagram, Based on Pure Sucrose-Water System.

In grained hard candies (such as grained mints), the sugar mass is "pulled" on rotary arms to induce crystallization. Powdered sugar is also added to provide seeds for crystallization, and the product is stored at warm temperatures to allow complete crystallization. In this case, the product contains many small crystals in a solution of highly concentrated sugar. It is similar to fondant in this respect, but since the water level is so much lower in these grained candies, they are much more solid than typical fondants.

The choice of corn syrup type and addition level is important to the characteristics of the hard candy and the processing steps. High-dextrose equivalent (DE) corn syrups (eg, 63 DE) contain substantial amounts of low-molecular-weight sugars and potentially can lower the T_g of a mixture with sucrose. Low-DE corn syrups (eg, 20 DE), on the other hand, contain few simple sugars and primarily the higher-order glucose polymers due to the lower extent of hydrolysis of starch. These increase the T_g of a sucrose mixture. In general, if the corn syrup increases the average molecular weight of the sugar mass (compared with pure sucrose), the sugar mass will have a higher T_g than sucrose. Conversely, reducing the average molecular weight causes a decrease in T_g. Typically, hard candy manufacturers use 42 DE corn syrup in their formulations for a combination of reasons. The level of corn syrup addition also impacts the T_g of the sugar mixture. Again, the magnitude of the effect depends primarily on the average molecular weight of the sugar mixture. In general, there is a direct relationship between T_g and the number-averaged molecular weight of the corn syrup component added to sucrose, as shown previously in Figure 4–31. Components with higher molecular weight have higher T_g. In general, higher corn syrup addition levels in hard candies increase the overall T_g of the mixture, but this depends on the average molecular weight of the final mixture. Addition of high-molecular-weight corn syrup components (lowest DE) cause the greatest increase in T_g. However, other factors must also be considered since differences in T_g arise from branched- versus straight-chain polysaccharides of equivalent molecular weight.[59]

There also are chemical changes that may occur during processing that impact the stability of hard candies. Under acidic conditions and at elevated temperatures, as might be encountered in hard candy processing, sucrose hydrolysis or inversion occurs. The breakdown of sucrose into glucose and fructose results in a decrease in average molecular weight of the system and causes a decrease in T_g. Thus, extensive inversion during processing may result in a hard candy with lower stability, although inversion generally results in a slightly sweeter candy. Reversion is a reaction that also occurs under acidic conditions where monosaccharides polymerize to form compounds such as polydextrose and polyfructose. These polymers cause an increase in average molecular weight and increase T_g, resulting in a more stable hard candy, but with reduced sweetness.

Sugar glasses are notoriously hygroscopic and sorption of water can have significant detrimental effect on the shelf life of hard candy. In general, the shelf life of a hard candy is limited by moisture sorption, although temperature also may play a role in some cases. When a sugar glass picks up moisture, the T_g decreases in accordance with the amount of water absorbed. Thus, a product that was initially stable at room temperature ($< T_g$) may become unstable as T_g falls below room temperature. Also, if the storage temperature of the hard candy increases above T_g, with or without moisture uptake, the product will become unstable. Once the product is no longer in the glassy state, there are numerous processes that can now occur at sufficient rate to end shelf life prematurely. Diffusion rates of flavorants increase dramatically once the glass transition condition is exceeded. This results in diffusional losses of flavors through the piece, and loss of flavor may be a problem in some cases. Also, sufficiently high moisture content or increased temperature can cause a syrupy layer on the surface to become sticky. The "sticky point" for sugars is often defined as the condition in an amorphous mass where the viscosity is between 10^6 to 10^8 Pa-s,[60–62] as shown schematically on the phase/state diagram in Figure 7–22. Any combination of temperature increase and moisture sorption that results in a decrease in viscosity to this level will cause the product to become sticky. In hard candies made with higher levels of corn syrup, this is often the main concern for end of shelf life. Sugar crystallization is less of a concern in these products because the high levels of corn syrup effectively prevent sucrose crystallization.

Even hard candies made with primarily sucrose become sticky when the T_g of the piece drops below ambient conditions (or if ambient temperature exceeds T_g). However, in this case, the amorphous sugar mass is prone to rapid crystallization or graining so stickiness is not usually considered to cause the end of shelf life. At some combination of temperature and moisture uptake, the sugar mass gains sufficient mobility that the sugar

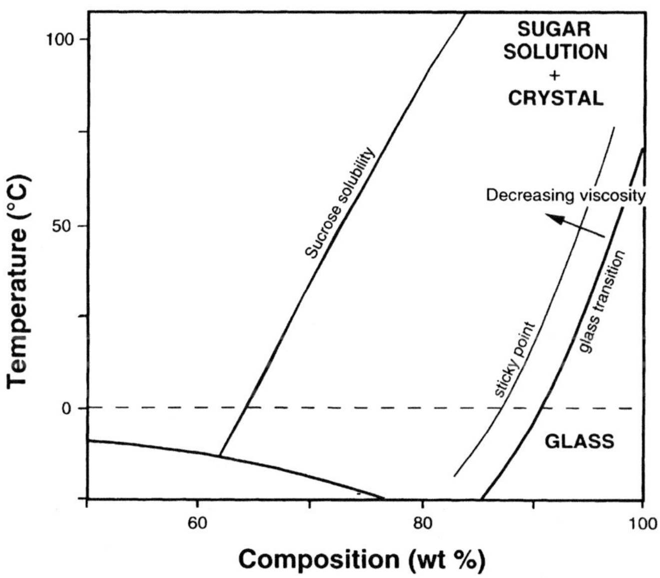

Figure 7–22 Sticky Point Line as Shown (Approximately) on the Phase/State Diagram for the Pure Sucrose-Water System.

molecules can rearrange and form into a crystal lattice. In fact, there are two potential causes of graining in hard candies: internal and external. Sometimes processing conditions are not controlled perfectly and a few seed crystals are present in the candy glass. These crystals are stable as long as the hard candy remains below its glass transition point. Once the temperature increases above T_g or the candy picks up sufficient moisture, these crystals can begin to grow. As they grow, they exclude colors and flavors resulting in mottled appearance and reduced flavor. Internal graining is shown schematically in Figure 7–23A. External graining occurs when the surface of the hard candy picks up sufficient moisture to initiate nucleation, as shown schematically in Figure 7–23B. Once crystallization begins, moisture in the amorphous mass migrates into the rest of the candy and eventually the candy is crystallized throughout. This also results in loss of flavors and a mottled appearance as dyes are distributed nonuniformly at the surface of the sugar crystals. After crystallization has occurred, the total moisture content of the piece begins to decrease since the equilibrium relative humidity of the crystallized piece is substantially lower than that of the glass. Figure 7–24 shows the change in moisture and onset of crystallization as a sugar glass picks up moisture and grains.[63]

One of the primary reasons for adding corn syrup to a hard candy formulation is to inhibit or prevent graining. Addition of 20% corn syrup reduces the likelihood of graining by increasing the induction time for nucleation (as discussed in Chapter 5). Addition of at least 50% corn syrup to a hard candy generally prevents sucrose crystallization for sufficient time that shelf life is well over 1 year. At this level, there is not enough sucrose remaining to crystallize (lower supersaturation), and the corn syrup provides a major

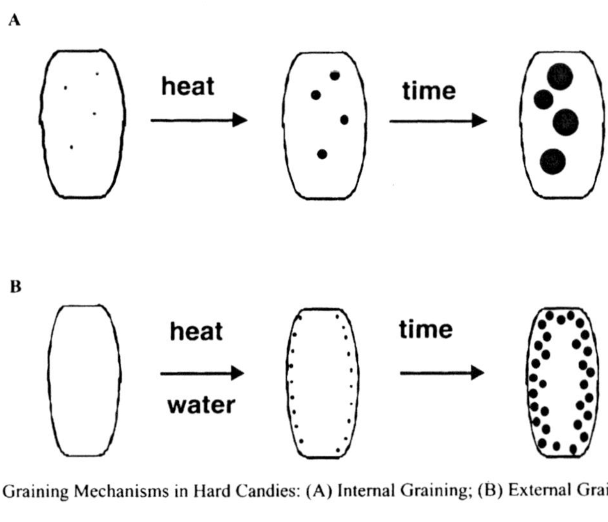

Figure 7–23 Graining Mechanisms in Hard Candies: (A) Internal Graining; (B) External Graining.

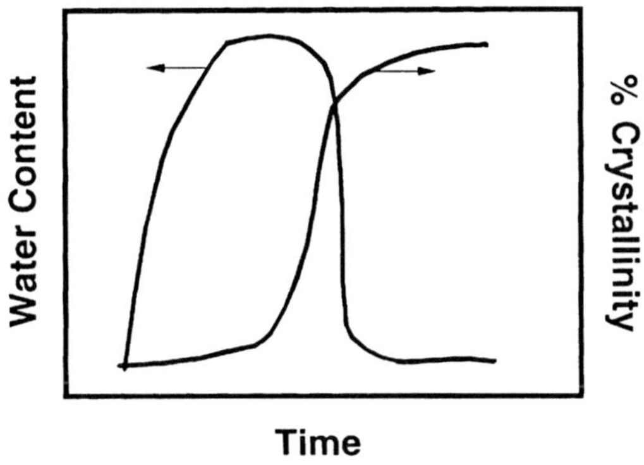

Figure 7–24 Moisture Sorption Followed by Crystallization and Subsequent Drying.

impediment to nuclei formation. Recent studies suggest that there may be some differences in inhibition of sucrose crystallization of different corn syrup components.[59] Corn syrup components with a wide range of molecular species, including branched-chain oligosaccharides, had the greatest inhibition of sucrose crystallization. However, further work is necessary to document the effects of specific polysaccharides on crystallization of sucrose from the amorphous state.

Graining in Caramels and Fudge

As mentioned previously, caramels can be either grained (containing crystals) or ungrained (not containing crystals), depending on the desired textural characteristics. The presence of sugar crystals in a caramel breaks up the protein network structure and causes a "short" texture in the product. Biting into a grained caramel does not result in a stretchy piece, but rather gives a clean break. However, the crystals must be sufficiently small to not induce a coarse or grainy texture to the piece. Controlling crystallization in caramel formulations is necessary to produce the desired texture. If a grained caramel is desired, crystallization must be induced (either through seeding or controlled nucleation) during processing to give the desired crystal size distribution and product texture. However, if crystallization is undesired, the caramel must be formulated and processed to prevent nucleation either during the process or during storage. Often, the end of shelf life for an ungrained caramel is when it crystallizes and changes texture. Uncontrolled crystallization can lead to a grainy product, where crystals are large enough to be detected in the mouth. Depending on the formulation, either lactose or sucrose may crystallize in a caramel. Formulations containing significant quantities of dairy ingredients (milk solids, whey, etc) may become supersaturated in lactose and crystallization of lactose must be controlled.

Quantification of the crystalline microstructure in caramel and fudge can be accomplished most easily using the optical microscope. A sample of caramel can be dispersed on a microscope slide using an appropriate organic solvent (adequately saturated in lactose and sucrose), such as acetone or petroleum ether, and stabilized for image analysis. However, it is difficult to distinguish between lactose and sucrose crystals based on their appearance on a microscope slide. It is possible to separate the sugar crystals from the caramel matrix completely by dissolution in saturated organic solvent. Filtration of this slurry leaves sugar crystals on the filter that can be weighed to give the total mass of crystals present in the caramel and analyzed for particle size distribution using either light-scattering techniques or the Coulter Counter. To distinguish between lactose and sucrose, these crystals must be dissolved and the solution passed through a chromatography column (eg, high pressure liquid chromatography) to identify specific peaks for each sugar. In this way, the crystalline microstructure of a caramel can be completely characterized.

Caramel is manufactured according to the general process shown in Figure 7–25. After mixing the ingredients together, the caramel mass is concentrated in an evaporator and then held at high temperature for a sufficient period of time to allow browning to occur. The extent of browning depends on the nature of the sugars present (reducing sugars) and the time spent at the elevated temperature. The cooked caramel mass is then cooled to the appropriate temperature for crystallization in the case of grained caramels. In some processes, granulated sugar that has been ground to a fine distribution is mixed into the mass to initiate crystallization. In other cases, the mass is simply worked (agitated) to a sufficient extent that nucleation is induced. The optimal temperature for nucleation is again based on the relative rates of nucleation and growth, as shown schematically in Figures 7–1 and 7–2. Typically, this optimal temperature for crystallization of caramel is found by trial and error for any given formulation. After crystallization is completed, the caramel mass is cooled further and formed into the desired shape for packaging. If the caramel is to be ungrained, the uncrystallized mass is cooled directly to forming temperature, followed by packaging.

Choice of ingredients is a major concern in the manufacture of caramel; it can directly influence the crystallization processes. The ratio of sucrose to other ingredients must be selected carefully to either promote or prevent crystallization. In particular, the ratio of sucrose to corn syrup must be adjusted to control sucrose crystallization. Higher levels of corn syrup must be used, in general, to produce an ungrained caramel that is stable for long periods of time. The final water content also is important to control crystallization during storage in an ungrained caramel since viscosity is another key element to inhibiting crystallization. Very low water content results in a highly viscous caramel and long shelf life. In many formulations, whey solids are added at substantial levels for economic reasons. This can create a substantial supersaturation in lactose that must be removed either during processing or storage. In a grained caramel with high lactose levels, it is necessary to control the crystallization of lactose during processing carefully so that many small crystals are formed. The basic principle to achieve this control is the same one operative in fondant manufacture, where nucleation should be induced at high supersaturation and over a short time period to result in a small average crystal size. Ungrained caramel

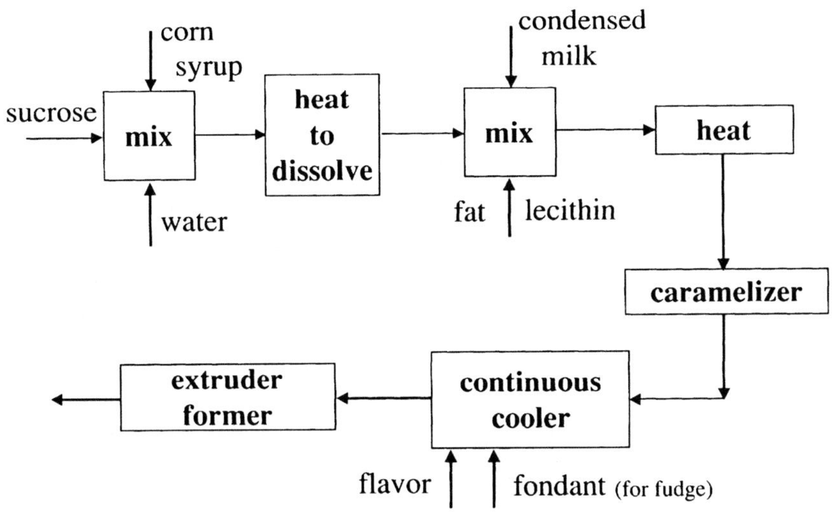

Figure 7–25 Process for Manufacture of Caramel.

with formulations that are high in lactose are very unstable due to the low solubility of lactose. The presence of sucrose also decreases the solubility of lactose, resulting in an even higher supersaturation. Thus, graining is likely during storage of caramels with high lactose content.

In fudge manufacture, either agitation is utilized to induce crystallization of the sugars during processing or fondant sugar is added as a seed to generate the desired crystalline microstructure. The texture of fudge is directly related to the amount of crystalline sugar,[14] so this is a critical element in the quality of fudge. However, little quantitative work has been published to document the relationships between processing conditions, formulation, and crystalline microstructure.

Starch Retrogradation

In many baked products, crystallization of starches over time (during storage) can cause changes in product texture. The changes in starch characteristics during retrogradation occur due to the rearrangement of the starch molecules into a more ordered state.[64] For example, staling of bread has been, at least in part, related to starch retrogradation as the starch molecules crystallize from the amorphous state.[62] The texture and quality of cereals and other extruded products also may be influenced by crystallization of starch.

Starch granules are composed of amorphous regions interspersed with crystalline regions. The crystalline regions are made up of the amylopectin, while the amorphous regions are made up of the branch points of amylopectin and the straight-chain amylose molecules.[65–67] During processing, the starch is gelatinized, which involves melting of the semi-crystalline region in the presence of water. This amorphous, gelatinized starch can be maintained in the fully disordered state by drying and cooling to a temperature below T_g. However, under certain storage conditions, the starch can slowly crystallize. When T_g of the product goes below the storage temperature, T, through either an increase in storage temperature or a decrease in T_g due to water sorption, crystallization is likely to occur. The rate of crystallization (or retrogradation) is dependent on the temperature difference ($T - T_g$).[62,68]

Crystallization in starch gels has been studied with a variety of techniques. Thermal methods (ie, differential scanning calorimetry, differential thermal analysis) have been used to monitor the change in melting enthalpy of the crystalline component.[69] Crystallization of starch gels also has been studied by monitoring changes in X-ray scattering patterns[70] to indicate the transition from amorphous to crystalline. Other methods that have been used include Raman, infrared, and nuclear magnetic resonance (NMR) spectroscopies. Chinachoti[71] reviews the work done with NMR spectroscopy to monitor changes in water and starch characteristics during storage of bread. Changes in rheological properties (eg, firmness) also have been used as an indirect measure of crystallization.

The main parameters that affect crystallization of gelatinized starch are type of starch, water content, and storage temperature.[62,67,68] Amylose has been found to crystallize more rapidly than amylopectin after cooling of gelatinized starch.[62,72,73] Water content has a major impact on the glass transition temperature of starch, as seen previously in Figure 4–30. T_g decreases dramatically (over 200°C) as water content increases from 0 to 20%.[62] Since crystallization only occurs when storage temperature is greater than T_g, moisture content and storage temperature are critical factors in determining stability of starch to crystallization.

Jouppila et al[70] followed the changes in crystallinity, by using X-ray spectroscopy, of gelatinized corn starch equilibrated at different water contents during storage at different temperatures. Their results were analyzed according to a polymer crystallization model developed by Avrami.[74,75] Results of this Avrami analysis give values of an Avrami exponent, which has been related to the nature of the crystallization process (nucleation and growth mechanisms), and a half-life of crystallization, which relates the rate kinetics to operating conditions. Their results showed that, at different conditions of water content and temperature, the nature of the crystalline regions were slightly different and the kinetics of crystallization were different. At lower storage temperatures, smaller and/or less perfect crystalline regions were formed, which correlated with slightly higher Avrami exponent (n) at lower temperatures. The rate of starch crystallization (Avrami half-life) decreased as the difference between storage temperature and T_g increased. However, rates of crystallization were lower, at equivalent ($T - T_g$), for samples with higher water content. Jouppila and Roos[69] also used DSC to monitor the change in melting enthalpy of starch during storage. The rate of increase in the melting enthalpy of corn starch was dependent on water content and temperature of storage. The final melting enthalpy (representative of the amount of starch crystallized) generally increased as $(T - T_g)$ increased. However, a slight decrease in melting enthalpy was observed at very high values of $(T - T_g)$, above 90°C.

The importance of water content and storage temperature is especially critical in controlling crystallization during storage. Dried (or low-moisture-content) food products will pick up moisture from the atmosphere when exposed to relative humidities above the equilibrium relative humidity for that product. The amount of water sorbed in air at different relative humidities is given by the sorption isotherm for a given product. The sorption isotherm gives the moisture content after the product has equilibrated with the water in the surrounding environment. Thus, during storage, a dried food product will pick up a certain amount of moisture dependent on its sorption isotherm. A product in the glassy state may pick up sufficient moisture, causing a decrease in T_g, so that storage temperature eventually exceeds T_g and the product is unstable. Jouppila and Roos[69] compare the moisture sorption isotherm with the plasticizing effect of water on the T_g of corn starch, as shown schematically in Figure 7–26. Selecting a critical temperature

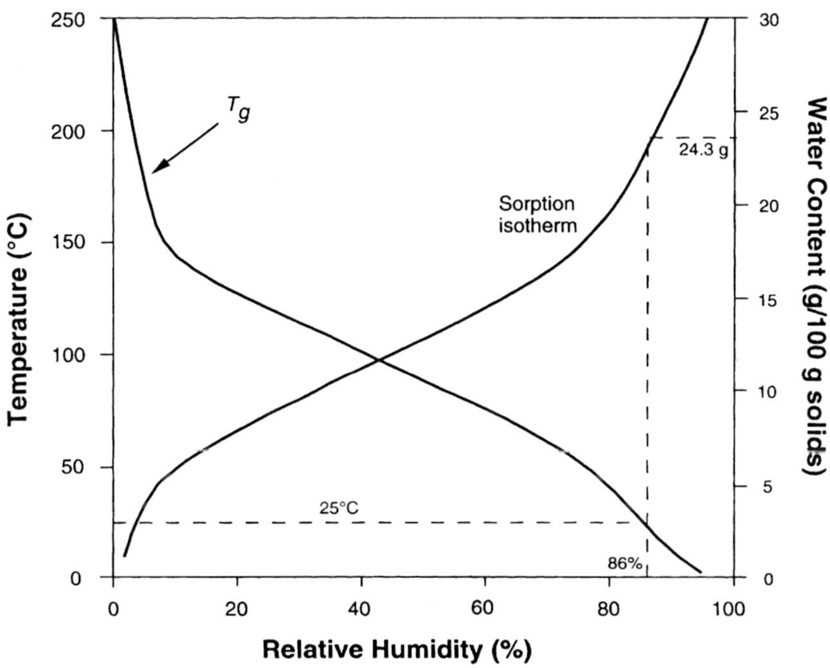

Figure 7–26 Overlay of Sorption Isotherm and Plasticizing Effect of Water on Glass Transition Temperature (T_g) of Starch.

of 25°C, approximately room temperature, allows one to determine the critical moisture content at which the product T_g has decreased to the point where crystallization can now occur.[62] If this critical moisture content is exceeded during storage at room temperature, the product is likely to undergo physical changes (eg, crystallization) and loss of product quality. A similar approach to stability of milk powders, where lactose could crystallize from the amorphous state, was taken by Jouppila and Roos.[76,77]

SUMMARY

There are numerous examples of food products where controlling crystallization is critical to the desired quality, texture, or appearance. An understanding of the principles of crystallization as applied to these processes allows food manufacturers to better control their process for the desired product characteristics. It is necessary to understand the influence of both processing factors and process operating conditions on crystallization phenomena. These factors influence the thermodynamic driving force for crystallization and the specific nucleation and growth rates. However, it often is difficult to apply these principles in commercial processes because of the complexity of food products. For example, the complex interactions of ingredients often leads to uncertainty in the thermodynamic driving force for crystallization in a food product. Clearly, further advances in our understanding of the complex physical chemistry will be required. In addition, our understanding of the relationship between crystalline microstructure and product attributes is not always well understood. For example, the effect of a specific lipid crystalline network on the physical characteristics of butter or shortening is only poorly understood. Thus, further advances in the relationship between crystalline microstructure and food texture are necessary so that the target for controlling crystallization can be accurately identified.

The principles outlined in this chapter provide a guideline for controlling crystallization in foods. However, the necessary depth of understanding of the key elements is lacking for many food products. That is, the thermodynamic behavior and crystallization kinetics have not been defined under the range of conditions needed for optimal control of crystallization. Thus, in practice, controlling crystallization is still very much an art rather than a science.

REFERENCES

1. Johnson ME, Riesterer BA, Olson NF. Influence of nonstarter bacteria on calcium lactate crystallization on the surface of Cheddar cheese. *J Dairy Sci.* 1990;73:1145–1149.
2. Dybing ST, Wiegand JA, Brudvig SA, Huang EA, Chandan RC. Effect of processing variables on the formation of calcium lactate crystals on Cheddar cheese. *J Dairy Sci.* 1988;71:1701.
3. Pearce KN, Creamer LK, Gilles J. Calcium lactate deposits on rindless Cheddar cheese. *NZ J Dairy Sci Technol.* 1973;8:3–7.
4. Deshpande SS, Cheryan M, Sathe SK, Salunkhe DK. Freeze concentration of fruit juices. *CRC Crit Rev Food Sci Nutr.* 1984;20(3):173–248.
5. Hartel RW. Concentration of liquid foods: evaporation and freeze concentration. In: Lund DB, Heldman DR, eds. *Handbook of Food Engineering.* New York: Marcel-Dekker; 1992:341–392.
6. Muller JG. Freeze concentration of liquid foods: theory, practice, and economics. *Food Technol.* 1967;21:49–61.

7. van der Malen BGM, van Pelt WHJM. Multi-stage freeze concentration economics and potentials. In: Cantarelli C, Peri C, eds. *Progress in Food Engineering*. Küsnacht, Switzerland: Foster-Verlag; 1983:413.
8. van Hook A. Nucleation in supersaturated solutions. In: Honig P, ed. *Principles of Sugar Technology*. New York: Elsevier; 1959:2–113.
9. Doner LW. The sugars of honey—a review. *J Science and Food Agriculture*. 1977,28:443–456.
10. Crane E, ed. *Honey: A Comprehensive Survey*. New York: Crane, Russak and Co.; 1975.
11. Assil HI, Sterling R, Sporns P. Crystal control in processed liquid honey. *J Food Sci*. 1991;56(4):1034.
12. Fennema OR, Powrie WD, Marth EH. *Low-Temperature Preservation of Foods and Living Matter*. New York: Marcel Dekker; 1973.
13. Ben-Yoseph E, Hartel RW, Howling D. Three-dimensional model of phase transition of thin sucrose films during drying. *J Food Eng*. 2000;44(1):13–22.
14. Lees R. *Factors Affecting Crystallization in Boiled Sweets, Fondants and Other Confectionery*. Leatherhead, UK: The British Food Manufacturing Industries Research Association; 1965. Scientific and Technical Survey, No. 42.
15. Cakebread SH. Candy chemistry: grained confections. III. *Manufacturing Confectioner*. 1971;51:25.
16. Hartel RW, Shastry AV. Sugar crystallization in food products. *CRC Crit Rev Food Sci Nutr*. 1991;1(1):49–112.
17. Desmarais AJ, Ganz AJ. Effect of cellulose gum on sugar crystallization and its utility in confections. *Manufacturing Confectioner*. 1962;42(10):33.
18. DuRoss J, Knightly WH. The effect of sorbitol on the crystallization of sucrose in selected confectionery products. *Manufacturing Confectioner*. 1963;43(7):26.
19. Hernqvist L. Chocolate temper. In: Beckett ST, ed. *Industrial Chocolate Manufacture and Use*. Glasgow: Blackie; 1988:159–171.
20. Jovanovic O, Karlovic D, Jakovljevic J. Chocolate pre-crystallization: a review. *Acta Alimentaria*. 1995;24(3):225–239.
21. Hartel RW. Phase transitions in chocolates and coatings. In: Rao MA, Hartel RW, eds. *Phase/State Transitions in Foods*. New York: Marcel Dekker; 1998:217–252.
22. Loisel C, Lecq G, Keller G, Ollivon M. Dynamic crystallization of dark chocolate as affected by temperature and lipid additives. *J Food Sci*. 1998;63(1):73–79.
23. Nelson RB. Tempcrers, enrobers, moulding equipment and coolers. In: Beckett ST, ed. *Industrial Chocolate Manufacture and Use*. Glasgow: Blackie; 1988:172–226.
24. Chasieri S, Dimick PS. Cocoa butter—its composition and properties. *Manufacturing Confectioner*. 1987;67(9): 115–122.
25. Kattenberg HR. The effect of cocoa butter on chocolate tempering and bloom. *Manufacturing Confectioner*. 1989;61(5): 77–81.
26. Jeffery M. The effect of cocoa butter origin, milk fat and lecithin levels on the temperability of cocoa butter systems. In: *Proceedings of the 45th Annual Production Conference, Pennsylvania Manufacturing Confectioners' Association*. Center Valley, PA:PMCA;1991:30–36.
27. Savage CM, Dimick PS. Influence of phospholipids during crystallization of hard and soft cocoa butters. *Proceedings of the Pennsylvania Manufacturing Confectioners' Association (PMCA) Production Conference*. Center Valley, PA: PMCA; 1995:54–57.
28. Dimick PS. Compositional impact on crystallization of cocoa butter. In: Widlak N, ed. *Proceedings of the International Conference on Physical Properties of Fats, Oil and Emulsifiers with Applications to Foods*. Champaign, IL: American Oil Chemist's Society Press; 2000:138–161.
29. Precht D. Fat crystal structure in cream and butter. In: Garti N, Sato K, eds. *Crystallization and Polymorphism of Fats and Fatty Acids* New York: Marcel Dekker; 1988:305–361.
30. Kubitz C. *The Effects of Season and Source on the Crystallization Behavior of Milk Fat*. Madison, WI: University of Wisconsin; 1999. MS thesis.
31. Walstra P. On the crystallization habit in fat globules. *Neth Milk Dairy J*. 1967;21:166–191.
32. Buchheim W. Aspects of sample preparation for freeze-fracture/freeze-etch studies of proteins and lipids in foods. A review. *Food Microstructure*. 1982;1:189–208.
33. Precht D, Buchheim W. Elektronmikroskopische untersüchung uber die physikalische struktur von streichfetten. I. Die mikrostruktur der fettkügelchen in butter. *Milchwissenschaft*. 1979;34:745–749

34. Van Boekel MAJS, Walstra P. Stability of oil-in-water emulsions with crystals in the disperse phase. *Colloids Surfaces*. 1981;3:109–118.

35. Mortensen BK, Pedersen HC. Verbutterung in kontinuierlichen butterungsmaschinen *Milchwissenschaft*. 1978;33: 204–208.

36. Mulder H, Walstra P. *The Milk Fat Globule*. Wageningen, The Netherlands: Pudoc; 1974.

37. Precht D, Buchheim W. Elektronmikroskopische untersüchung uber die physikalische struktur von streichfetten. II. Die mikrostruktur der zwischenglobulären fettphase in butter. *Milchwissenschaft*. 1980;35.393–398.

38. DeMan JM, Wood FW. Hardness of butter. II. Influence of setting. *J. Dairy Sci*. 1959;42:56–61.

39. Russell AB, Cheney PE, Wantling SD. Influence of freezing conditions on ice crystallization in ice cream. *J Food Eng* 1999;39:179–191.

40. Arbuckle WS. *Ice Cream*. 4th ed. Westport, CT: AVI; 1986.

41. Hartel RW. Phase transitions during manufacture and storage of ice cream. In: Rao MA, Hartel RW, eds. *Phase/State Transitions in Foods*. New York: Marcel Dekker; 1998:327–368.

42. Berger KG. Ice cream. In: Larrson K, Friberg SE, eds. *Food Emulsions*. 2nd ed. New York: Marcel Dekker; 1990: 367–444.

43. Caldwell KB, Goff HD, Stanley DW. A low-temperature scanning electron microscopy study of ice cream. I. Techniques and general microstructure. *Food Structure*. 1992;11:1–9.

44. Donhowe DP, Hartel RW, Bradley RL. Determination of ice crystal size distributions in frozen desserts. *J Dairy Sci*. 1991;74:3334–3344.

45. Hartel RW. Ice crystallization during manufacture of ice cream. *Trends Food Sci Technol*. 1996;7(10):315–320.

46. Frazeur DR. Body and texture characteristics of low temperature frozen product. *Ice Cream Trade J*. 1964;60(11):72.

47. Frazeur DR, Harrington RB. Low temperature and conventionally frozen ice cream. *Food Technol*. 1968;22:98.

48. Schwartzberg HG. Food freeze concentration. In: Schwartzberg HG, Rao MA, eds. *Biotechnology and Food Process Engineering*. New York: Marcel Dekker; 1990.127–202.

49. Jacobsen J. *Ice Crystallization in a Scraped-Surface Slush Freezer*. Madison, WI: University of Wisconsin-Madison; 1999. MS thesis.

50. Grall DS, Hartel RW. Kinetics of butterfat crystallization. *J Am Oil Chem Soc*. 1992;69(8):741.

51. Ng WL. Nucleation behaviour of tripalmitin from a triolein solution. *J Am Oil Chem Soc*. 1989;66(8):1103–1106.

52. Dibildox-Alvarado E, Toro-Vazquez JF. Isothermal crystallization of tripalmitin in sesame oil. *J Am Oil Chem Soc*. 1997;74(2):69–76.

53. Dibildox-Alvarado E, Toro-Vazquez JF. Evaluation of tripalmitin crystallization in sesame oil through a modified Avrami equation. *J Am Oil Chem Soc*. 1998;75(1):73–76.

54. Kloek W, Walstra P, Van Vliet T. Crystallization kinetics of fully hydrogenated palm oil in sunflower oil mixtures. *J Am Oil Chem Soc*. 2000;77(4):389–398.

55. Herrera ML, de Leon Gatti M, Hartel RW. A kinetic analysis of crystallization of a milk fat model system. *Food Res Int*. 1999;32:289–298.

56. Timms RE. The phase behavior and polymorphism of milk fat, milk fat fractions and fully hardened milk fat. *Aust J Dairy Technol* 1980;35(2):47–53.

57. Hartel RW, Kaylegian KE. Advances in milk fat fractionation: technology and applications. In: Garti N, Sato K, eds. *Crystallization Processes in Fats and Lipid Systems*. In press.

58. Patience D, Hartel RW, Illingworth D. Crystallization and pressure filtration of anhydrous milk fat: mixing effects. *J Am Oil Chem Soc*. 1999;76(5):585–594.

59. Gabarra P, Hartel RW. Corn syrup solids and their saccharide fractions affect crystallization of amorphous sucrose. *J Food Sci*. 1998;63(3):523–528.

60. Downton GE, Flores-Luna JL, King CJ. Mechanisms of stickiness in hygroscopic, amorphous powders. *Industrial Engineering Chemistry Fundamentals* 1982;21:447–451.

61. Wallack DA, King CJ. Sticking and agglomeration of hygroscopic, amorphous carbohydrate and food powders. *Biotechnol Prog*. 1988;4:32–35.

62. Roos YH. *Phase Transitions in Foods*. San Diego: Academic Press; 1995.

63. Makower B, Dye WB. Equilibrium moisture content and crystallization of amorphous sucrose and glucose. *J Agric Food Chem.* 1956;4:72–77.

64. Atwell WA, Hood LF, Lineback DR, Varriano-Marston E, Zobel HF. The terminology and methodology associated with basic starch phenomena. *Cereal Foods World* 1988;33:306–311.

65. Morris VJ. Starch gelation and retrogradation. *Trends Food Sci Technol.* 1990;1:2–6.

66. Oates CG. Towards an understanding of starch granule structure and hydrolysis. *Trends Food Sci Technol.* 1997; 8:375–382.

67. Buleon A, Le Bail P, Colonna P, Bizot H. Phase and polymorphic transitions of starches at low and intermediate water contents. In: Reid DS, ed. *The Properties of Water in Foods: ISOPOW 6.* Glasgow: Blackie; 1998:160–178.

68. Slade L, Levine H. Beyond water activity: recent advances based on an alternative approach to the assessment of food quality and safety. *CRC Crit Rev Food Sci Nutr.* 1991;30:115–360.

69. Jouppila K, Roos YH. The physical state of amorphous corn starch and its impact on crystallization. *Carbohydr Polymers.* 1997;32:95–104.

70. Jouppila K, Kansikas J, Roos YH. Factors affecting crystallization and crystallization kinetics in amorphous corn starch. *Carbohydr Polymers.* 1998;36:143–149.

71. Chinachoti P. NMR dynamics properties of water in relation to the thermal characteristics of bread. In: Reid DS, ed. *The Properties of Water in Foods: ISOPOW 6.* Glasgow: Blackie; 1998:139–159

72. Miles MJ, Morris VJ, Orford PD, Ring SG. The roles of amylose and amylopectin in the gelation and retrogradation of starch. *Carbohydr Res.* 1985;135:271–281.

73. Miles MJ, Morris VJ, Ring SG. Some recent observations on the retrogradation of amylose. *Carbohydr Polymers.* 1984;4:73–77.

74. Avrami M. Kinetics of phase change. I. General theory. *J Chem Phys.* 1939;7:1103–1112.

75. Avrami M. Kinetics of phase change. II. Transformation-time relations for random distribution of nuclei. *J Chem Phys.* 1940;8:212–224.

76. Jouppila K, Roos YH. Water sorption and time-dependent phenomena of milk powders. *J Dairy Sci.* 1994;77: 1798–1808.

77. Jouppila K, Roos YH. Glass transitions and crystallization in milk powders. *J Dairy Sci* 1994;77:2907–2915.

Chapter 8

Recrystallization Processes

Even though a distribution of crystals has reached a phase equilibrium with its surrounding environment, changes to this crystalline microstructure may still occur. The crystals may change in size, number, shape, orientation, and degree of internal perfection[1] as the system approaches a different energy minimum (surface free energy, internal energy, etc). These changes typically occur during packaging, storage, and distribution, although they also may occur during processing.[2] The result of recrystallization is usually a product that starts out initially with high quality and ends up with undesired characteristics. Some typical cases where recrystallization occurs in foods are given in Table 8–1.

A classic example of recrystallization is the growth of ice crystals in ice cream (and other frozen foods) during storage. In frozen desserts, larger ice crystals indicate a coarser product, whereas larger ice crystals in frozen foods can disrupt cell structures and lead to loss of structural integrity upon thawing. Recrystallization also occurs in sugar crystals during storage of fondant, and it may occur to some extent in grained caramels and nougats as well. Development of grainy texture in margarine due to polymorphic transformation[3] also is a form of recrystallization. Fat bloom in chocolates is due to recrystallization of cocoa butter into a more stable polymorph at the surface[4,5]

Typically, recrystallization is promoted when a product is stored at conditions where the mobility of molecules is high. In frozen foods, storage at temperatures close to the melting point causes the most rapid recrystallization. Mobility of water and solute molecules is highest at this point and diffusion-based processes can occur rapidly. In products containing lipids, recrystallization is promoted at temperatures close to the final

Table 8–1 Food Product Examples Where Changes In Crystalline Structure during Storage Is Important

	Control to Prevent Changes in Crystalline Structure	
Frozen foods	Ice	Prevent coarsening
Frozen desserts	Ice	Prevent coarsening
Chocolates	Cocoa butter	Prevent fat bloom
Margarines	Vegetable fat	Prevent graininess
Fondants	Sucrose	Prevent coarsening
Baked goods	Starch	Retrogradation and staling

melting point of the fat. This usually occurs at temperatures above room temperature but it depends on the type of fat used in the product. Temperature fluctuations also are important to the rate of recrystallization during storage. In general, larger thermal fluctuations lead to more rapid recrystallization, although this finding has not been verified in every circumstance. More important, perhaps, recrystallization may still occur, although at reduced rates, even when temperature is held constant. Other factors that influence recrystallization include the state of the original crystal size distribution (CSD) at the onset of recrystallization and the composition of the product. These factors also influence the formation of crystalline microstructure during processing, and here their effect on subsequent changes in crystallinity also must be considered.

Recrystallization is generally evidenced by an increase in the mean size and spread in the width of a CSD, as seen in Figure 8–1. The increase in mean size is accompanied by a decrease in the number of crystals (Figure 8–2) so that the total phase volume of crystalline matter remains the same. In this case, the surface free energy is being minimized. However, recrystallization also may result in a change in the shape of crystals and a transformation to a more stable polymorph as the internal energy of the crystals is minimized. In lipid systems, both forms of recrystallization may be evident.

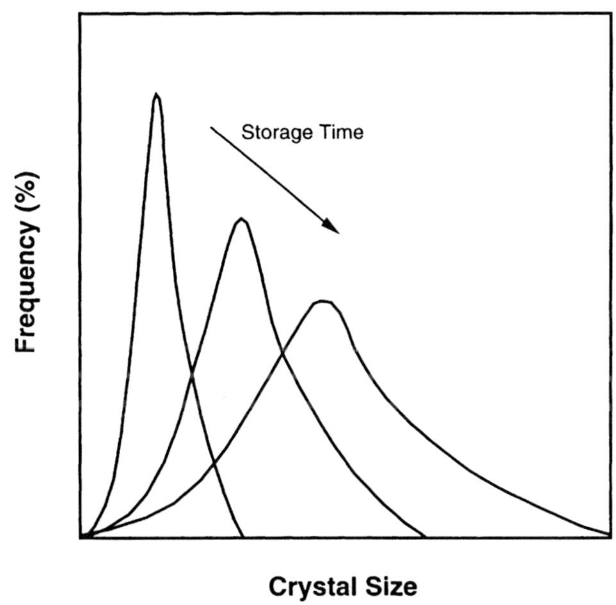

Figure 8–1 Typical Changes in Crystal Size Distribution in Ice Cream during Storage.

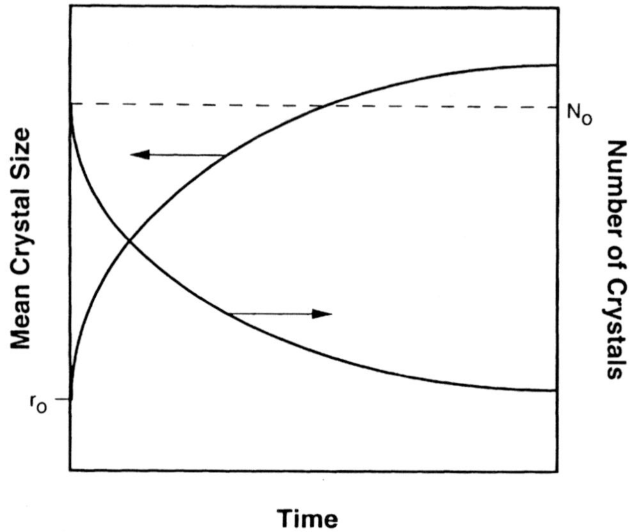

Figure 8–2 Schematic Representation of Increase in Mean Size in Decrease in Total Crystal Numbers with Time of Recrystallization.

MECHANISMS OF RECRYSTALLIZATION

There are a number of mechanisms that act together to cause recrystallization.[1,6] Although most of these mechanisms occur with constant phase volume, some changes in crystalline microstructure in food products may be due to a change in the phase volume. Mechanisms of recrystallization that involve no net change in crystal phase volume are often categorized according to whether temperature is constant (static) or whether it is fluctuating (dynamic). One form of recrystallization does involve a change in crystalline phase volume. This is irruptive recrystallization, which occurs when the full amount of crystalline material has not been formed due to some type of constraint (molecular mobility) during processing, but then subsequent rewarming causes additional crystallization. Although irruptive crystallization involves a change in crystalline mass, it will be included in this discussion as a mechanism of recrystallization important in foods.

Ostwald Ripening

The underlying physical phenomenon responsible for most recrystallization is the thermodynamic ripening phenomenon, often attributed to Ostwald.[7] In many cases, it is the thermodynamic difference in equilibrium between large and small crystals that leads to recrystallization (sometimes also called ripening). Thus, Ostwald ripening occurs due to the small difference in equilibrium between large and small crystals. In this case, the small crystals disappear and the large ones grow.

The difference in thermodynamic equilibrium between crystals of different size is defined by the Kelvin or the Gibbs-Thomson equation.[7] The original Kelvin equation was written to describe droplets of liquid in a vapor in terms of vapor pressures around droplets of different size. However, in crystallization studies, the vapor pressure may be related, through the solution activity, to the equilibrium concentration in solution for crystals of different size. Making the assumption that concentration ratio (or actually the ratio of activities) can be used directly for vapor pressure ratio gives:

$$kT \ln\left(\frac{p}{p_\infty}\right) = \frac{2\sigma v}{r} \approx kT \ln\left\{\frac{C(r)}{C_s}\right\} \tag{8.1}$$

where $C(r)$ is the concentration in equilibrium with a crystal of size r, C_s is the concentration in equilibrium with an infinitely large crystal (or flat plane), k is the Boltzman constant, T is absolute temperature, v is molecular volume, and σ is the interfacial tension between crystal and solution. Equation 8.1 assumes that the interfacial tension of a crystal in a solution is independent of crystal size and predicts that solubility concentration approaches infinity as particle size goes to zero. Since this cannot be true, Equation 8.1 has been modified[7] to account for surface electrical charge as particles get increasingly smaller. Dadyburjor and Ruckenstein[8,9] further develop the Gibbs-Thomson equation for particles in solution.

The Gibbs-Thomson equation also can be written for melt systems[7] as

$$\Delta T = T_0 - T(r) = \frac{2\sigma v T_0}{\Delta H_f r} \tag{8.2}$$

where ΔT is the difference in melting points between a crystal of infinite size, T_0, and a crystal of size, r, and ΔH_f is the latent heat of fusion for the crystallizing species.

Based on the Gibbs-Thomson equation, conditions may occur where small crystals are in undersaturated conditions at the same time that large crystals are in supersaturated conditions. Thus, small crystals dissolve or melt and large crystals grow. For example, as seen in Figure 8–3, a single small ice crystal can be seen to disappear in the vicinity of a larger crystal.[10] However, the difference in equilibrium condition between large and small crystals based on the Gibbs-Thomson equation is very small. In most systems, true Ostwald ripening is only important for crystals smaller than 1 to 2 μm. For ice, Ostwald ripening may be important in crystals as large as 5 μm. Based on approximations for the ice-sucrose system, the differences in melting point between ice crystals of different sizes at –5°C can be calculated,[6,11] and are shown in Figure 8–4. The line represents the difference in equilibrium melting temperature as size increases. Note that even ice crystals

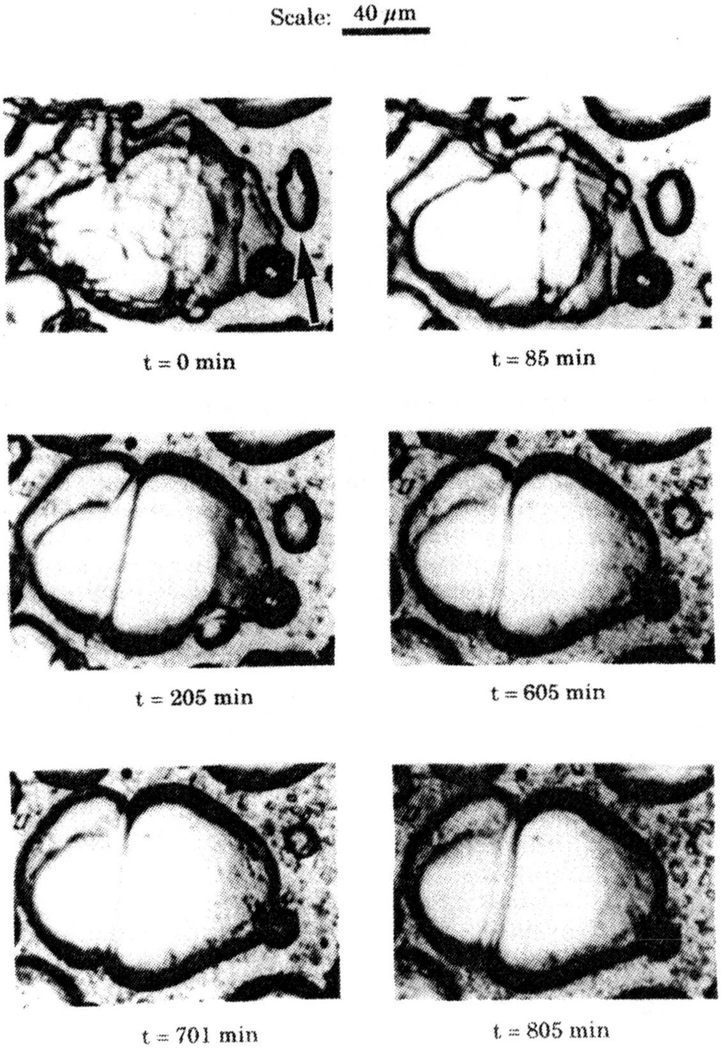

Figure 8–3 Ostwald Ripening of a Small Crystal during Recrystallization of an Ice Cream Sample on a Controlled Temperature Microscope Stage at −10 ± 0.01°C.

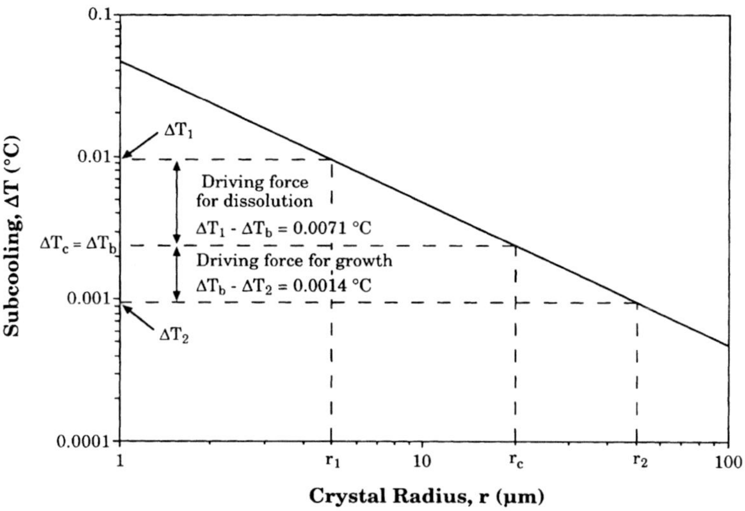

Figure 8–4 Effect of Radius of Curvature on Subcooling Relative to a Plane Ice Crystal Surface.

of 1-μm radius have only about 0.05°C difference in melting temperature as compared with an infinitely large crystal. Although these driving forces are quite small, during the long times of storage for many products, significant changes in CSD can still occur.

The rate of increase in mean crystal size depends on the mechanism that controls crystallization under these conditions. A general asymptotic solution[12] is given by

$$\bar{r}^n = \bar{r}_0^n + \frac{t}{\tau} \tag{8.3}$$

where r is the mean radius at any time, t; r_0 is initial mean crystal size; and n and τ are parameters that depend on the mechanism of recrystallization. When bulk diffusion of molecular species limits the rate of ripening, $n = 3$, and when surface integration process controls ripening, $n = 2$.[7,12] Equation 8.3 can be simplified for longer times to give

$$\bar{r} = \bar{r}_0 + Rt^{1/n} \tag{8.4}$$

where R is the rate of recrystallization. The general mechanism for recrystallization processes can be determined, in part, by determining the value of n from Equation 8.4 that matches the experimental results. For recrystallization of ice during storage of ice cream, it has been found that $n = 3$.[10,13] This indicates that it is the mobility of water molecules through the ice cream matrix that limits the rate of recrystallization in ice cream. In either model sugar systems or ice cream, mean size of ice crystals increased as a function of time to a power between 0.33 and 0.5.[14–18] Bevilacqua and Zaritzky[19] and Martino and Zaritzky[20,21] found that mean size increased with time to the power of 0.5 in frozen beef and 0.28 sodium chloride (NaCl) solutions. Differences in results for values of n in Equation 8.4 can be attributed to different experimental conditions that lead to the dominance of particular recrystallization mechanisms.

The total number of crystals must decrease to balance the increase in crystal size in order to maintain constant crystal phase volume during migratory recrystallization. For the case of diffusion-limited ripening, the decrease in number of crystals can be written as[22]

$$\frac{1}{N} = \frac{1}{N_0}\left(1 + \frac{t}{\tau}\right) \tag{8.5}$$

where N is the number of crystals at any time, t; N_0 is the initial number of crystals; and τ is the same time constant for recrystallization used in Equation 8.3.

Ostwald ripening has been studied to a great extent in materials other than foods, particularly in metals.[23] Dunning[22] and Kahlweit[24] provide further details on the mechanisms and kinetics of Ostwald ripening. However, application of these principles to ripening in foods has been limited due to the additional complexities caused by the nature of food structures.

Isomass Rounding

The ripening process may occur even within a single crystal of nonspherical shape. Rough edges with different radii of curvature have slightly different equilibrium energies than smooth flat surfaces; over time, these differences may be reduced as the system approaches a lower surface free energy. This process, often called isomass rounding,[1] may occur in a single isolated crystal, although often there are neighboring crystals that also are involved. If there are no interactions between adjacent crystals, isomass rounding does not lead to a change in crystal number. The phase volume of crystalline material again remains the same during isomass rounding, with the phase distribution changing to minimize surface free energy. When rounding is the primary mechanism of recrystallization, the crystals gradually become increasingly spherical in nature (Figure 8–5). Although this is an important mechanism of recrystallization under certain circumstances, particularly when temperature does not fluctuate substantially, there has been very little quantitative study of this mechanism of recrystallization.

Accretion

Another static mechanism of recrystallization is accretion or fusion. Crystals that are very close to each other tend to grow together to form one larger crystal (Figure 8–6).

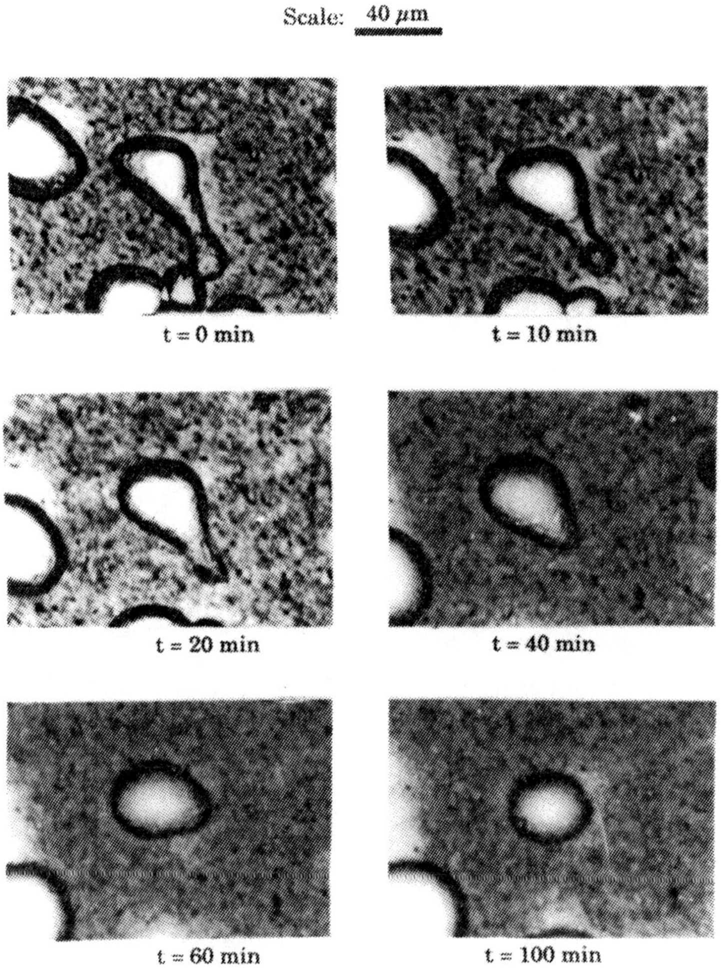

Figure 8–5 Isomass Rounding of Ice Crystals during Recrystallization of an Ice Cream Sample on a Controlled Temperature Microscope Stage at –5 ± 0.01°C.

Heat and mass transfer in the region directly between the two crystals cause a bridge of crystalline material to form between them until eventually the two crystals have grown into one.[25] This mechanism is very important in some foods, particularly those with high

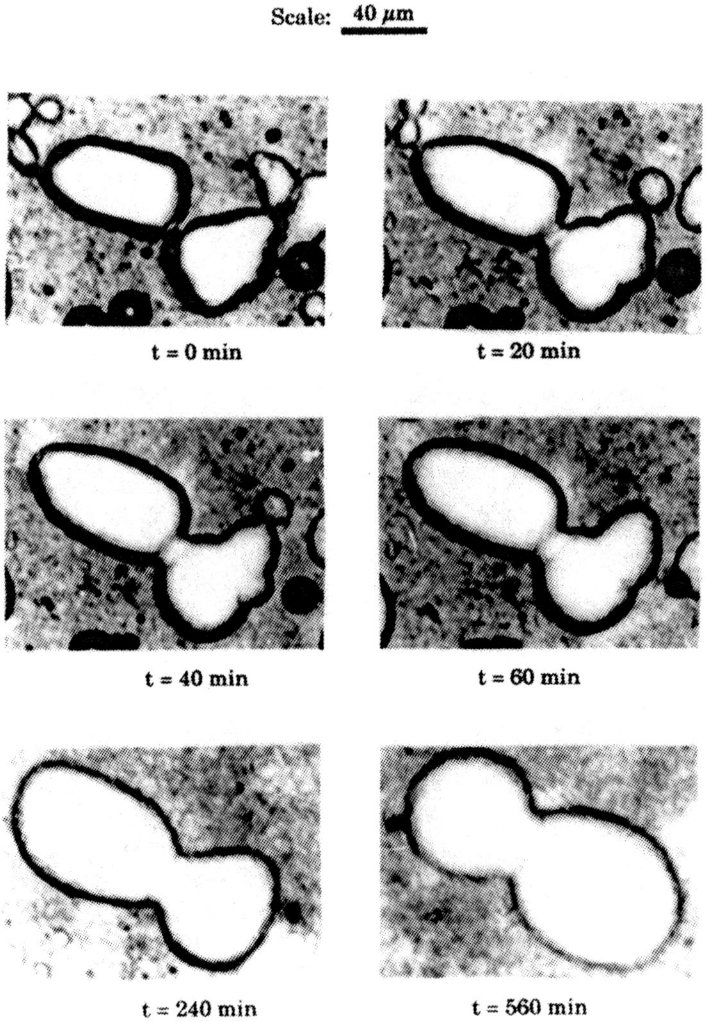

Figure 8–6 Accretion of Two Crystals, Initially Separated, during Recrystallization of an Ice Cream Sample on a Controlled Temperature Microscope Stage at $-5 \pm 0.01°C$.

numbers of very small crystals in close proximity. For example, it has been observed as a substantial contributing mechanism in ripening of ice crystals in ice cream.[10]

Studies on accretion have primarily focused on the joining of crystals surrounded by air, a process often called sintering.[25–28] Sintering is important, for example, as freshly fallen snow packs into denser structures over time. It also is important in materials processing, as the individual particles of a composite material that form initially in a process are sintered together to form a bulk structure. Local concentration (or vapor pressure in the case of air) gradients are higher in the region surrounding two adjacent crystals, and this eventually leads to formation of a neck or bridge between the two crystals.[9] Once this neck has formed, this region has high curvature and is unstable compared with the rest of the two crystals according to the Gibbs-Thomson equation. The rate at which the neck widens with time has been given as[25]

$$\frac{x^a}{r^b} = B(T)t \tag{8.6}$$

where x is half-width of the neck at time, t; r is crystal radius; $B(T)$ is a function of temperature that incorporates physical properties of the system; and the parameters a and b depend on the mechanism of mass transport to the neck. When bulk diffusion limits the rate of mass transport, a and $b = 1$, whereas when surface diffusion is limiting, $a = 7$ and $b = 3$.[29]

The rate of recrystallization by accretion also is dependent on the probability that two crystals are in the same region and can form a bridge. This is a function of how tightly packed the crystals are in the product. The "coalescence frequency" has been related to the average volume of the two crystals, although the surface area also has been used to refine sintering theory. However, these theories have not always met with great success and our understanding of the process of accretion recrystallization, especially in complex foods, remains quite poor.

Despite this lack of understanding of the fundamental mechanisms, accretion is a major factor in recrystallization in some food products. Through microscopic observations, it has been shown that accretion of ice crystals is the dominant process of recrystallization when ice crystals are very small and numerous.[30] A sugar solution rapidly frozen on a microscope slide contained many small ice crystals that initially recrystallized due to accretion of these small crystals. After some time, the remaining crystals were sufficiently separated in space that accretion was no longer the main mechanism of ripening. This study also clearly documented how the mechanism of recrystallization may change during the process (or during storage of a food product) as conditions within the product change.

Fusion or sintering of fat crystals also is an important process in many lipid-based foods. Butter, margarine and chocolate are a few examples of products where the nature of the fat crystalline microstructure influences product rheology and texture. Once fat crystals have formed during processing of these foods, additional changes in the crystalline microstructure (number, size, perfection, polymorphic form, etc) may still take place due to thermodynamic ripening processes. Johansson and Bergenstahl[31] studied sintering of fat crystal networks during post-crystallization (cooling, storage, etc) and found that numerous factors influenced the nature of the fat crystal microstructure. In particular, cooling rate after processing and the presence of various surface-active agents (eg, emulsifiers) had significant effect on the amount of sintering between fat crystals. These differences in fat crystal microstructure due to sintering also influenced the mechanical properties of the material, as measured by changes in the Bingham yield stress by Johansson and Bergenstahl.[31] The effects of fat crystals on stability of emulsions also was studied.[32–34]

Polymorphic Transformation

Changes in the internal structure of crystals also may occur during storage (both static and dynamic) and lead to negative changes in product quality. For example, polymorphic changes that occur in lipid crystals during storage of foods is an example of recrystallization. In this case, the particulate system is approaching a lower internal energy state, although this change in internal energy also may be combined with changes in the particle size distribution and surface energy of the crystals. Margarine and chocolate are products where the correct form of the lipid crystal is desired to maintain product quality. In both products, a more stable lipid polymorph often exists than what is formed in the commercial product, and transformation to this lower energy state brings about undesirable textural changes in the product. In both products, the lipids are tempered during processing (according to some desired temperature-time profile) to produce the desired crystalline phase. Subsequent changes in that crystalline phase due to recrystallization generally lead to negative changes in product quality.

As discussed in Chapter 2, lipids undergo a polymorphic change as less stable crystal forms change to more stable forms. In most cases, it is sufficient to consider only three different polymorphs in lipids: α, β', and β, (in order of increasing stability or lower internal energy level). However, in some situations it is necessary to consider subtypes of these main polymorphs. During manufacture of lipid-containing foods, typically the polymorphs of lower stability are formed first due to the faster nucleation kinetics for forming these "looser" crystals. However, the unstable polymorphs eventually convert to the more stable forms with a subsequent release of additional latent heat. If this polymorphic transition is uncontrolled, the additional latent heat may cause substantial problems in the product. The time scale for this transition depends on such factors as temperature of storage and solid fat content at that temperature, the difference in melting temperature between polymorphs, and the complexity of the lipid (triacylglycerol) structure.

Melt/Refreeze

Another important mechanism of recrystallization (dynamic) is due to thermal fluctuations that naturally occur during storage. A change in the phase volume of crystalline material occurs as temperatures fluctuate. These changes in phase volume are often accompanied by a change in crystalline structure (size, number, shape, etc) and decrease in product quality. For example, recrystallization in ice cream or frozen food is enhanced due to fluctuating temperatures during storage and leads to either a coarse product (ice cream) or loss of structural integrity (frozen food). As temperature goes up, crystals either melt or dissolve to maintain the phase equilibrium at the new product temperature. When temperature cycles back down, crystalline material grows back onto the remaining crystals. The smallest crystals are most sensitive to temperature increases (based on the Gibbs-Thomson equation) and may eventually disappear or dissolve/melt away. Once a crystal is gone, the mass contained in that crystal is distributed among the remaining crystals and they all grow slightly larger. This assumes that no nucleation occurs during the temperature-decreasing cycle. As long as the temperature fluctuations are reasonably slow and of limited amplitude, nucleation is unlikely to occur within the product because subcoolings remain low.[7] An extreme case of melting and then refreezing may occur when a frozen food is allowed to remain at room temperature for a sufficiently long time that the product at the package surface actually melts. Upon placing the package back in the freezer, ice crystals may nucleate in the melted region, although growth of the remaining crystals also will still occur.

The process of cycling in phase volume with fluctuating temperatures is shown schematically in Figures 8–7 and 8–8. As temperature increases in frozen systems (Figure 8–7), the amount of ice present is reduced as the system strives to maintain equilibrium conditions. In solution systems (Figure 8–8), the system strives to maintain an equilibrium amount of solute in solution as defined by the solubility curve. In this case too, the amount of crystalline phase volume decreases as temperature increases due to the higher solubility. When the temperature decreases again, an increase in phase volume for both melt and solution systems occurs. However, this phase volume may be redistributed among the remaining crystals and result in changes in the CSD and product quality. The difference in equilibrium conditions for small crystals as compared with larger ones enhances the likelihood that small crystals will disappear first. Because of this difference in equilibrium condition, smaller crystals melt or dissolve slightly faster than larger crystals. Eventually, the small ones will melt or dissolve away, causing the expected decrease in particle number and increase in mean size.[10] Figure 8–9 documents the effect of temperature cycling on ice crystals in ice cream, showing the rapid disappearance of a small ice crystal in the vicinity of a larger one. The actual change in ice crystal size distribution during accelerated recrystallization of ice cream is shown in Figure 8–10.

The discussion above assumes that temperature fluctuations are sufficiently slow that the system maintained an approximate equilibrium phase volume at all times. This is probably true, in most cases, during long-term storage of foods since temperatures generally change slowly in storage facilities. In storage freezers, temperature fluctuations generally have at least a 2-hour cycle time, and the amplitude may only be about a degree or two. In this case, the temperature within the frozen food changes according to the laws of unsteady state heat transfer and depends on the thermal diffusivity of the product. In addition, the change in latent heat caused by melting and regrowth during temperature cycling insulates the effects of external temperature fluctuations. Thus, large external

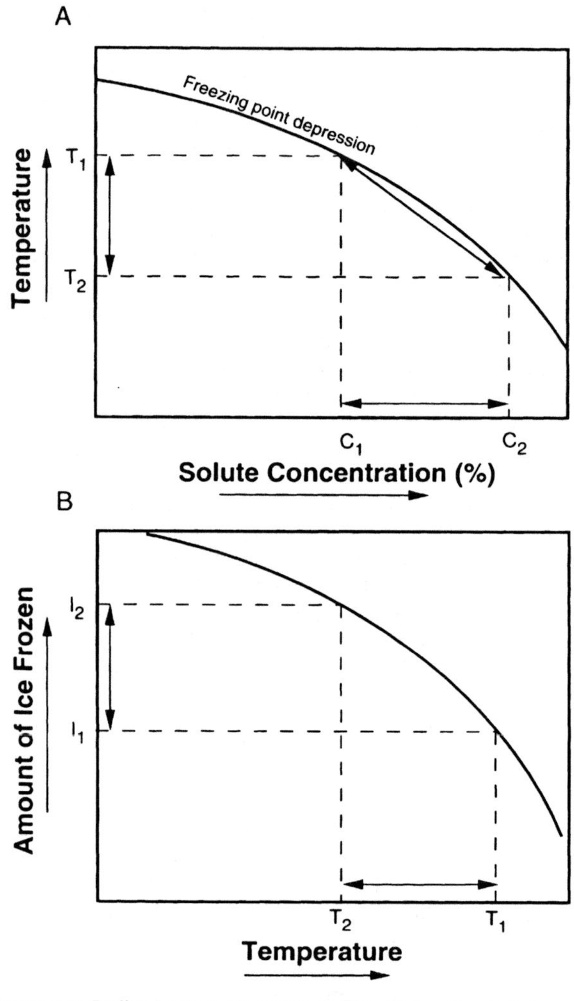

Figure 8–7 Temperature Cycling Leads to Phase Volume Changes in a Frozen System: (A) Freezing Point Depression Curve; (B) Amount of Ice Frozen.

temperature fluctuations often result in relatively small changes within the product,[35] and the crystalline structure maintains an approximate equilibrium at the temperature within the product. A similar argument can probably be made for products stored at room

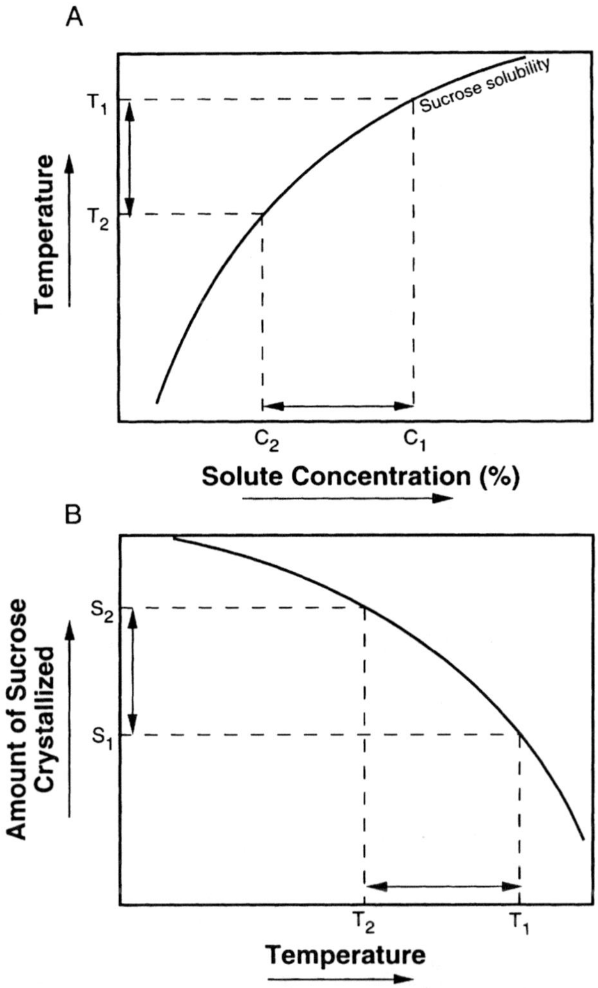

Figure 8–8 Temperature Cycling Leads to Phase Volume Changes in Solution System: (A) Solubility Curve; (B) Amount of Sugar Crystallized.

temperature and experiencing normal, daily temperature fluctuations. The magnitude of these changes may be substantial, but the rate at which they occur is generally slow enough that the product maintains approximate phase volume equilibrium.

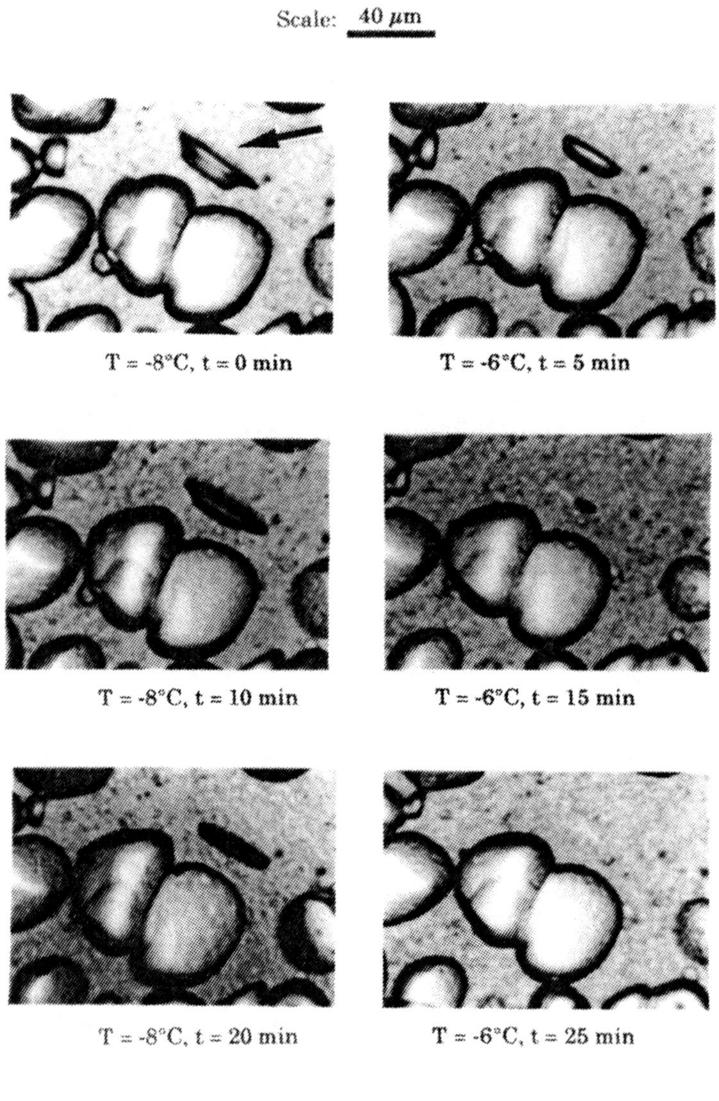

Figure 8–9 Disappearance of an Ice Crystal in Ice Cream due to Melting and Refreezing during Recrystallization of an Ice Cream Sample on a Controlled Temperature Microscope Stage (–7 ± 1°C with 10 min oscillations).

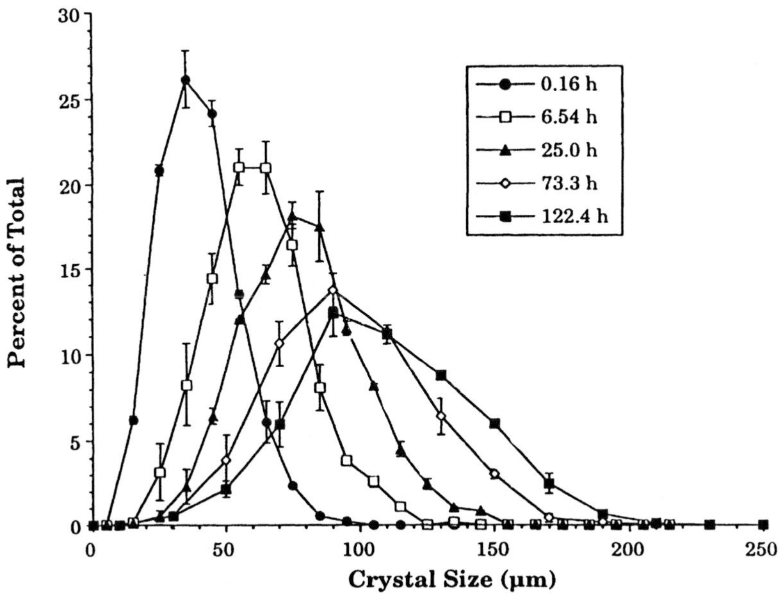

Figure 8–10 Change in Ice Crystal Size Distribution during Accelerated Recrystallization of Vanilla Ice Cream on a Cold-Stage Microscope at a Constant Temperature of −5°C (± 0.01°C)

However, there may be cases where temperature fluctuations are so fast and/or the rate of melting/crystallization is so slow that a phase volume equilibrium is not maintained. In this case, less than the maximum phase volume change occurs as temperature cycles. That is, before the system can equilibrate to a reduced phase volume at high temperature, the system is already cooling and causing an increase in phase volume. This may occur, for example, in frozen foods during heat shocking (large temperature swings). There are several differences in this case that may impact the CSD in a food product. If less than the equilibrium phase volume changes, then there is less crystalline mass to be redistributed and the chance for less change in the crystalline phase. However, if there are large and rapid temperature swings, as in the case of heat shock of frozen foods, the surface of the food may be completely melted. When the product is refrozen, uncontrolled nucleation may occur and result in an undesirable CSD and an unacceptable product. This also may occur in chocolates, for example, that have been melted accidentally. When resolidified as the temperature cools, cocoa butter nucleates into an unstable polymorphic form. This leads to a product that quickly develops fat bloom as the unstable polymorph transforms to the more stable forms.

Irruptive

Irruptive recrystallization has been described as the change in crystallinity of a food product when the equilibrium amount of crystalline material was not formed initially, but it then approaches the equilibrium phase volume during storage.[1,36] For example, a food product can be frozen so rapidly that the glassy state was approached before crystallization could be completed. For a given storage temperature, there is an equilibrium phase volume of ice as determined by the freezing point depression curve. During rapid cooling and freezing, mobility decreases and viscosity increases so dramatically that crystallization is inhibited and there may be less ice than expected. In this case, warming the product causes additional ice to freeze, and this would continue until the equilibrium phase volume for ice at that temperature was attained. This increase in crystal phase volume may have an impact on product texture. For example, a rapidly frozen fruit or vegetable would contain small ice crystals that could increase dramatically in size if the product were to be warmed up. The increase in ice crystal size may affect the cellular structure of the food and cause irreversible damage to texture.

In some cases, rapid cooling of a material results in formation of the glassy state with no crystalline structure. When this material is rewarmed, nucleation and growth must occur. This is slightly different from irruptive recrystallization, which is based on limited crystal formation during rapid cooling. In this case, nucleation already has occurred during the cooling step, and rewarming results in growth, release of additional latent heat, and recrystallization. The uncontrolled nucleation and growth that occur in this case may be particularly detrimental to product quality.

Luyet[36] discussed the phase transitions that occur during rapid freezing of aqueous solutions. Studies on biological materials (sugars, proteins, etc) using microscopy (optical and scanning electron), X-ray spectroscopy, and calorimetry showed that rapid freezing resulted in formation of very small ice crystal spherulites or dendrites that did not grow as long as the temperature remained below some critical temperature. Upon warming, however, these solutions exhibited rapid ice crystal growth with a subsequent release of latent heat. As these crystals grew and released latent heat, recrystallization of the ice crystals also occurred. Changes in crystalline structure, as measured by X-ray spectroscopy, also occurred during rewarming, which suggested that different crystalline lattice structures were formed.

FACTORS THAT INFLUENCE RECRYSTALLIZATION

Initial Crystalline State

Recrystallization rates depend on the nature of the initial crystalline material. Crystals that are smooth and rounded, are of about the same size, have stable internal energy (most stable polymorph), are relatively distant from each other, and experience uniform temperatures during storage are not likely to experience rapid recrystallization. However, crystals that have widely different sizes, take unusual shapes, are in unstable polymorphs, are compacted together, and undergo significant thermal fluctuations during storage are likely to undergo rapid recrystallization. Thus, it is extremely important to produce crystals of desired size, shape, and perfection during processing.

The formation of crystals during processing to control the desired amount, shape, size distribution, and degree of perfection is critical to extended shelf life in many food products. For example, the process for forming many small crystals during freezing of foods (fruits, vegetables, frozen desserts, etc) is critical to controlling recrystallization during storage. Formation of an improper size distribution during processing typically leads to more rapid recrystallization and degradation of product quality.

The effects of initial beating temperature during production of fondant on change in crystal size distribution is shown in Table 8–2. The lower beating temperature promotes many more and smaller nuclei than beating at higher temperatures. These smaller crystals are also more resistant to changes during storage, although recrystallization still takes place during storage; this causes the fondant to become coarser.[37]

In the same way, crystallization of lipid materials into the appropriate size, distribution of sizes, shape, and polymorph is critical to controlling product quality in many cases. Production of high-quality butter, chocolate, and margarine, to name just a few products, requires careful control of crystallization.

Subsequent Processing Steps

In some processes, crystals are produced during the process, but additional crystallization then occurs during packaging and storage. For example, ice crystals in ice cream continue to grow in the hardening tunnel. In the same way, tempered chocolate is cooled further in a cooling tunnel to set the desired crystalline structure. Butter and margarine are packaged while still in a fluid state, with less crystalline structure than desired in the final product, and then set into the desired shape for final crystallization. In fact, there is a wide range of products where subsequent processing causes continuing changes in

Table 8–2 Effects of Beating Temperature and Storage Time on Size Distribution of Sugar Crystals in Fondant

	Crystal Size					
	1 Day		*1 Week*		*1 Month*	
Beating Temperature (°C)	*Range (μm)*	*< 16 μm* (%)*	*Range (μm)*	*< 16 μm (%)*	*Range (μm)*	*< 16 μm (%)*
42	2–16	98	7–20	95	5–70	45
66	2–16	98	3–50	50	7–80	30
119	3–30	90	10–70	20	10–120	5

* Percentage of crystals smaller than 16 μm.

the CSD. In many products, the best quality occurs shortly after packaging and before the crystals have a chance to change due to recrystallization. This is certainly true in the case of frozen desserts.

During hardening of ice cream, the amount of water frozen increases from about 50% at the draw condition (about –5 to –6°C) to about 75 to 80% after hardening to –18°C. This increase in the amount of ice causes a significant increase in crystal size and width of the distribution. In fact, recrystallization occurs in ice cream during the hardening stage and, due to temperature differences between the inner and outer regions of the container during cooling, the average size of ice crystals in the core of the package is larger than that at the package surface.[6] Poor hardening techniques, where the ice cream container remains at warm temperatures for any substantial time period (even minutes) can cause significant increase in ice crystal size.[11]

When chocolate is cooled in a cooling tunnel, it is important to maintain the proper temperature profile to maintain the optimal cocoa butter crystallinity.[38] If chocolate (even well-tempered chocolate) is cooled too rapidly, for example, problems can occur. Cocoa butter at the surface of the chocolate may nucleate during rapid cooling and, since these crystals will be in an unstable polymorphic form, bloom will appear as the polymorphic transformation to the more stable form occurs. Rapid cooling of large blocks of chocolate in a cooling tunnel can cause another form of bloom (E Seguine, personal communication). Here, the rapid cooling causes rapid solidification of the cocoa butter at the surface, which delays removal of heat from the center. The heat being generated as crystallization continues during cooling, builds up in the center, and melts the tempered crystal seed left in the fluid chocolate at the center of the bar. When the center finally does cool sufficiently to solidify the chocolate, it will form an unstable polymorph. When this polymorph transforms to the more stable form, a form of internal bloom can occur that appears as striations or even slip planes of different crystalline structures within the solidified block.

Storage Conditions

Most recrystallization occurs during storage, and adverse storage conditions increase the rate of recrystallization. Some of the recrystallization mechanisms noted previously occur even at constant temperature (migratory recrystallization mechanisms), but all are enhanced when the temperature fluctuates. The rate of recrystallization during storage depends primarily on the temperature of storage and the extent of thermal fluctuations, although the interactions of ingredients also impact the kinetic rates.

At temperatures close to an equilibrium condition (melting point or solubility line), recrystallization occurs most rapidly because any slight variations in temperature result in the maximum change in the amount of crystalline material. For example, recrystallization in ice cream or frozen foods occurs rapidly at temperatures close to the melting point of the mix. At this condition, the viscosity of the continuous phase is relatively low, and molecules are free to take part in migratory recrystallization. At very low temperatures, especially below the glass transition temperature of the frozen product, recrystallization occurs at extremely slow rates due to the limited mobility of water molecules. Thus, to maintain optimal product quality, as related to crystal microstructure, storage temperatures as low as possible are desired.

In ice cream, the rate of change of mean size with time follows Equation 8.4 with $n = 3$.[11] This indicates that diffusion mechanisms limit the rate of ice recrystallization during storage of ice cream. At low storage temperatures, diffusion of water molecules is inhibited, and the amount of water unfrozen is reduced. Thus, the limited mobility of water molecules, due to low storage temperature, keeps the recrystallization rates low.

Recrystallization occurs even under constant temperature storage, although the rate of recrystallization is lower than when temperature fluctuations occur. In ice cream, recrystallization rates increased slowly as temperature fluctuations increased from near 0 to about ± 1.0°C.[10] Above this level of thermal fluctuation, however, recrystallization rate was hardly affected by the magnitude of thermal fluctuations. The primary mechanism of recrystallization during storage of ice cream is likely to be different under different storage conditions. During constant temperature storage (if this can ever be truly achieved), the melt-refreeze mechanism of recrystallization does not occur. This leaves the migratory mechanisms as the main contributors to recrystallization and this typically leads to smooth, spherical crystals.[10] Since there are very few crystals less than 5 μm or so in hardened ice cream, true Ostwald ripening is unlikely to be a major contributor. Thus, isomass rounding and accretion are the main mechanisms of recrystallization under constant temperature conditions. However, when thermal fluctuations occur, melt-refreeze also can occur. In this case, recrystallization may be dominated by melt-refreeze and accretion, which can result in formation of irregular shaped ice crystals.[10]

Similar principles of storage apply for recrystallization in chocolates. At temperatures close to the melting point of the lipid component, relatively large amounts of lipid crystals melt with any temperature fluctuations. Also, at temperatures close to the melting point, relatively high amounts of fat are in the liquid state, allowing rapid migration of lipid molecules during recrystallization. Temperature fluctuations lead to an increased rate of recrystallization in chocolate, leading to fat bloom formation. This is due to the amount of liquid lipid formed as temperature increases and the subsequent recrystallization of that lipid as temperature cools down again. Under the correct circumstances, the cocoa butter triacylglycerols form into crystalline structures that lead to visual bloom.[4] This is accompanied by the polymorphic transformation to the most stable polymorph (β VI).

Thus, conditions that lead to high mobility of the molecules of the crystalline species generally lead to rapid recrystallization. Storage of food products at low temperatures and with minimal temperature fluctuations is typically required to achieve the longest shelf stability.

Product Formulation

The composition of the product also plays an important role in recrystallization rates. The other ingredients in a food, besides the crystallizing species, may have several effects on recrystallization. They include an influence on phase equilibrium, inhibition of crystal growth rates, and rates of polymorphic transformation.

Probably the most important formulation factor is the overall content of the crystallizing species. This is especially true in frozen foods where water content can be quite high. Both Hagiwara and Hartel[39] and Livney-Miller and Hartel[40] found that recrystallization rate in ice cream made with no added stabilizer and stored in bulk containers (as opposed to studies on microscope slides) was directly proportional to the amount of ice formed during storage. Ice phase volume was varied by a combination of storage temperature and formulation (varied freezing point). Higher ice phase volumes led to reduced recrystallization rates in these frozen desserts. However, Sutton et al[41] found that increased ice phase volume in a fructose-water system led to faster recrystallization rates at higher temperatures (–10°C) and had little effect at lower temperatures (–20°C). Sutton and Wilcox[42,43] also found that ice phase volume had no effect on recrystallization rates in ice cream. In the first case,[39,40] ice phase volume was changed by changing storage temperature and/or sweetener type, whereas in the latter case[41] the ice phase volume was changed by processing conditions. These two situations lead to different thermodynamic situations and apparently lead to different recrystallization rates. More work in this area is warranted to understand fully the effects of phase volume on recrystallization.

The phase equilibrium condition may be altered by addition of certain ingredients, as with addition of small-molecular-weight solutes in aqueous solution that cause a decrease in freezing point. A change in phase equilibrium in a product may make that product more stable at a given storage condition. For example, addition of low-molecular-weight sugars to ice cream mix results in a product with a very low freezing point. When frozen to normal storage temperatures, such a product has more water unfrozen than a similar product with higher freezing point. Less water frozen, with all other conditions being equal, results in more rapid recrystallization since the water molecules have greater mobility.

However, more liquid fat does not always lead to reduced storage stability as evidenced by the effect of milk fat in chocolate. Interestingly, when milk fat is added to chocolate, it causes a substantial softening (decrease in solid fat content at equilibrium at any temperature) of the mixture of fats compared with the original cocoa butter. In this case, there is a significant effect on product texture, due to softening, but not necessarily an effect on recrystallization. That is, milk chocolate blooms more slowly than dark chocolate stored at the same temperature despite this decrease in the phase equilibrium temperature. In this case, there is more liquid lipid in a milk chocolate than a dark chocolate stored at the same temperature, but this does not necessarily lead to more rapid recrystallization. Apparently, other effects that influence recrystallization of cocoa butter (specific inhibition of crystallization) are more important than the influence of phase volume.

The specific inhibition of certain food ingredients on recrystallization is widely recognized, although we do not always understand the mechanisms for this inhibition. For example, we add stabilizers (gums and proteins) to ice cream to help reduce the rates of recrystallization and the subsequent coarsening of the ice crystals. However, the effectiveness of stabilizers to inhibit recrystallization depends on the type of stabilizer used and the storage temperature.[15,40] According to Livney-Miller and Hartel,[40] stabilizers work best under conditions of intermediate water mobility. When water molecules had high mobility (higher storage temperature and/or low freezing point of the mix), stabilizers had little effect. In this case, recrystallization rates were very rapid, and no effect of stabilizers was observed. Also, when the ice cream was stored at low temperatures and/or had high freezing point, where the water had low mobility, recrystallization was very slow and no effects of stabilizers were observed. In between these conditions, at conditions of intermediate mobility, recrystallization occurred at intermediate rates and was inhibited by the addition of stabilizer. However, not all of the stabilizers studied had the same effect. Some stabilizers seemed to work better under different conditions. Sutton and Wilcox[42,43] also studied the effect of stabilizers on recrystallization rates in ice cream and found that different stabilizers had different effects. Interestingly, their results also showed the effects of changing stabilizer concentration. For guar gum and pectin, the rate of recrystallization initially decreased as stabilizer concentration increased; it then began to increase again at higher levels of stabilizer addition. This effect was not observed for an enzyme-modified guar gum or for locust bean gum at the levels of addition used.

Several mechanisms have been proposed to explain the inhibition of ice recrystallization due to the addition of stabilizers,[1,40,44] although the exact mechanisms remain unknown. It has been suggested that stabilizers increase the local viscosity and thereby decrease the mobility of water molecules, which results in an inhibition of recrystallization. If the molecules do not move very rapidly, then recrystallization cannot occur very rapidly. It also has been suggested that a weak stabilizer gel structure is formed under frozen conditions and that this inhibits the rate of ice crystal growth during storage.[45] Another potential mechanism is that stabilizer molecules provide some specific inhibition of ice crystal growth and dissolution rates and that this inhibition reduces the rate of recrystallization. However, not all stabilizers have been found to reduce linear ice crystal growth rates. Martin et al[46] used nuclear magnetic resonance (NMR) spectroscopy to show that the molecular mobility of water and solutes in frozen ice cream were not influenced by the addition of stabilizers. They proposed that stabilizers inhibited recrystallization of ice crystals in ice cream by modifying the ice crystal interface with the unfrozen solution. Based on a combination of differential scanning calorimeter (DSC), mechanical analysis, and NMR spectroscopy applied to frozen sucrose solutions, Sahagian and Goff[44] suggested that the influence of stabilizers on recrystallization of ice involves a modification of kinetic properties of the unfrozen phase rather than an influence on thermodynamic properties of the system. Further work is needed to verify which mechanism of recrystallization applies and under what conditions.

In chocolates, the inhibitory effect of milk fat on recrystallization rate in chocolate has been related to a specific inhibition effect. In some way, the milk fat must inhibit recrystallization of cocoa butter during storage and prevent the formation of crystalline structures at the surface of the chocolate that lead to visual bloom. It has been shown that milk fat reduces the rate of cocoa butter crystallization (both nucleation and growth), but how this influences recrystallization is not clear. Milk fat also has been shown to inhibit the polymorphic transformation from β V to β VI,[47] and this is thought to influence the rate of bloom formation during storage of chocolate.

In fondants, the presence of invert sugar has been shown to both decrease the initial size of the distribution and inhibit further changes during storage. Table 8–3 shows the differences in initial size distribution and subsequent changes during storage of fondants. A similar effect on initial crystal size was seen upon addition of glucose syrup, as seen previously in Table 7–3. The presence of invert sugar reduces the amount of sucrose crystalline phase volume (yield), which results in increased separation of the small crystals in fondant. This increased separation reduces the likelihood of accretion during storage. In addition, both corn syrup and invert sugar adsorb to the sucrose crystal surface and inhibit crystal growth rates. Both effects are undoubtedly important in the inhibition of recrystallization in fondants.

SUMMARY

Controlling formation of crystalline microstructure in many food products is critical to obtaining the desired texture and quality. However, subsequent changes in the crystalline microstructure can lead to loss of quality and end of shelf life. These changes may occur even when the total amount of crystalline material does not change (constant phase volume). However, the most dramatic changes occur when storage conditions cause a fluctuation in crystal phase volume due either to temperature fluctuations or changes in moisture content. Thus, recrystallization is typically a combination of several mechanisms, including thermodynamic ripening, accretion, and melt/refreeze.

Control of recrystallization during storage can be accomplished through control of storage conditions. Typically, cold temperatures with only small fluctuations provide the best protection against loss of product quality due to recrystallization. However, certain formulation agents can be used to inhibit the kinetics of these recrystallization processes. Examples include the addition of stabilizers in ice cream to minimize ice recrystallization and the addition of milk fat to chocolate to inhibit fat bloom formation. However, our understanding of how and when these formulation factors work is still somewhat limited, particularly in the case of stabilizers in ice cream.

Table 8–3 Changes in Crystal Size Distribution during Storage of Fondants Made with Sucrose and Invert Sugar

Invert Sugar (%)	*Range of Crystal Sizes (μm)*		
	Initial	*1 Week*	*3 Weeks*
0.42	5.0–25.5	5.8–28.3	6.1–26.1
6.49	5.2–15.5	5.2–18.3	6.1–20.0
11.1	3.6–12.5	4.0–13.3	3.3–14.4
15.8	2.8–13.2	3.3–15.8	2.8–17.5

REFERENCES

1. Fennema O. The nature of the freezing process. In: Fennema O, Powrie WD, Marth EH, eds. *Low-Temperature Preservation of Foods and Living Matter*. New York: Marcel Dekker; 1973·151–239.
2. Russell AB, Cheney PE, Wantling SD. Influence of freezing conditions on ice crystallization in ice cream. *J Food Eng*. 1999;39:179–191.
3. deMan L, deMan JM, Blackman B. Polymorphic behavior of some fully hydrogenated oils and their mixtures with liquid oil. *J Am Oil Chem Soc*. 1989;66:1777–1780.
4. Hartel RW. Mechanism of bloom formation in chocolate. *Manufacturing Confectioner*. 1999;79(5):89–99.
5. Schlichter-Aronhime J, Garti N. Solidification and polymorphism in cocoa butter and the blooming problems In: Garti N, Sato K, eds. *Crystallization and Polymorphism of Fats and Fatty Acids* New York: Marcel Dekker; 1988:363–394.
6. Hartel RW. Mechanisms and kinetics of recrystallization in ice cream. In: Reid D, ed. *Properties of Water: ISOPOW VI*. London: Blackie; 1998,287–319.
7. Mullin JW. *Crystallization*. 2nd ed. Oxford: Butterworth; 1993.
8. Dadyburjor DB, Ruckenstein E. Kinetics of Ostwald ripening. *J Crystal Growth*. 1977;40:279–290.
9. Dadyburjor DB, Ruckenstein E. Crowded ripening in unsaturated solution. *J Crystal Growth*. 1977;38:28–287.
10. Donhowe DP, Hartel RW. Recrystallization of ice in ice cream during accelerated storage. *Int Dairy J*. 1996;6: 1191–1208.
11. Donhowe DP. *Ice Recrystallization in Ice Cream and Ice Milk*. Madison, WI: University of Wisconsin–Madison; 1993. PhD dissertation.
12. Jain SC, Hughes AE. Ostwald ripening and its application to precipitates and colloids in ionic crystals and glasses. *J Material Sci*. 1978;13:1611–1631.
13. Donhowe DP, Hartel, RW. Influence of temperature on ice recrystallization in frozen desserts II. Bulk storage studies. *Int Dairy J* 1996;6:1209–1221.
14. Harper EK, Shoemaker CF. Effect of locust bean gum and selected sweetening agents on ice recrystallization rates. *J Food Sci*. 1983;48:1801–1803, 1806.
15. Lim MH. *The Effect of Polymers on the Growth and Recrystallization Kinetics of Ice*. Davis, CA: University of California–Davis; 1983. MS thesis.
16. Buyong N, Fennema O. Amount and size of ice crystals in frozen samples as influenced by hydrocolloids. *J Dairy Sci*. 1988;71(10):2630–2639.
17. Caldwell KB, Goff, HD, Stanley DW. A low-temperature scanning electron microscopy study of ice cream. II. Influence of selected ingredients and processes. *Food Structure*. 1992;11:11–23.
18. Sutton RL, Lips A, Piccirillo G, Sztehlo A. Kinetics of ice recrystallization in aqueous fructose solutions. *J Food Sci*. 1996;61(4):741–745.
19. Bevilacqua AE, Zaritzky NE. Ice recrystallization in frozen beef. *J Food Sci*. 1982;47.1410–1414.
20. Martino M, Zaritzky N. Effects of temperature on recrystallization in polycrystalline ice. *Sci Aliments*. 1987;7(1): 147–166.
21. Martino M, Zaritzky N. Ice recrystallization in a model system and in frozen muscle tissue. *Cryobiology*. 1989;26:138.
22. Dunning WJ. Ripening and ageing processes in precipitates. In: Smith AL, ed. *Particle Growth in Suspensions*. London: Academic Press; 1973:3.
23. Van Suchtelen J. Coarsening of eutectic structure during and after unidirectional growth. *J Crystal Growth* 1978;43:28–46.
24. Kahlweit M. Ostwald ripening of precipitates *Adv Colloid Interface Sci*. 1975;5:1–5.
25. Kuczynski GC. Towards the understanding of the process of sintering. In: Kuczynski GC, Uskokovic DP, Palmour III H, Ristic MM, eds. *Sintering '85*. New York: Plenum Press; 1989:3–16.
26. Kingery WD. Regelation, surface diffusion and ice sintering. *J Appl Phys*. 1960;31(5):833–838.
27. Kingery WD. Sintering in the presence of a liquid phase. In: Kingery WD, ed. *Kinetics of High-Temperature Processes*. New York: John Wiley & Sons; 1960:31.
28. Hobbs PV. *Ice Physics*. Oxford: Clarendon Press; 1974.

29. Kuczynski GC. Self-diffusion in sintering of metallic particles. *J Metals.* 1949;1:69.

30. Sutton RL, Evans ID, Crilly JF. Modeling of ice crystals coarsening in concentrated disperse food systems. *J Food Sci.* 1994;59(6):1227.

31. Johansson D, Bergenstahl B. Sintering of fat crystal networks in oil during post-crystallization processes. *J Am Oil Chem Soc.* 1995;72(8):911–920.

32. Johansson D, Bergenstahl B, Lundgren E. Wetting of fat crystals by triglyceride oil and water. 1. The effect of additives. *J Am Oil Chem Soc.* 1995;72(8):921–931.

33. Johansson D, Bergenstahl B. Wetting of fat crystals by triglyceride oil and water. 2. Adhesion to the oil/water interface. *J Am Oil Chem Soc.* 1995;72(8):933–938.

34. Johansson D, Bergenstahl B, Lundgren E. Water-in-triglyceride oil emulsions. Effect of fat crystals on stability. *J Am Oil Chem Soc.* 1995;72(8):939–950.

35. Ben-Yoseph E, Hartel RW. Computer simulation of ice recrystallization in ice cream during storage. *J Food Eng.* 1999;38(3):309–331.

36. Luyet BJ. Phase transitions encountered in the rapid freezing of aqueous solutions. *Ann NY Acad Sci.* 1965;125(2): 502–521.

37. Lees R. *Factors Affecting Crystallization in Boiled Sweets, Fondants and Other Confectionery.* Leatherhead, UK: The British Food Manufacturing Industries Research Association; 1965. Scientific and Technical Survey, No. 42.

38. Hartel RW. Phase transitions in chocolates and coatings. In: Rao MA, Hartel RW, eds. *Phase/State Transitions in Foods.* New York: Marcel Dekker; 1998:217–252.

39. Hagiwara T, Hartel RW. Effect of sweetener, stabilizer and storage temperature on ice recrystallization in ice cream. *J Dairy Sci.* 1996;79(5):735–744.

40. Livney-Miller T, Hartel RW. Ice recrystallization in ice cream: sweetener-stabilizer interactions. *J Dairy Sci.* 1997; 80:447–456.

41. Sutton RL, Lips A, Piccirillo G. Recrystallization in aqueous fructose solutions as affected by locust bean gum. *J Food Sci.* 1996;61(4):746–748.

42. Sutton RL, Wilcox J. Recrystallization in ice cream as affected by stabilizers. *J Food Sci.* 1998;63(1):104–107.

43. Sutton RL, Wilcox J. Recrystallization in model ice cream solutions as affected by stabilizer concentration. *J Food Sci.* 1998;63(1):9–11.

44. Sahagian ME, Goff HD. Thermal, mechanical and molecular relaxation properties of stabilized sucrose solutions at sub-zero temperatures. *Food Res Int* 1995;28(1):1–8.

45. Muhr AH, Blanshard JMV. Effect of polysaccharide stabilizers on the rate of growth of ice. *J Food Technol.* 1986; 21:683–710.

46. Martin DR, Ablett S, Darke A, Sutton RL, Sahagian M. Diffusion of aqueous sugar solutions as affected by locust bean gum studied by NMR. *J Food Sci.* 1999;64(1):46–49.

47 Bricknell JP, Hartel RW. Relation of fat bloom in chocolate to polymorphic transition of cocoa butter. *J American Oil Chemists Society.* 1998;75(11):1609–1616.

Sources

CHAPTER 2

Figure 2–3 Reprinted with permission from B.M. Smythe, *Australian Journal of Chemistry*, Vol. 20, p. 1116, © 1967, CSIRO Publishing.

Figure 2–4 From J.W. Mullin, *Crystallization, 3rd Edition*, p. 12, © 1993, adapted with permission of Butterworth Heinemann Publishers, a division of Reed Educational & Professional Publishing Ltd.

Table 2–2 Data from K.J. Klacik, Continuous Production of Sugar-Free Hard Candies, *Manufacturing Confectioner*, Vol. 69, No. 8, pp. 61–67, © 1989.

Table 2–3 Data from J. Hagemann, Thermal Behavior and Polymorphism of Acylglycerides, in *Crystallization and Polymorphism of Fats and Fatty Acids*, N. Garti and K. Sato, eds., pp. 9–96, © 1988, Marcel Dekker, Inc.

Table 2–4 Data from T.R. Davis and P.S. Dimick, Solidification of Cocoa Butter, *Proceedings of the Pennsylvania Manufacturers Confectionary Association Production Conference*, Vol. 40, pp. 104–108, © 1986 and R.L. Wille and E.S. Lutton, Polymorphism of Cocoa Butter, *Journal of the American Oil Chemists' Society*, Vol. 43, pp. 491–496, © 1966, AOCS Press.

Table 2–5 Reprinted from R.D. O'Brien, *Fats and Oils*, p. 268, with permission from Technomic Publishing Co., Inc., copyright 1998.

Figure 2–12 Reprinted with permission from W. Si, *Mechanisms of Ice Crystallization in Scraped Surface Heat Exchanger*, MS Thesis, © 2000, University of Wisconsin-Madison.

Figure 2–13 Reprinted with permission from D.S. Grall, *Kinetics of Butterfat Crystallization from the Melt*, MS Thesis, © 1991, University of Wisconsin-Madison.

CHAPTER 3

Figure 3–5 Reprinted with permission from D. Donhowe, R.W. Hartel, and R.C. Bradley, Jr., Ice Crystallization Processes During Manufacture and Storage of Ice Cream, *Journal of Dairy Science*, Vol. 74, No. 10, p. 3336, © 1991, American Dairy Science Association.

Figure 3–11 Reprinted with permission from P. Gabarra, *The Effects of Corn Syrup Solids and Their Oligosaccharide Fractions on the Glass Transition and* istics of Aqueous Citric Acid Solutions, pp. 75–76, Copyright 1969, with permission from Elsevier Science.

Figure 5–7 Reprinted from R.W. Hartel and A.V. Shastry, Sugar Crystallization in Food Products, *Critical Reviews in Food Science and Nutrition*, Vol. 30, p. 75, © 1991, CRC Press, Inc.

Figure 5–8 Reprinted with permission from Y.D. Livney, D.P. Donhowe, and R.W. Hartel, Influence of Temperature on Crystallization of Lactose in Ice Cream, *International Journal of Food Science and Technology*, Vol. 30, No. 3, p. 316, © 1995, Blackwell Science Ltd.

Figure 5–9 Reprinted with permission from B. Makower and W.B. Dye, Equilibrium Moisture Content and Crystallization of Amorphous Sucrose and Glucose, *Journal of Agricultural and Food Chemistry*, Vol. 4, p. 73. Copyright 1956 American Chemical Society.

Figure 5–10 Adapted with permission from Y.H. Roos, *Phase Transitions in Foods*, p. 169, © 1995, Academic Press.

Figure 5–11 Reprinted with permission from L. Hernqvist, On the Structure of Triglycerides in the Liquid State and Fat Crystallization, *Fette Seifen Anstrichmittel*, Vol. 86, p. 300, © 1984, Wiley VCH.

Figure 5–12 Reprinted from R.W. Hartel, *Phase Transitions in Chocolates and Coatings, in Phase/State Transitions in Foods*, M.A. Rao and R.W. Hartel, eds., p. 230, © 1998, by courtesy of Marcel Dekker Inc.

Figure 5–13 Data from K. Sato, Crystallization of Fats and Fatty Acids, in *Crystallization and Polymorphism of Fats and Fatty Acids*, N. Garti and K. Sato, eds., pp. 227–266, © 1988, Marcel Dekker Inc.

Figure 5–14 Reprinted from R.W. Hartel, Phase Transitions in Chocolates and Coatings, in *Phase/State Transitions in Foods*, M.A. Rao and R.W. Hartel, eds., p. 230, © 1998, by courtesy of Marcel Dekker Inc.

Table 5–1 Reprinted with permission from S. Metin, *Crystallization Behavior and Kinetics of Blends of Cocoa Butter and Milk Fat or Milk Fat Fractions*, Ph.D. Dissertation, © 1997, University of Wisconsin-Madison.

Figure 5–15 Reprinted with permission from R.W. Hartel and A.V. Shastry, Sugar Crystallization in Food Products, *Critical Reviews in Food Science and Nutrition*, Vol. 30, p. 60, © 1991, CRC Press, Inc.

Figure 5–18 Reprinted with permission from M.L. Herrera and R.W. Hartel, Effect of Processing Conditions on Physical Properties of a Milk Fat Model System Microstructure, *Journal of the American Oil Chemists' Society*, Vol. 77, No. 11, pp. 1197–1204, © 2000, AOCS Press.

Figure 5–19 Reprinted with permission from M.L. Herrera and R.W. Hartel, Effect of Processing Conditions on Physical Properties of a Milk Fat Model System Microstructure, *Journal of the American Oil Chemists' Society*, Vol. 77, No. 11, pp. 1197–1204, © 2000, AOCS Press.

Figure 5–21 Reprinted from *Food Research International*, Vol. 32, M.L. Herrera, et al., A Kinetic Analysis of Crystallization of a Milk Fat Model System, pp. 289–298, Copyright 1999, with permission from Elsevier Science.

CHAPTER 6

Exhibit 6–1 Reprinted with permission from R.W. Hartel and A.V. Shastry, Sugar Crystallization in Food Products, *Critical Reviews in Food Science Nutrition*, Vol. 30, pp. 49–112, © 1991, CRC Press, Inc.

Figure 6–4 Reprinted with permission from R.W. Hartel and A.V. Shastry, Sugar Crystallization in Food Products, *Critical Reviews in Food Science Nutrition*, Vol. 30, p. 85, © 1991, CRC Press, Inc.

Figure 6–5 Adapted with permission from W.A. Van Hook, *Crystallization: Theory and Practice*, p. 93, © 1961.

Table 6–1 Data from R.A. Visser, A Natural Crystal Growth Retarder in Lactose, *Journal of Dairy Science*, Vol. 34, p. 255, © 1980.

Figure 6–11 Data from Y. Shi, R.W. Hartel, and B.M. Liang, Formation and Growth Phenomena of Lactose Nuclei Under Contact Nucleation Conditions, *Journal of Dairy Science*, Vol. 72, No. 11, p. 2903, © 1989.

Figure 6–13 Data from J. Fabian, *Untersuchung zur Auflosungskinetik in der Losungkristallisation*, Ph.D. Dissertation, © 1996, Universitat Bremen, Germany.

Figure 6–15 Reprinted with permission from E. Ben-Yoseph, *Computer Modeling of Sucrose Crystallization During Drying of Thin Sucrose Films*, Ph.D. Dissertation, © 1999, University of Wisconsin-Madison.

Figure 6–16 Data from B.M. Smythe, Sucrose Crystal Growth Rate: I. Rate of Crystal Growth in Pure Solutions, *Australian Journal of Chemistry*, Vol. 20, pp. 1087–1096, © 1967, CSIRO Publishing.

Figure 6–17 Data from B.M. Smythe, Sucrose Crystal Growth Rate: I. Rate of Crystal Growth in Pure Solutions, *Australian Journal of Chemistry*, Vol. 20, pp. 1087–1096, © 1967, CSIRO Publishing.

Figure 6–19 Reprinted with permission from B.M. Smythe, Sucrose Crystal Growth Rate: I. Rate of Crystal Growth in Pure Solutions, *Australian Journal of Chemistry*, Vol. 20, p. 1091, © 1967, CSIRO Publishing.

Figure 6–20 Data from B.M. Smythe, Sucrose Crystal Growth Rate: II. Rate of Crystal Growth in the Presence of Impurities, *Australian Journal of Chemistry*, Vol. 20, pp. 1097–1114, © 1967, CSIRO Publishing.

Figure 6–21 Data from P. Tjuradi and R.W. Hartel, Effects of Corn Syrup Oligosaccharides on Sucrose Crystallization, *Journal of Food Science*, Vol. 60, No. 6, pp. 1353–1356, © 1995.

CHAPTER 7

Figure 7–2 Reprinted from R.W. Hartel, Phase Transitions During Manufacture and Storage of Ice Cream, in *Phase/State Transitions in Foods*, M.A. Rao and R.W. Hartel, eds., p. 346, © 1998, by courtesy of Marcel Dekker Inc.

Figure 7–6 Reprinted with permission from R.W. Hartel, Controlling Sugar

Index

R

S

Printed in the United States
115889LV00001B/55-66/A

9 780834 216341